Simon Singh received his PhD [illegible] University of Cambridge. A former BBC producer, he directed the BAFTA award-winning documentary film *Fermat's Last Theorem* and wrote the bestselling book of the same name, followed by *The Code Book*, also a bestseller, which was the basis for the Channel 4 series, *The Science of Secrecy*.

For automatic updates on Simon Singh visit harperperennial.co.uk and register for AuthorTracker

From the reviews of *Big Bang*:

'Singh is a very gifted storyteller who never misses the chance to make his subject clearer or more entertaining'

SCARLETT THOMAS, *Independent on Sunday*

'This very well-written book conveys the ideas underpinning cosmological theory with great clarity' *Nature*

'Singh uses beautifully simple analogies and clearly explained diagrams to enable even the most mathematically hobbled of us to recapitulate the history of man's intellectual engagement with the dark spaces around him' *Sunday Telegraph*

'If you are intrigued by the story but wary of mathematics, do not worry; Simon Singh spares us most of the maths, and he juggles big ideas with tact and care' *Daily Mail*

'A model of clarity' *Economist*

'Singh tells his tale well, with chatty anecdotes leavening the astrophysics' *Guardian*

'An epic tale brilliantly told, packed with courage and tragedy, heroes and martyrs'
Daily Telegraph

'Even if the cosmologists don't know where the universe is going, at least they have found out where it has come from. Anybody who wants to understand this wonderful achievement will not do better than start with Singh's book'
Mail on Sunday

'An excellent introduction to the way modern science works'
The Times Higher Education Supplement

By the same author

Fermat's Last Theorem
The Code Book

BIG BANG

The Most Important Scientific Discovery of All Time and Why You Need to Know About it

Simon Singh

HARPER PERENNIAL

Harper Perennial
An imprint of HarperCollins*Publishers*
77–85 Fulham Palace Road
Hammersmith
London W6 8JB

www.harperperennial.co.uk

This edition published by Harper Perennial 2005
5

First published by Fourth Estate 2004

A catalogue record for this book
is available from the British Library

ISBN 0 00 715252 3

Set in Linotype Bembo

Printed and bound in Great Britain by
Clays Ltd, St Ives plc

This book would not have been possible without Carl Sagan, James Burke, Magnus Pyke, Heinz Wolff, Patrick Moore, Johnny Ball, Rob Buckman, Miriam Stoppard, Raymond Baxter, and all the science TV producers and directors who inspired my interest in science.

Place three grains of sand inside a vast cathedral, and the cathedral will be more closely packed with sand than space is with stars.
JAMES JEANS

The effort to understand the universe is one of the very few things that lifts human life a little above the level of farce, and gives it some of the grace of tragedy.
STEVEN WEINBERG

In science one tries to tell people, in such a way as to be understood by everyone, something that no one ever knew before. But in poetry, it's the exact opposite.
PAUL DIRAC

The most incomprehensible thing about the universe is that it is comprehensible.
ALBERT EINSTEIN

CONTENTS

Chapter 1

IN THE BEGINNING

Science must begin with myths, and with the criticism of myths.
KARL POPPER

I do not feel obliged to believe that the same God who has endowed us with sense, reason and intellect has intended us to forgo their use. **GALILEO GALILEI**

Living on Earth may be expensive, but it includes an annual free trip around the Sun. **ANONYMOUS**

Physics is not a religion. If it were, we'd have a much easier time raising money. **LEON LEDERMAN**

Our universe is dotted with over 100 billion galaxies, and each one contains roughly 100 billion stars. It is unclear how many planets are orbiting these stars, but it is certain that at least one of them has evolved life. In particular, there is a life form that has had the capacity and audacity to speculate about the origin of this vast universe.

Humans have been staring up into space for thousands of generations, but we are privileged to be part of the first generation who can claim to have a respectable, rational and coherent description for the creation and evolution of the universe. The Big Bang model offers an elegant explanation of the origin of everything we see in the night sky, making it one of the greatest achievements of the human intellect and spirit. It is the consequence of an insatiable curiosity, a fabulous imagination, acute observation and ruthless logic.

Even more wonderful is that the Big Bang model can be understood by everyone. When I first learned about the Big Bang as a teenager, I was astonished by its simplicity and beauty, and by the fact that it was built on principles which, to a very large extent, did not go beyond the physics I was already learning at school. Just as Charles Darwin's theory of natural selection is both fundamental and comprehensible to most intelligent people, the Big Bang model can be explained in terms that will make sense to non-specialists, without having to water down the key concepts within the theory.

But before encountering the earliest stirrings of the Big Bang model, it is necessary to lay some groundwork. The Big Bang model of the universe was developed over the last hundred years, and this was only possible because twentieth-century breakthroughs were built upon a foundation of astronomy constructed in previous centuries. In turn, these theories and observations of the sky were set within a scientific framework that had been assiduously crafted over two millennia. Going back even further, the scientific method as a path to objective truth about the material world could start to blossom only when the role of myths and folklore had begun to decline. All in all, the roots of the Big Bang model and the desire for a scientific theory of the universe can be traced right back to the decline of the ancient mythological view of the world.

From Giant Creators to Greek Philosophers

According to a Chinese creation myth that dates to 600 BC, Phan Ku the Giant Creator emerged from an egg and proceeded to create the world by using a chisel to carve valleys and mountains from the landscape. Next, he set the Sun, Moon and stars in the sky; he died as soon as these tasks were finished. The death of the Giant Creator was an essential part of the creation process, because fragments of his own body helped to complete the world. Phan Ku's skull formed the dome of sky, his flesh formed the soil, his bones became rocks and his blood created rivers and seas. The last of his breath forged the wind and clouds, while his sweat became rain. His hair fell to Earth, creating plant life, and the fleas that had lodged in his hair provided the basis for the human race. As our birth required the death of our creator, we were to be cursed with sorrow forever after.

In contrast, in the Icelandic epic myth *Prose Edda* creation started not with an egg, but within the Yawning Gap. This void separated the contrasting realms of Muspell and Niflheim, until one day the fiery, bright heat of Muspell melted the freezing snow and ice of Niflheim, and the moisture fell into the Yawning Gap, sparking life in the form of Imir, the giant. Only then could the creation of the world begin.

The Krachi people of Togo in West Africa speak of another giant, the vast blue god Wulbari, more familiar to us as the sky. There was a time when he lay just above the Earth, but a woman pounding grain with a long timber kept prodding and poking him until he raised himself above the nuisance. However, Wulbari was still within reach of humans, who used his belly as a towel and snatched bits of his blue body to add spice to their soup. Gradually, Wulbari moved higher and higher until the blue sky was out of reach, where it has remained ever since.

For the Yoruba, also of West Africa, Olorun was Owner of the Sky. When he looked down upon the lifeless marsh, he asked another divine being to take a snail shell down to the primeval Earth. The shell contained a pigeon, a hen and a tiny amount of soil. The soil was sprinkled on the marshes of the Earth, whereupon the hen and pigeon began scratching and picking at it, until the marsh became solid ground. To test the world, Olorun sent down the Chameleon, which turned from blue to brown as it moved from sky to land, signalling that the hen and pigeon had completed their task successfully.

Throughout the world, every culture has developed its own myths about the origin of the universe and how it was shaped. These creation myths differ magnificently, each reflecting the environment and society from which it originated. In Iceland, it is the volcanic and meteorological forces that form the backdrop to the birth of Imir, but according to the Yoruba of West Africa it is the familiar hen and pigeon that give rise to solid land. Nevertheless, all these unique creation myths have some features in common. Whether it is the big, blue, bruised Wulbari or the dying giant of China, these myths inevitably invoke at least one supernatural being to play a crucial role in explaining the creation of the universe. Also, every myth represents the absolute truth within its society. The word 'myth' is derived from the Greek word *mythos*, which can mean 'story', but also means 'word', in the sense of 'the final word'. Indeed, anybody who dared to question these explanations would have laid themselves open to accusations of heresy.

Nothing much changed until the sixth century BC, when there was a sudden outbreak of tolerance among the intelligentsia. For the very first time, philosophers were free to abandon accepted mythological explanations of the universe and develop their own theories. For example, Anaximander of Miletus argued that the Sun was a hole in a fire-filled ring that encircled the Earth and revolved around it. Similarly, he believed that the Moon and stars were nothing more than holes in the firmament, revealing otherwise hidden fires. Alternatively, Xenophanes of Colophon believed that the Earth exuded combustible gases that accumulated at night until they reached a critical mass and ignited, thereby creating the Sun. Night fell again when the ball of gas had burned out, leaving behind just the few sparks that we call stars. He explained the Moon in a similar way, with gases developing and burning over a twenty-eight-day cycle.

The fact that Xenophanes and Anaximander were not very close to the truth is unimportant, because the real point is that they were developing theories that explained the natural world without resorting to supernatural devices or deities. Theories that say that the Sun is a celestial fire seen through a hole in the firmament or a ball of burning gas are qualitatively different from the Greek myth that explained the Sun by invoking a fiery chariot driven across the sky by the god Helios. This is not to say that the new wave of philosophers necessarily wanted to deny the existence of the gods, rather that they merely refused to believe that it was divine meddling that was responsible for natural phenomena.

These philosophers were the first *cosmologists*, inasmuch as they were interested in the scientific study of the physical universe and its origins. The word 'cosmology' is derived from the ancient Greek word *kosmeo*, which means 'to order' or 'to organise', reflecting the belief that the universe could be understood and is worthy of analytical study. The cosmos had patterns, and it was the ambition of the Greeks to recognise these patterns, to scrutinise them and to understand what was behind them.

It would be a great exaggeration to call Xenophanes and Anaximander scientists in the modern sense of the term, and it would flatter them to consider their ideas as full blown scientific theories. Nevertheless, they were certainly contributing to the birth of scientific thinking, and their ethos had much in common with modern science. For example, just like ideas in modern science, the ideas of the Greek cosmologists could be criticised and compared, refined or abandoned. The Greeks loved a good argument, so a community of philosophers would examine theories, question the reasoning behind them and ultimately choose which was the most convincing. In contrast, individuals in many other cultures would not dare to question their own mythology. Each mythology was an article of faith within its own society.

Pythagoras of Samos helped to reinforce the foundations of this new rationalist movement from around 540 BC. As part of his philosophy, he developed a passion for mathematics and demonstrated how numbers and equations could be used to help formulate scientific theories. One of his first breakthroughs was to explain the harmony of music via the harmony of numbers. The most important instrument in early Hellenic music was the tetrachord, or four-stringed lyre, but Pythagoras developed his theory by experimenting with the single-stringed monochord. The string was kept under a fixed tension, but the length of the string could be altered. Plucking a particular length of string generated a particular note, and Pythagoras realised that halving the length of the same string created a note that was one octave higher and in harmony with the note from the plucking of the original string. In fact, changing the string's length by any simple fraction or ratio would create a note harmonious with the first (e.g. a ratio of 3:2, now called a musical fifth), but changing the length by an awkward ratio (e.g. 15:37) would lead to a discord.

Once Pythagoras had shown that mathematics could be used to help explain and describe music, subsequent generations of scientists used numbers to explore everything from the trajectory of a cannonball to chaotic weather patterns. Wilhelm Röntgen, who discovered X-rays in 1895, was a firm believer in the Pythagorean philosophy of mathematical science, and once pointed out: 'The physicist in preparing for his work needs three things: mathematics, mathematics and mathematics.'

Pythagoras' own mantra was 'Everything is number.' Fuelled by this belief, he tried to find the mathematical rules that governed the heavenly bodies. He argued that the movement of the Sun, Moon and planets across the sky generated particular musical notes, which were determined by the lengths of their orbits. Therefore, Pythagoras concluded, these orbits and notes had to have specific numerical proportions for the universe to be in harmony. This became a popular theory in its time. We can re-examine it from a modern perspective and see how it stands up to the rigours of today's scientific method. On the positive side, Pythagoras' claim that the universe is filled with music does not rely on any supernatural force. Also, the theory is rather simple and quite elegant, two qualities that are highly valued in science. In general, a theory founded on a single short, beautiful equation is preferred to a theory that relies on several awkward, ugly equations qualified by lots of complicated and spurious caveats. As the physicist Berndt Matthias put it: 'If you see a formula in the *Physical Review* that extends over a quarter of a page, forget it. It's wrong. Nature isn't that complicated.' However, simplicity and elegance are secondary to the most important feature of any scientific theory, which is that it must match reality and it must be open to testing, and this is where the theory of celestial music fails completely. According to Pythagoras, we are constantly bathed in his hypothetical heavenly music, but we cannot perceive it because we have been hearing it since birth and have become habituated to it. Ultimately, any theory that predicts a music that could never be heard, or anything else that could never be detected, is a poor scientific theory.

Every genuine scientific theory must make a prediction about the universe that can be observed or measured. If the results of an experiment or observation match the theoretical prediction, this is a good reason why the theory might become accepted and then incorporated into the grander scientific framework. On the other hand, if the theoretical prediction is inaccurate and conflicts with an experiment or observation, then the theory must be rejected, or at least adapted, regardless of how well the theory does in terms of beauty or simplicity. It is the supreme challenge, and a brutal one, but every scientific theory must be testable and compatible with reality. The nineteenth-century naturalist Thomas Huxley stated it thus: 'The great tragedy of Science – the slaying of a beautiful hypothesis by an ugly fact.'

Fortunately, Pythagoras' successors built on his ideas and improved on his methodology. Science gradually became an increasingly sophisticated and powerful discipline, capable of staggering achievements such as measuring the actual diameters of the

Sun, Moon and Earth, and the distances between them. These measurements were milestones in the history of astronomy, representing as they do the first tentative steps on the road to understanding the entire universe. As such, these measurements deserve to be described in a little detail.

Before any celestial distances or sizes could be calculated, the ancient Greeks first had to establish that the Earth is a sphere. This view gained acceptance in ancient Greece as philosophers became familiar with the notion that ships gradually disappear over the horizon until only the tip of the mast could be seen. This made sense only if the surface of the sea curves and falls away. If the sea has a curved surface, then presumably so too does the Earth, which means it is probably a sphere. This view was reinforced by observing lunar eclipses, when the Earth casts a disc-shaped shadow upon the Moon, exactly the shape you would expect from a spherical object. Of equal significance was the fact that everyone could see that the Moon itself was round, suggesting that the sphere was the natural state of being, adding even more ammunition to the round Earth hypothesis. Everything began to make sense, including the writings of the Greek historian and traveller Herodotus, who told of people in the far north who slept for half the year. If the Earth was spherical, then different parts of the globe would be illuminated in different ways according to their latitude, which naturally gave rise to a polar winter and nights that lasted for six months.

But a spherical Earth raised a question that still bothers children today – what stops people in the southern hemisphere from falling off? The Greek solution to this puzzle was based on the belief that the universe had a centre and that everything was attracted to this centre. The centre of the Earth supposedly coincided with the hypothetical universal centre, so the Earth itself was static and everything on its surface was pulled towards the centre. Hence, the Greeks would be held on the ground by this force, as would everybody else on the globe, even if they lived down under.

The feat of measuring the size of the Earth was first accomplished by Eratosthenes, born in about 276 BC in Cyrene, in modern-day Libya. Even when he was a little boy it was clear that Eratosthenes had a brilliant mind, one that he could turn to any discipline, from poetry to geography. He was even nicknamed Pentathlos, meaning an athlete who participates in the five events of the pentathlon, hinting at the breadth of his talents. Eratosthenes spent many years as the chief librarian at Alexandria, arguably the most prestigious academic post in the ancient world. Cosmopolitan Alexandria had taken over from Athens as the intellectual hub of the Mediterranean, and the city's library was the most respected institution of learning in the world. Forget any notion of strait-laced librarians stamping books and whispering to each other, because this was a vibrant and exciting place, full of inspiring scholars and dazzling students.

While at the library, Eratosthenes learned of a well with remarkable properties, situated near the town of Syene in southern Egypt, near modern-day Aswan. At noon on 21 June each year, the day of the summer solstice, the Sun shone directly into the well and illuminated it all the way to the bottom. Eratosthenes realised that on that particular day the Sun must be directly overhead, something that never happened in Alexandria, which was several hundred kilometres north of Syene. Today we know that Syene lies close to the Tropic of Cancer, the most northerly latitude from which the Sun can appear overhead.

Aware that the Earth's curvature was the reason why the Sun could not be overhead at both Syene and Alexandria simultaneously, Eratosthenes wondered if he could exploit this to measure the circumference of the Earth. He would not necessarily have thought about the problem in the same way we would, as his interpretation

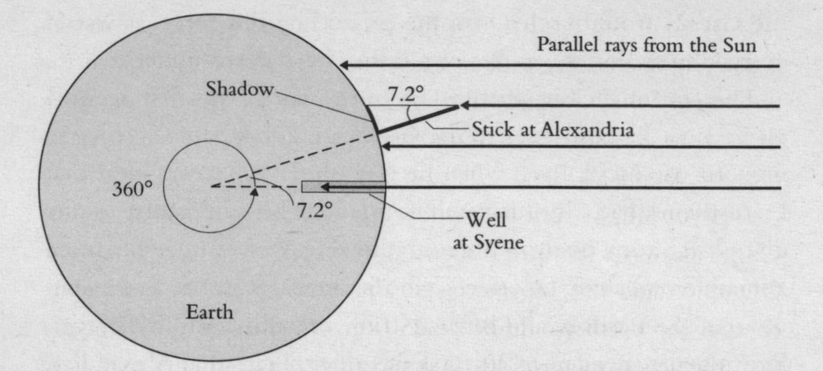

Figure 1 Eratosthenes used the shadow cast by a stick at Alexandria to calculate the circumference of the Earth. He conducted the experiment at the summer solstice, when the Earth was at its maximum tilt and when towns lying along the Tropic of Cancer were closest to the Sun. This meant that the Sun was directly overhead at noon at those towns. For reasons of clarity, the distances in this and other diagrams are not drawn to scale. Similarly, angles may be exaggerated.

of geometry and his notation would have been different, but here is a modern explanation of his approach. Figure 1 shows how parallel rays of light from the Sun hit the Earth at noon on 21 June. At exactly the same moment that sunlight was plunging straight down the well at Syene, Eratosthenes stuck a stick vertically in the ground at Alexandria and measured the angle between the Sun's rays and the stick. Crucially, this angle is equivalent to the angle between two radial lines drawn from Alexandria and Syene to the centre of the Earth. He measured the angle to be 7.2°.

Next, imagine somebody at Syene who decides to walk in a straight line towards Alexandria, and who carries on walking until they circumnavigate the globe and return to Syene. This person would go right round the Earth, traversing a complete circle and covering 360°. So, if the angle between Syene and Alexandria is only 7.2°, then the distance between Syene and Alexandria represents

$^{7.2}/_{360}$, or $^{1}/_{50}$ of the Earth's circumference. The rest of the calculation is straightforward. Eratosthenes measured the distance between the two towns, which turned out to be 5,000 stades. If this represents $^{1}/_{50}$ of the total circumference of the Earth, then the total circumference must be 250,000 stades.

But you might well be wondering, how far is 250,000 stades? One stade was a standard distance over which races were held. The Olympic stade was 185 metres, so the estimate for the circumference of the Earth would be 46,250 km, which is only 15% bigger than the actual value of 40,100 km. In fact, Eratosthenes may have been even more accurate. The Egyptian stade differed from the Olympic stade and was equal to just 157 metres, which gives a circumference of 39,250 km, accurate to 2%.

Whether he was accurate to 2% or 15% is irrelevant. The important point is that Eratosthenes had worked out how to reckon the size of the Earth scientifically. Any inaccuracy was merely the result of poor angular measurement, an error in the Syene–Alexandria distance, the timing of noon on the solstice, and the fact that Alexandria was not quite due north of Syene. Before Eratosthenes, nobody knew if the circumference was 4,000 km or 4,000,000,000 km, so nailing it down to roughly 40,000 km was a huge achievement. It proved that all that was required to measure the planet was a man with a stick and a brain. In other words, couple an intellect with some experimental apparatus and almost anything seems achievable.

It was now possible for Eratosthenes to deduce the size of the Moon and the Sun, and their distances from the Earth. Much of the groundwork had already been laid by earlier natural philosophers, but their calculations were incomplete until the size of the Earth had been established, and now Eratosthenes had the missing value. For example, by comparing the size of the Earth's shadow cast upon the Moon during a lunar eclipse, as shown in Figure 2, it was possible to deduce that the Moon's diameter was about one-quarter

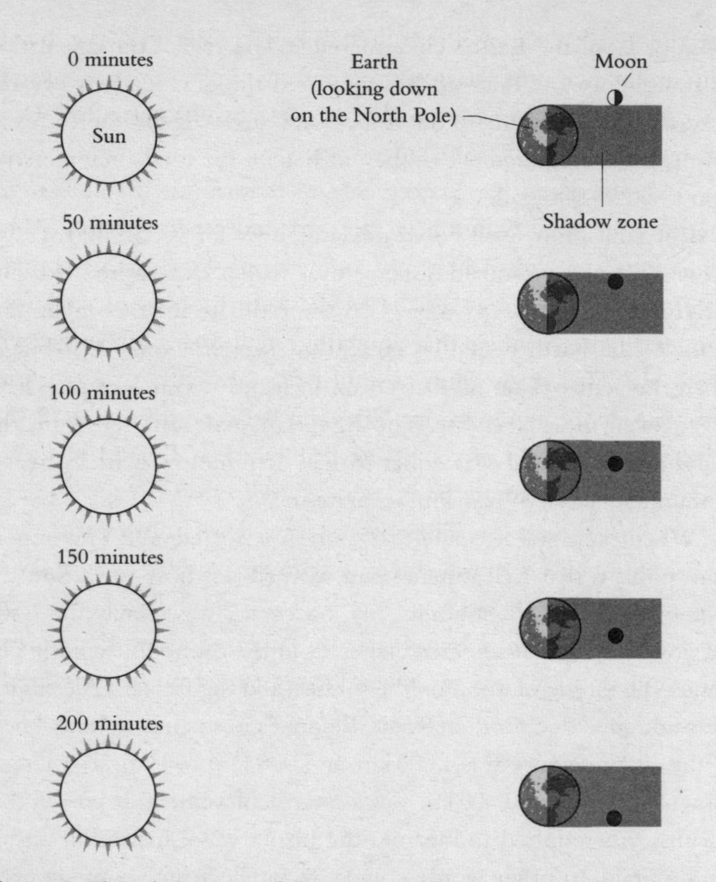

Figure 2 The relative sizes of the Earth and the Moon can be estimated by observing the Moon's passage through the Earth's shadow during a lunar eclipse. The Earth and Moon are very far from the Sun compared with the distance from the Earth to the Moon, so the size of the Earth's shadow is much the same as the size of the Earth itself.

The diagram shows the Moon passing through the Earth's shadow. In this particular eclipse – when the Moon passes roughly through the centre of the Earth's shadow – it takes 50 minutes for the Moon to go from touching the shadow to being fully covered, so 50 minutes is an indication of the Moon's own diameter. The time required for the front of the Moon to cross the entire Earth's shadow is 200 minutes, which is an indication of the Earth's diameter. The Earth's diameter is therefore roughly four times the Moon's diameter.

of the Earth's. Once Eratosthenes had shown that the Earth's circumference was 40,000 km, then its diameter was roughly (40,000 ÷ π) km, which is roughly 12,700 km. Therefore the Moon's diameter was (¼ × 12,700) km, or nearly 3,200 km.

It was then easy for Eratosthenes to estimate the distance to the Moon. One way would have been to stare up at the full Moon, close one eye and stretch out your arm. If you try this you will notice that you can cover the Moon with the end of your forefinger. Figure 3 shows that your fingernail forms a triangle with your eye. The Moon forms a similar triangle, with a vastly greater size but identical proportions. The ratio between the length of your arm and the height of your fingernail, which is about 100:1, must be the same as the ratio between the distance to the Moon and the Moon's own diameter. This means that the distance to the Moon must be roughly 100 times greater than its diameter, which gives a distance of 320,000 km.

Next, thanks to a hypothesis by Anaxagoras of Clazomenae and a clever argument by Aristarchus of Samos, it was possible for

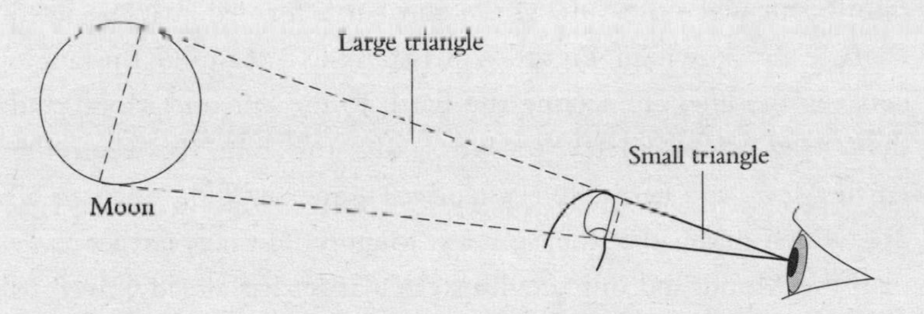

Figure 3 Having estimated the size of the Moon, it is relatively easy to work out the distance to the Moon. First, you will notice that you can just block out the Moon with a fingertip at arm's length. Therefore, it becomes clear that the ratio of a fingernail's height to an arm's length is roughly the same as the ratio of the Moon's diameter to its distance from the Earth. An arm's length is roughly a hundred times longer than a fingernail, so the distance to the Moon is roughly a hundred times its diameter.

Eratosthenes to calculate the size of the Sun and how far away it was. Anaxagoras was a radical thinker in the fifth century BC who deemed the purpose of life to be 'the investigation of the Sun, the Moon and the heavens'. He believed that the Sun was a white-hot stone and not a divinity, and similarly he believed that the stars were also hot stones, but too far away to warm the Earth. In contrast, the Moon was supposed to be a cold stone that did not emit light, and Anaxagoras argued that moonshine was nothing more than reflected sunlight. Despite the increasingly tolerant intellectual climate in Athens, where Anaxagoras lived, it was still controversial to claim that the Sun and Moon were rocks and not gods, so much so that jealous rivals accused Anaxagoras of heresy and organised a campaign that resulted in his exile to Lampsacus, in Asia Minor. The Athenians had a penchant for adorning their city with idols, which is why in 1638 Bishop John Wilkins pointed out the irony of a man who turned gods into stones being persecuted by people who turned stones into gods.

In the third century BC, Aristarchus built on Anaxagoras' idea. If moonshine was reflected sunshine, he argued, then the half Moon must occur when the Sun, Moon and Earth formed a right-angled triangle, as shown in Figure 4. Aristarchus measured the angle between the lines connecting the Earth to the Sun and Moon, and then used trigonometry to work out the ratio between the Earth–Moon and Earth–Sun distances. He measured the angle to be 87°, which meant that the Sun was roughly 20 times farther away than the Moon, and our previous calculation has already given us the distance to the Moon. In fact, the correct angle is 89.85°, and the Sun is 400 times further away than the Moon, so Aristarchus had clearly struggled to measure this angle accurately. Once again, accuracy is not the point: the Greeks had come up with a valid method, which was the key breakthrough, and better measuring tools would take future scientists closer to the true answer.

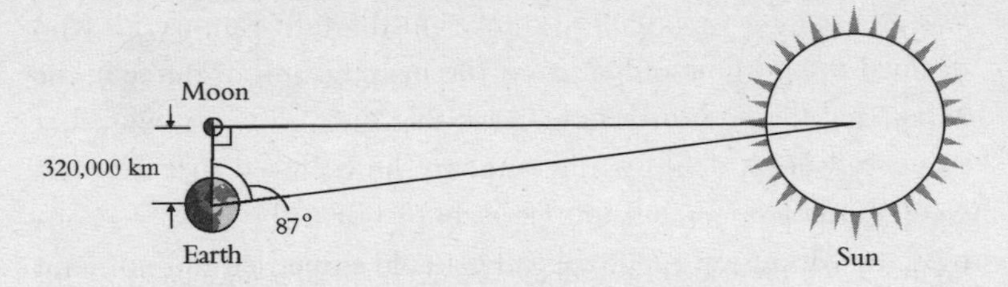

Figure 4 Aristarchus argued that it was possible to estimate the distance to the Sun using the fact that the Earth, Moon and Sun form a right-angled triangle when the Moon is at its half phase. At half Moon he measured the angle shown in the diagram. Simple trigonometry and the known Earth–Moon distance can then be used to determine the Earth–Sun distance.

Finally, deducing the size of the Sun is obvious, because it is a well-established fact that the Moon fits almost perfectly over the Sun during a solar eclipse. Therefore, the ratio of the Sun's diameter to the Sun's distance from the Earth must be the same as the ratio of the Moon's diameter to the Moon's distance from the Earth, as shown in Figure 5. We already know the Moon's diameter and its distance from the Earth, and we also know the Sun's distance from the Earth, so the Sun's diameter is easy to calculate. This method is identical to the one illustrated in Figure 3, whereby the distance to and height of our fingernail was used to measure the distance to the Moon, except that now the Moon has taken the place of our fingernail as an object of known size and distance.

The amazing achievements of Eratosthenes, Aristarchus and Anaxagoras illustrate the advances in scientific thinking that were taking place in ancient Greece, because their measurements of the universe relied on logic, mathematics, observation and measurement. But do the Greeks really deserve all the credit for laying the foundations of science? After all, what about the Babylonians, who were great practical astronomers, making thousands of detailed

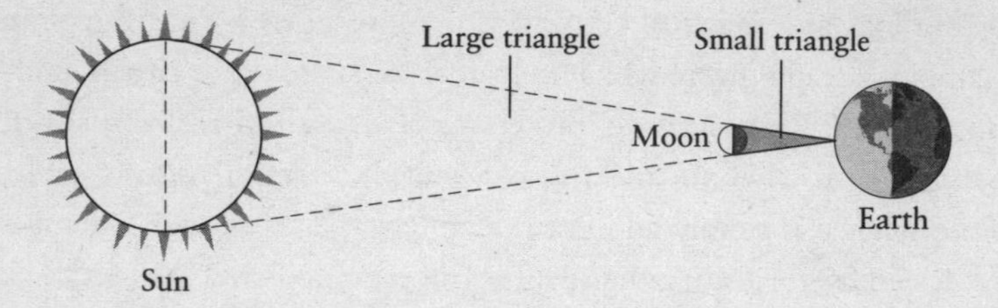

Figure 5 It is possible to estimate the size of the Sun, once we know its distance. One approach is to use a total solar eclipse and our knowledge of the Moon's distance and diameter. A total solar eclipse is visible only from a small patch on the Earth's surface at any given time, because the Sun and the Moon appear almost the same size when viewed from the Earth. This diagram (not to scale) shows how an eclipse observer on the Earth is at the apex of two similar triangles. The first triangle stretches to the Moon, and the second triangle to the Sun. Knowing the distances to the Moon and to the Sun and knowing the diameter of the Moon is enough to deduce the diameter of the Sun.

---

observations? It is generally agreed by philosophers and historians of science that the Babylonians were not true scientists, because they were still content with a universe guided by gods and explained with myths. In any case, collecting hundreds of measurements and listing endless stellar and planetary positions was trivial compared with genuine science, which has the glorious ambition of trying to explain such observations by understanding the underlying nature of the universe. As the French mathematician and philosopher of science Henri Poincaré rightly declared: 'Science is built up with facts, as a house is with stones. But a collection of facts is no more a science than a heap of stones is a house.'

If the Babylonians were not the first proto-scientists, then what about the Egyptians? The Great Pyramid of Cheops predates the Parthenon by two thousand years, and the Egyptians were certainly far in advance of the Greeks in terms of their development of weighing scales, cosmetics, inks, wooden locks, candles and many other inventions. These, however, are examples of technology, not science.

Technology is a practical activity, as demonstrated by the Egyptian examples already given, which helped to facilitate death rituals, trading, beautification, writing, protection and illumination. In short, technology is all about making life (and death) more comfortable, while science is simply an effort to understand the world. Scientists are driven by curiosity, rather than comfort or utility.

Although scientists and technologists have very different goals, science and technology are frequently confused as being one and the same, probably because scientific discoveries often lead to technological breakthroughs. For example, scientists spent decades making discoveries about electricity, which technologists then used to invent light bulbs and many other devices. In ancient times, however, technology grew without the benefit of science, so the Egyptians could be successful technologists without having any grasp of science. When they brewed beer, they were interested in the technological methods and the results, but not why or how one material was being transformed into another. They had no inkling of the underlying chemical or biochemical mechanisms at work.

So, the Egyptians were technologists, not scientists, whereas Eratosthenes and his colleagues were scientists, not technologists. The intentions of the Greek scientists were identical to those described two thousand years later by Henri Poincaré:

> The scientist does not study nature because it is useful; he studies it because he delights in it, and he delights in it because it is beautiful. If nature were not beautiful, it would not be worth knowing, and if nature were not worth knowing, life would not be worth living. Of course I do not here speak of that beauty that strikes the senses, the beauty of qualities and appearances; not that I undervalue such beauty, far from it, but it has nothing to do with science; I mean that profounder beauty which comes from the harmonious order of the parts, and which a pure intelligence can grasp.

In summary, the Greeks had shown how knowing the diameter of the Sun depends on knowing the distance to the Sun, which depends on knowing the distance to the Moon, which depends on knowing the diameter of the Moon, which depends on knowing the diameter of the Earth, and that was Eratosthenes' great breakthrough. These distance and diameter stepping stones were made possible by exploiting a deep vertical well on the Tropic of Cancer, the Earth's shadow cast upon the Moon, the fact that the Sun, Earth and Moon form a right angle at half Moon, and the observation that the Moon fits perfectly over the Sun during a solar eclipse. Throw in some assumptions, such as moonlight being nothing more than reflected sunlight, and a framework of scientific logic takes shape. This architecture of scientific logic has an inherent beauty which emerges from how various arguments fit together, how several measurements interlock with one another, and how different theories are suddenly introduced to add strength to the edifice.

Having completed their initial phase of measurement, the astronomers of ancient Greece were now ready to examine the motions of the Sun, Moon and planets. They were about to create a dynamic model of the universe in an attempt to discern the interplay between the various celestial bodies. It would be the next step on the road to a deeper understanding of the universe.

Circles within Circles

Our most distant ancestors studied the sky in detail, whether it was to predict changes in the weather, keep track of time or measure direction. Every day they watched the Sun cross the sky, and every night they watched the procession of stars that followed in its wake. The land on which they stood was firm and fixed, so it was only natural to assume that it was the heavenly bodies that moved relative to a static Earth, not vice versa. Consequently, the ancient

Table 1

The measurements made by Eratosthenes, Aristarchus and Anaxagoras were inaccurate, so the table below corrects previously quoted figures by providing modern values for the various distances and diameters.

Earth's circumference	40,100 km	$= 4.01 \times 10^4$ km
Earth's diameter	12,750 km	$= 1.275 \times 10^4$ km
Moon's diameter	3,480 km	$= 3.48 \times 10^3$ km
Sun's diameter	1,390,000 km	$= 1.39 \times 10^6$ km
Earth–Moon distance	384,000 km	$= 3.84 \times 10^5$ km
Earth–Sun distance	150,000,000 km	$= 1.50 \times 10^8$ km

This table also serves as an introduction to *exponential notation*, a way of expressing very large numbers – and in cosmology there are some very, very large numbers:

10^1 means 10	= 10
10^2 means 10×10	= 100
10^3 means $10 \times 10 \times 10$	= 1,000
10^4 means $10 \times 10 \times 10 \times 10$	= 10,000
etc.	

The Earth's circumference, for example, can be expressed as:

$40{,}100 \text{ km} = 4.01 \times 10{,}000 \text{ km} = 4.01 \times 10^4 \text{ km}.$

Exponential notation is an excellent way of concisely expressing numbers that would otherwise be full of zeros. Another way to think of 10^N is as 1 followed by N zeros, so that 10^3 is 1 followed by three zeros, which is 1,000.

Exponential notation is also used for writing very small numbers:

10^{-1} means $1 \div 10$	= 0.1
10^{-2} means $1 \div (10 \times 10)$	= 0.01
10^{-3} means $1 \div (10 \times 10 \times 10)$	= 0.001
10^{-4} means $1 \div (10 \times 10 \times 10 \times 10)$	= 0.0001
etc.	

astronomers developed a view of the world in which the Earth was a central static globe with the universe revolving around it.

In reality, it is of course the Earth that moves around the Sun, and not the Sun moving around the Earth, but nobody considered this possibility until Philolaus of Croton entered the debate. A pupil of the Pythagorean school in the fifth century BC, he was the first to suggest that the Earth orbited the Sun, not vice versa. In the following century, Heracleides of Pontus built on Philolaus' ideas, even though his friends thought he was crazy, nicknaming him *Paradoxolog*, 'the maker of paradoxes'. And the final touches to this vision of the universe were added by Aristarchus, who was born in 310 BC, the same year that Heracleides died.

Although Aristarchus contributed to measuring the distance to the Sun, this was a minor accomplishment compared with his stunningly accurate overview of the universe. He was trying to dislodge the instinctive (though incorrect) picture of the universe, in which the Earth is at the centre of everything, as shown in Figure 6(a). In contrast, Aristarchus' less obvious (though correct) picture has the Earth dashing around a more dominant Sun, as shown in Figure 6(b). Aristarchus was also right when he stated that the Earth spins on its own axis every 24 hours, which explained why each day we face towards the Sun and each night we face away from it.

Aristarchus was a highly respected philosopher, and his ideas on astronomy were well known. Indeed, his belief in a Sun-centred universe was documented by Archimedes, who wrote: 'He hypothesises that the fixed stars and the Sun remain unmoved; that the Earth is borne around the Sun on the circumference of a circle.' Yet philosophers completely abandoned this largely accurate vision of the Solar System, and the idea of a Sun-centred world disappeared for the next fifteen hundred years. The ancient Greeks were supposed to be smart, so why did they reject Aristarchus' insightful world-view and stick to an Earth-centred universe?

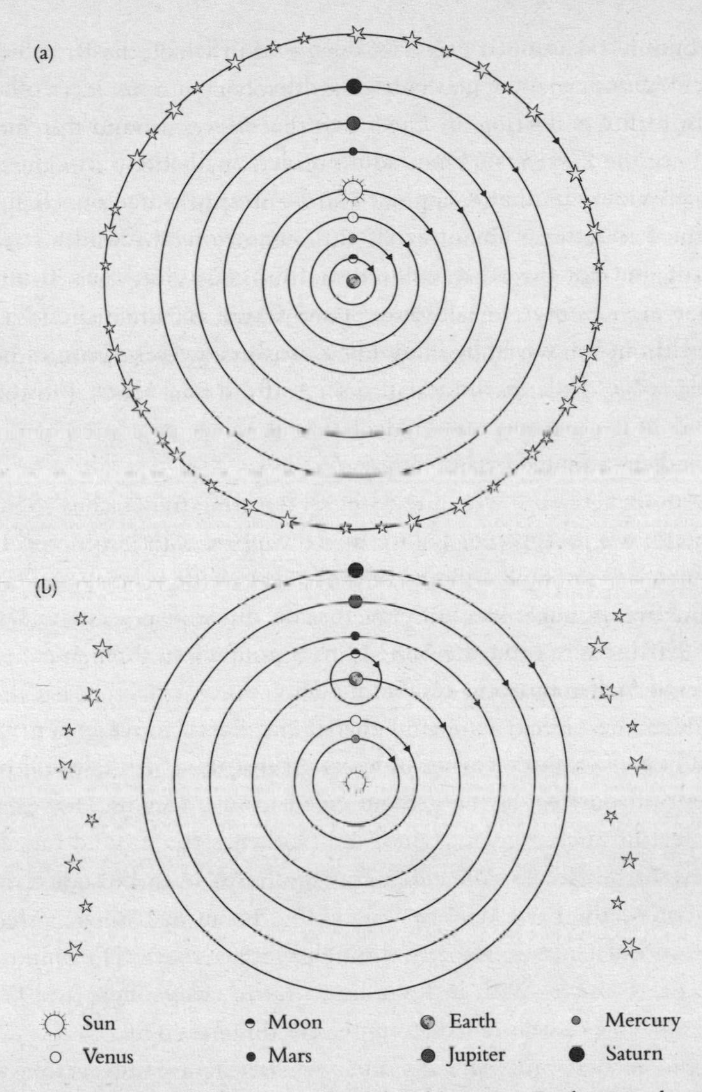

Figure 6 Diagram (a) shows the classical and incorrect Earth-centred model of the universe, in which the Moon, Sun and other planets orbit the Earth. Even the thousands of stars orbit the Earth. Diagram (b) shows Aristarchus' Sun-centred view of the universe, with only the Moon orbiting the Earth. In this case, the stars form a static backdrop to the universe.

Egocentric attitudes may have been a contributory factor behind the dominance of the geocentric world-view, but there were other reasons for preferring an Earth-centred universe to Aristarchus' Sun-centred universe. One basic problem with the Sun-centred world-view was that it appeared to be simply ridiculous. It just seemed so utterly obvious that the Sun revolved round a static Earth, and not the other way round. In short, a Sun-centred universe ran counter to common sense. Good scientists, however, should not be swayed by common sense, because it sometimes has little to do with the underlying scientific truth. Albert Einstein condemned common sense, declaring it to be 'the collection of prejudices acquired by age eighteen'.

Another reason why the Greeks rejected Aristarchus' Solar System was its apparent failure to stand up to scientific scrutiny. Aristarchus had built a model of the universe that was supposed to match reality, but it was not clear that his model was accurate. Did the Earth really orbit the Sun? Critics pointed to three apparent flaws in Aristarchus' Sun-centred model.

First, the Greeks expected that if the Earth moved then we would feel a constant wind blowing against us, and we would be swept off our feet as the ground raced from under us. However, we feel no such constant wind, and neither is the ground tugged away, so the Greeks concluded that the Earth must be stationary. Of course, the Earth does move, and the reason that we are oblivious to our fantastic velocity through space is that everything on the Earth moves with it, including us, the atmosphere and the ground. The Greeks failed to appreciate this argument.

The second problematic point was that a moving Earth was incompatible with the Greek understanding of gravity. As mentioned earlier, the traditional view was that everything tended to move towards the centre of the universe, and the Earth was already at the centre, so it did not move. This theory made perfect sense, because it explained that apples fell from trees and headed towards the centre of the Earth because they were being attracted to the centre of the universe. But if the Sun were at the centre of the universe, then why would objects fall towards the Earth? Instead, apples should not fall down from trees, but should be sucked up towards the Sun – indeed, everything on Earth should fall towards the Sun. Today we have a clearer understanding of gravity, which makes a Sun-centred Solar System much more sensible. The modern theory of gravity describes how objects close to the massive Earth are attracted to the Earth, and in turn the planets are held in orbit by the attraction of the even more massive Sun. Once again, however, this explanation was beyond the limited scientific framework of the Greeks.

The third reason why philosophers rejected Aristarchus' Sun-centred universe was the apparent lack of any shift in the positions of the stars. If the Earth were travelling huge distances around the Sun, then we would see the universe from different positions during the course of the year. Our changing vantage point should mean a changing perspective on the universe, and the stars should move relative to one another, which is known as *stellar parallax*. You can see parallax in action at a local level by simply holding one finger in the air just a few centimetres in front of your face. Close your left eye and use your right eye to line your finger up with a nearby object, perhaps the edge of a window. Next, close your right eye and open your left one, and you will see that your finger has shifted to the right relative to the edge of the window. Switch between your eyes quickly and your finger will jump to and fro. So shifting your vantage point from one eye to the other, a distance of just a few centimetres, moves the apparent position of your finger relative to another object. This is illustrated in Figure 7(a).

The distance from the Earth to the Sun is 150 million km, so if the Earth orbited the Sun then it would be 300 million km away

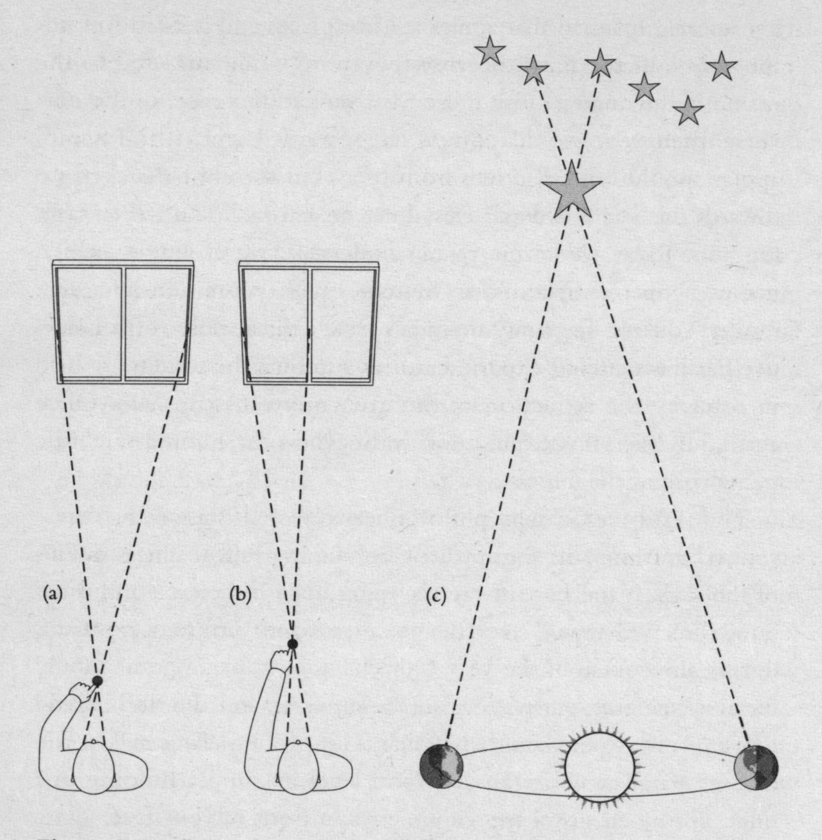

Figure 7 Parallax is the apparent shift in the position of an object due to a change in an observer's vantage point. Diagram (a) shows how a marker finger lines up with the left window edge when viewed with the right eye, but shifts when viewed with the other eye. Diagram (b) shows that the parallax shift caused by switching between eyes is significantly reduced if the marker finger is more distant. Because the Earth orbits the Sun, our vantage point changes, so if one star is used as a marker then it should shift relative to more distant stars over the course of a year. Diagram (c) shows how the marker star lines up with two different background stars depending on the position of the Earth. However, if diagram (c) were drawn to scale, then the stars would be over 1 km off the top of the page! Therefore the parallax shift would be minuscule and imperceptible to the ancient Greeks. The Greeks assumed that the stars were much closer, so to them a lack of parallax shift implied a static Earth.

from its original position after six months. The Greeks found it impossible to detect any shift in the positions of the stars relative to one another over the course of the year, despite the enormous shift in Earthly perspective that would happen if we orbited the Sun. Once more, the evidence seemed to point to the conclusion that the Earth did not move and was at the centre of the universe. Of course, the Earth does orbit the Sun, and stellar parallax does exist, but it was imperceptible to the Greeks because the stars are so very far away. You can see how distance reduces the parallax effect by repeating the winking experiment, this time fully extending your arm so that your finger is almost a metre away. Again, use your right eye to line up your finger with the edge of the window. This time, when you switch to your left eye the parallax shift should be much less significant than before because your finger is farther away, as illustrated in Figure 7(b). In summary, the Earth does move, but the parallax shift rapidly reduces with distance and the stars are very far away, so stellar parallax could not be detected with primitive equipment.

At the time, the evidence against Aristarchus' Sun-centred model of the universe seemed overwhelming, so it is quite understandable why all his philosopher friends stayed loyal to the Earth-centred model. Their traditional model was perfectly sensible, rational and self-consistent. They were content with their vision of the universe and their place within it. However, there was one outstanding problem. Sure enough, the Sun, Moon and stars all seemed to march obediently around the Earth, but there were five heavenly bodies that dawdled across the heavens in a rather haphazard manner. Occasionally, some of them even dared to stop momentarily before temporarily reversing their motion in a volte-face known as *retrograde motion*. These wandering rebels were the five other known planets: Mercury, Venus, Mars, Jupiter and Saturn. Indeed, the word 'planet' derives from the Greek *planetes*, meaning 'wanderer'.

Similarly, the Babylonian word for planet was *bibbu*, literally 'wild sheep' – because the planets seemed to stray all over the place. And the ancient Egyptians called Mars *sekded-ef em khetkhet*, meaning 'one who travels backwards'.

From our modern Earth-orbits-Sun perspective, it is easy enough to understand the behaviour of these heavenly vagabonds. In reality, the planets orbit the Sun in a steady manner, but we view them from a moving platform, the Earth, which is why their motion appears to be irregular. In particular, the retrograde motions exhibited by Mars, Saturn and Jupiter are easy to explain. Figure 8(a) shows a stripped-down Solar System containing just the Sun, Earth and Mars. Earth orbits the Sun more quickly than Mars, and as we catch up to Mars and pass it, our line of sight to Mars shifts back and forth. However, from the old Earth-centred perspective, in which we sit at the centre of the universe and everything revolves around us, the orbit of Mars was a riddle. It appeared that Mars, as shown in Figure 8(b), looped the loop in a most peculiar manner as it orbited the Earth. Saturn and Jupiter displayed similar retrograde motions, which the Greeks also put down to looping orbits.

These loopy planetary orbits were hugely problematic for the ancient Greeks, because all the orbits were supposed to be circular according to Plato and his pupil Aristotle. They declared that the circle, with its simplicity, beauty and lack of beginning or end, was the perfect shape, and since the heavens were the realm of perfection then celestial bodies had to travel in circles. Several astronomers and mathematicians looked into the problem and, over the course of several centuries, they developed a cunning solution – a way to describe looping planetary orbits in terms of combinations of circles, which was in keeping with Plato and Aristotle's edict of circular perfection. The solution became associated with the name of one astronomer, Ptolemy, who lived in Alexandria in the second century AD.

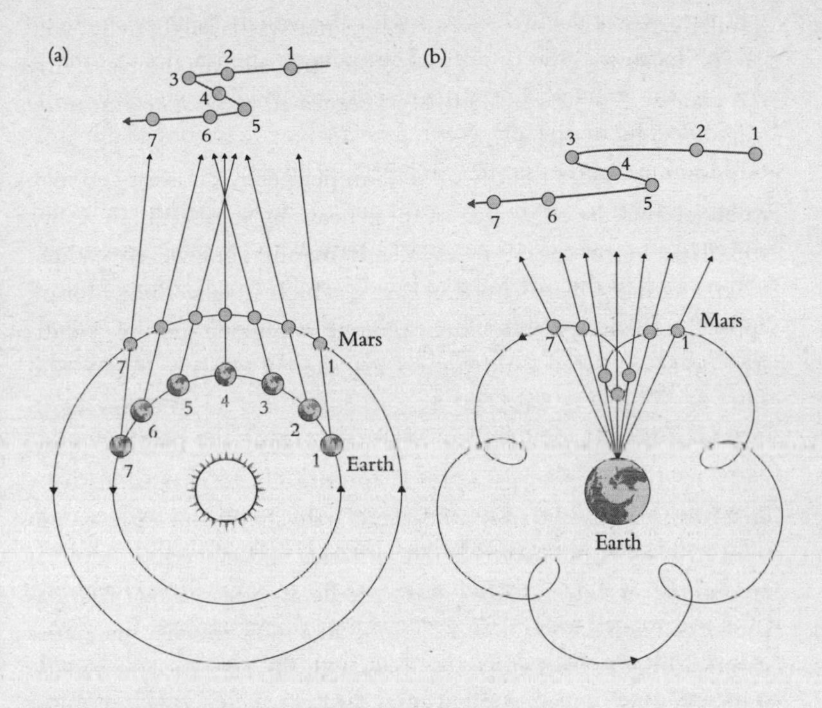

Figure 8 Planets such as Mars, Jupiter and Saturn exhibit so-called retrograde motion when viewed from Earth. Diagram (a) shows a stripped-down Solar System with just the Earth and Mars orbiting (anticlockwise) around the Sun. From position 1, we would see Mars move increasingly ahead of us, which continues as we observe Mars from position 2. But Mars pauses at position 3, and by position 4 is now moving to the right, and even further to the right when Earth arrives at position 5. There it pauses once more, before resuming its original direction of travel, as seen from positions 6 and 7. Of course, Mars is continually moving anticlockwise around the Sun, but it appears to us that Mars is zigzagging because of the relative motions of the Earth and Mars. Retrograde motion makes perfect sense in a Sun-centred model of the universe.

Diagram (b) shows how believers in an Earth-centred model perceived the orbit of Mars. The zigzag of Mars was interpreted as an actual looping orbit. In other words, traditionalists believed that the static Earth sat at the centre of the universe, while Mars looped its way around the Earth.

Ptolemy's world-view started with the widely held assumption that the Earth is at the centre of the universe and stationary, otherwise 'all the animals and all the separate weights would be left behind floating on the air'. Next, he explained the orbits of the Sun and Moon in terms of simple circles. Then, in order to explain retrograde motions, he developed a theory of circles within circles, as illustrated in Figure 9. To generate a path with periodic retrograde motion, such as the one followed by Mars, Ptolemy proposed starting with a single circle (known as the *deferent*), with a rod attached to the circle so that it pivoted. The planet then occupied a position at the end of this pivoted rod. If the main deferent circle remained fixed and the rod rotated around its pivot, then the planet would follow a circular path with a short radius (known as the *epicycle*), as shown in Figure 9(a). Alternatively, if the main deferent circle rotated and the rod remained fixed, then the planet would follow a circular path with a larger radius, as shown in Figure 9(b). However, if the rod rotated around its pivot and at the same time the pivot rotated with the main deferent circle, then the planet's path would be a composite of its motion around the two circles, which mimics a retrograde loop, as shown in Figure 9(c).

Although this description of circles and pivots conveys the central idea of Ptolemy's model, it was actually far more complicated. To start with, Ptolemy thought of his model in three dimensions and constructed it from crystal spheres, but for simplicity we will continue to think in terms of two-dimensional circles. Also, in order to accurately explain the retrogrades of different planets, Ptolemy had to carefully tune the radius of the deferent and the radius of the epicycle for each planet, and select the speed at which each rotated. For even greater accuracy he introduced two other variable elements. The *eccentric* defined a point to the side of the Earth which acted as a slightly displaced centre for the deferent circle, while the *equant* defined another point close to the Earth,

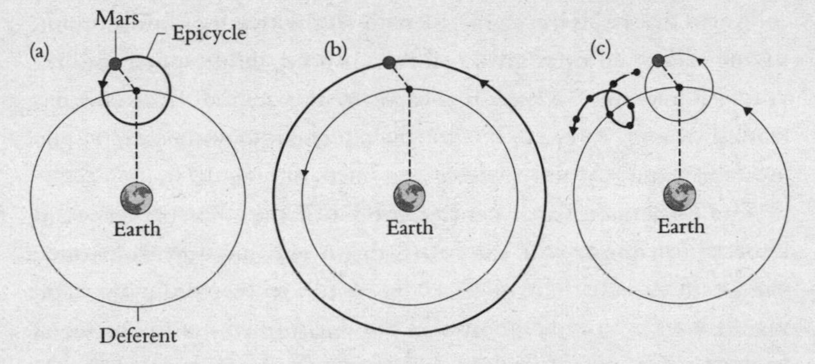

Figure 9 The Ptolemaic model of the universe explained the loopy orbits of planets such as Mars using combinations of circles. Diagram (a) shows the main circle, called the deferent, and a pivoted rod with a planet on the end. If the deferent does not rotate, but the rod does rotate, then the planet follows the smaller, bold circle mapped out by the end of the rod, which is called an epicycle.

Diagram (b) shows what happens if the pivoted rod remains fixed and the deferent is allowed to rotate. The planet follows a circle with a large radius.

Diagram (c) shows what happens when both the rod rotates around its pivot, and the pivot rotates with the deferent. This time the epicycle is superimposed on the deferent, and the planet's orbit is the combination of two circular paths, which results in the loopy retrograde orbit associated with a planet such as Mars. The radii of the deferent and epicycle can be adjusted and both speeds of rotation can be tuned to mimic the path of any planet.

---

whose influence contributed to the variable speed of the planet. It is hard to imagine this increasingly complicated explanation for planetary orbits, but essentially it consisted of nothing more than circles on top of more circles within yet more circles.

The best analogy for Ptolemy's model of the universe is to be found in a fairground. The Moon follows a simple path, a bit like a horse on a rather tame merry-go-round for young children. But the path of Mars is more like a wild waltzer ride, which locks the rider in a cradle that pivots at the end of a long rotating arm. The rider follows a circular path while spinning in the cradle, but at the same time he is following another, much larger, circular path at the end of the long arm that holds the cradle. Sometimes the two motions combine, giving rise to an even greater forward speed, while sometimes the cradle is moving backwards relative to the arm and the speed is slowed or even reversed. In Ptolemaic terminology, the cradle spins around an epicycle and the long arm traces out the deferent.

The Ptolemaic Earth-centred model of the universe was constructed to comply with the beliefs that everything revolves around the Earth and that all celestial objects follow circular paths. This resulted in a horribly complex model, replete with epicycles heaped upon deferents, upon equants, upon eccentrics. In *The Sleepwalkers*, Arthur Koestler's history of early astronomy, the Ptolemaic model is described as 'the product of tired philosophy and decadent science'. But despite being fundamentally wrong, the Ptolemaic system satisfied one of the basic requirements of a scientific model, which is that it predicted the position and movement of every planet to a higher degree of accuracy than any previous model. Even Aristarchus' Sun-centred model of the universe, which happens to be basically correct, could not predict the motion of the planets with such precision. So, all in all, it is not surprising that Ptolemy's model endured while Aristarchus' disappeared. Table 2 summarises the key strengths and weaknesses of the two models, as understood by the ancient Greeks, and it serves only to reinforce the apparent superiority of the Earth-centred model.

Ptolemy's Earth-centred model was enshrined in his *Hè megalè syntaxis* ('The Great Collection'), written in about AD 150, which became the most authoritative text on astronomy for centuries to come. In fact, every astronomer in Europe for the next millennium was influenced by the *Syntaxis*, and none of them seriously questioned its Earth-centred picture of the universe. *Syntaxis* reached an even wider audience in AD 827, when it was translated into Arabic and retitled the *Almagest* ('The Greatest'). So, during the lull in scholasticism during the European Middle Ages, Ptolemy's ideas were kept alive and studied by the great Islamic scholars in the Middle East. During the golden age of the Islamic empire, Arab astronomers invented many new astronomical instruments, made significant celestial observations and built several major observatories, such as the al-Shammasiyyah observatory in Baghdad, but they never doubted Ptolemy's Earth-centred universe with its planetary orbits defined by circles within circles within circles.

As Europe finally began to emerge from its intellectual slumber, the ancient knowledge of the Greeks was exported back to the West via the Moorish city of Toledo in Spain, where there was a magnificent Islamic library. When the city was captured from the Moors by the Spanish King Alfonso VI in 1085, scholars all over Europe were given an unprecedented opportunity to gain access to one of the world's most important repositories of knowledge. Most of the library's contents were written in Arabic, so the first priority was to establish an industrial-scale bureau of translation. Most translators worked with the aid of an intermediary to translate from Arabic into the Spanish vernacular, which they then translated into Latin, but one of the most prolific and brilliant translators was Gerard of Cremona, who learned Arabic so that he could achieve a more direct and accurate interpretation. He had been drawn to Toledo by rumours that Ptolemy's masterpiece was to be found at the library and, of the seventy-six seminal books that he translated from Arabic into Latin, the *Almagest* was his most significant achievement.

Thanks to the efforts of Gerard and other translators, European scholars were able to reacquaint themselves with the writings of the past, and astronomical research in Europe was reinvigorated. Paradoxically, progress became stifled, because there was such reverence for the writings of the ancient Greeks that nobody dared to question their work. It was assumed that the classical scholars had mastered everything that could ever be understood, so books such

Table 2

This table lists various criteria against which the Earth-centred and Sun-centred models could be judged, based on what was known in the first millennium AD. The ticks and crosses give crude indications of how well each theory fared in relation to the seven criteria, and a question mark

Criterion	Earth-centred model	Success
1. Common sense	It seems obvious that everything revolves around the Earth	✓
2. Awareness of motion	We do not detect any motion, therefore the Earth cannot be moving	✓
3. Falling to the ground	The centrality of the Earth explains why objects appear to fall downwards, i.e. objects are being attracted to the centre of the universe	✓
4. Stellar parallax	There is no detection of stellar parallax, absence of which is compatible with a static Earth and a stationary observer	✓
5. Predicting planetary orbits	Very close agreement – the best yet	✓
6. Retrograde paths of planets	Explained with epicycles and deferents	✓
7. Simplicity	Very complicated – epicycles, deferents, equants and eccentrics	✗

indicates a lack of data or a mixture of agreement and disagreement. From an ancient point of view, the Sun-centred model does better than its rival in only one area (simplicity), even though we now know it to be closer to reality.

Criterion	Sun-centred model	Success
1. Common sense	It requires a leap of imagination and logic to see that the Earth might circle the Sun	✗
2. Awareness of motion	We do not detect any motion, which is not easy to explain if the Earth is moving	✗
3. Falling to the ground	There is no obvious explanation for why objects fall to the ground in a model where the Earth is not centrally located	✗
4. Stellar parallax	The Earth moves, so the apparent lack of stellar parallax must be due to huge stellar distances; hopefully parallax would be detected with better equipment	?
5. Predicting planetary orbits	Good agreement, but not as good as in the Earth-centred model	?
6. Retrograde paths of planets	A natural consequence of the motion of the Earth and our changing vantage point	✓
7. Simplicity	Very simple – everything follows circles	✓

as the *Almagest* were taken as gospel. This was despite the fact that the ancients had made some of the biggest blunders imaginable. For example, the writings of Aristotle were considered sacred, even though he had stated that men have more teeth than women, a generalisation based on the observation that stallions have more teeth than mares. Although he was married twice, Aristotle apparently never bothered to look into the mouth of either of his wives. He might have been a superlative logician, but he failed to grasp the concepts of observation and experimentation. The irony is that scholars had waited for centuries to recover the wisdom of the ancients – and then they had to spend centuries unlearning all the ancients' mistakes. Indeed, after Gerard's translation of the *Almagest* in 1175, Ptolemy's Earth-centred model of the universe continued to survive intact for another four hundred years.

In the meantime, however, a few minor criticisms did emerge from such figures as Alfonso X, King of Castile and León (1221–84). Having made Toledo his capital, he instructed his astronomers to draw up what became known as the *Alphonsine Tables* of planetary motion, based partly on their own observations and partly on translated Arabic tables. Although he was a strong patron of astronomy, Alfonso remained resolutely unimpressed with Ptolemy's intricate system of deferents, epicycles, equants and eccentrics: 'If the Lord Almighty had consulted me before embarking upon Creation, I should have recommended something simpler.'

Then, in the fourteenth century, Nicole d'Oresme, chaplain to Charles V of France, openly stated that the case for an Earth-centred universe had not been fully proved, although he did not go as far as saying that he believed it to be wrong. And in fifteenth-century Germany, Cardinal Nicholas of Cusa suggested that the Earth is not the hub of the universe, but he stopped short of suggesting that the Sun should occupy the vacated throne.

The world would have to wait until the sixteenth century before an astronomer would have the courage to rearrange the universe and seriously challenge the cosmology of the Greeks. The man who would eventually reinvent Aristarchus' Sun-centred universe was christened Mikolaj Kopernik, but he is better known by his Latinised name of Nicholas Copernicus.

The Revolution

Born in 1473 into a prosperous family in Torun, on the banks of the Vistula in modern-day Poland, Copernicus was elected a canon at the cathedral chapter of Frauenburg, largely thanks to the influence of his uncle Lucas, who was Bishop of Ermland. Having studied law and medicine in Italy, his main duty as canon was to act as physician and secretary to Lucas. These were not onerous responsibilities, and Copernicus was free to dabble in various activities in his spare time. He became an expert economist and advisor on currency reform, and even published his own Latin translations of the obscure Greek poet Theophylactus Simocattes.

However, Copernicus's greatest passion was astronomy, which had interested him ever since he had bought a copy of the *Alphonsine Tables* as a student. This amateur astronomer would grow increasingly obsessed with studying the motion of the planets, and his ideas would eventually make him one of the most important figures in the history of science.

Surprisingly, all Copernicus's astronomical research was contained in just 1½ publications. Even more surprising, these 1½ publications were hardly read during his lifetime. The ½ refers to his first work, the *Commentariolus* ('Little Commentary'), which was handwritten, never formally published and circulated only among a few people in roughly 1514. Nevertheless, in just twenty pages Copernicus shook the cosmos with the most radical idea in astronomy for over one thousand years. At the heart of his pamphlet were the seven axioms upon which he based his view of the universe:

1. The heavenly bodies do not share a common centre.
2. The centre of the Earth is not the centre of the universe.
3. The centre of the universe is near the Sun.
4. The distance from the Earth to the Sun is insignificant compared with the distance to the stars.
5. The apparent daily motion of the stars is a result of the Earth's rotation on its own axis.
6. The apparent annual sequence of movements of the Sun is a result of the Earth's revolution around it. All the planets revolve around the Sun.
7. The apparent retrograde motion of some of the planets is merely the result of our position as observer on a moving Earth.

Copernicus's axioms were spot on in every respect. The Earth does spin, the Earth and the other planets do go around the Sun, this does explain the retrograde planetary orbits, and failure to detect any stellar parallax was due to the remoteness of the stars. It is not clear what motivated Copernicus to formulate these axioms and break with the traditional world-view, but perhaps he was influenced by Domenico Maria de Novara, one of his professors in Italy. Novara was sympathetic to the Pythagorean tradition, which was at the root of Aristarchus' philosophy, and it was Aristarchus who had first posited the Sun-centred model 1,700 years earlier.

The *Commentariolus* was a manifesto for an astronomical mutiny, an expression of Copernicus's frustration and disillusionment with the ugly complexity of the ancient Ptolemaic model. Later he would condemn the makeshift nature of the Earth-centred model: 'It is as though an artist were to gather the hands, feet, head and other members for his images from diverse models, each part excellently drawn, but not related to a single body, and since they in no way match each other, the result would be a monster rather than a man.' Nevertheless, despite its radical contents, the pamphlet caused no ripples among the intellectuals of Europe, partly because it was read by so few people and partly because its author was a minor canon working on the fringes of Europe.

Copernicus was not dismayed, for this was only the start of his efforts to transform astronomy. After his uncle Lucas died in 1512 (having quite possibly been poisoned by the Teutonic Knights, who had described him as 'the devil in human shape'), he had even more time to pursue his studies. He moved to Frauenburg Castle, set up a small observatory and concentrated on fleshing out his argument, adding in all the mathematical detail that was missing in the *Commentariolus*.

Copernicus spent the next thirty years reworking his *Commentariolus*, expanding it into an authoritative two-hundred-page manuscript. Throughout this prolonged period of research, he spent a great deal of time worrying about how other astronomers would react to his model of the universe, which was fundamentally at odds with accepted wisdom. There were often days when he even considered abandoning plans to publish his work for fear that he would be mocked far and wide. Moreover, he suspected that theologians would be wholly intolerant to what they would perceive as sacrilegious scientific speculation.

He was right to be concerned. The Church later demonstrated its intolerance by persecuting the Italian philosopher Giordano Bruno, who was part of the generation of dissenters that followed Copernicus. The Inquisition accused Bruno of eight heresies, but the existing records do not specify them. Historians think that it is likely that Bruno had offended the Church by writing *On the Infinite Universe and Worlds*, which argued that the universe is infinite, that stars have their own planets and that life flourishes on these other planets. When condemned to death for his crimes, he responded: 'Perchance you who pronounce my sentence are in greater fear than I who receive it.' On 17 February 1600, he was taken to Rome's Campo dei Fiori (Field of Flowers), stripped naked, gagged, tied to a stake and burned to death.

Copernicus's fear of persecution could have meant a premature end to his research, but fortunately a young German scholar from Wittenberg intervened. In 1539, Georg Joachim von Lauchen, known as Rheticus, travelled to Frauenburg to seek out Copernicus and find out more about his cosmological model. It was a brave move, because not only was the young Lutheran scholar facing an uncertain welcome in Catholic Frauenburg, but also his own colleagues were not sympathetic to his mission. The mood was typified by Martin Luther, who kept a record of dinner-table conversation about Copernicus: 'There is talk of a new astronomer who wants to prove that the Earth moves and goes around instead of the sky, the Sun and the Moon, just as if somebody moving in a carriage or ship might hold that he was sitting still and at rest while the ground and the trees walked and moved . . . The fool wants to turn the whole art of astronomy upside-down.'

Luther called Copernicus 'a fool who went against Holy Writ', but Rheticus shared Copernicus's unshakeable confidence that the route to celestial truth lay with science rather than Scripture. The sixty-six-year-old Copernicus was flattered by the attentions of the twenty-five-year-old Rheticus, who spent three years at Frauenburg reading Copernicus's manuscript, providing him with feedback and reassurance in equal measure.

By 1541, Rheticus's combination of diplomatic and astronomical skills was sufficient for him to obtain Copernicus's blessing to take the manuscript to the printing house of Johannes Petreius in Nuremberg for publication. He had planned to stay to oversee the entire printing process, but was suddenly called away to Leipzig on urgent business, and so handed responsibility for supervising publication to a clergyman by the name of Andreas Osiander. At last, in the spring of 1543, *De revolutionibus orbium cœlestium* ('On the Revolutions of the Heavenly Spheres') was finally published and several hundred copies were on their way to Copernicus.

Meanwhile, Copernicus had suffered a cerebral haemorrhage at the end of 1542, and was lying in bed, fighting to stay alive long enough to set eyes on the finished book that contained his life's work. Copies of his treatise reached him just in time. His friend Canon Giese wrote a letter to Rheticus describing Copernicus's plight: 'For many days he had been deprived of his memory and mental vigour; he only saw his completed book at the last moment, on the day he died.'

Copernicus had completed his duty. His book offered the world a convincing argument in favour of Aristarchus' Sun-centred model. *De revolutionibus* was a formidable treatise, but before discussing its contents it is important to address two perplexing mysteries surrounding its publication. The first of these relates to Copernicus's incomplete acknowledgements. The introduction to *De revolutionibus* mentioned several people, such as Pope Paul III, the Cardinal of Capua and the Bishop of Kulm, yet there was no mention of Rheticus, the brilliant apprentice who had played the vital role of midwife to the birth of the Copernican model. Historians are baffled as to why his name was omitted and can only speculate that crediting a Protestant might have been looked upon unfavourably by the Catholic hierarchy which Copernicus was trying to impress. One consequence of this lack of acknowledgement was that Rheticus felt snubbed and would have nothing more to do with *De revolutionibus* after its publication.

The second mystery concerns the preface to *De revolutionibus*, which was added to the book without Copernicus's consent and

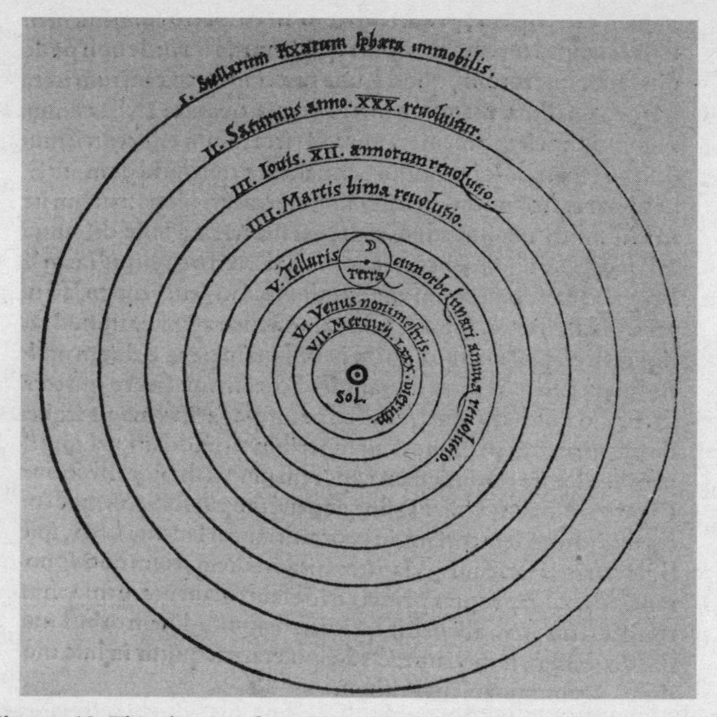

Figure 10 This diagram from Copernicus's *De revolutionibus* illustrates his revolutionary view of the universe. The Sun is firmly at the hub and is orbited by the planets. Earth itself is orbited by the Moon and is correctly located between the orbits of Venus and Mars.

---

which effectively retracted the substance of his claims. In short, the preface undermined the rest of the book by stating that Copernicus's hypotheses 'need not be true or even probable'. It emphasised 'absurdities' within the Sun-centred model, implying that Copernicus's own detailed and carefully argued mathematical description was nothing more than a fiction. The preface does admit that the Copernican system is compatible with observations to a reasonable degree of accuracy, but it emasculates the theory by stating that it is merely a convenient way to do calculations, rather than an attempt to represent reality. Copernicus's original handwritten manuscript still exists, so we know that the original opening was quite different in tone from the printed preface that trivialised his work. The new preface must therefore have been inserted after Rheticus had left Frauenburg with the manuscript. This would mean that Copernicus was on his deathbed when he first read it, by which time the book had been printed and it was too late to make any changes. Perhaps it was the very sight of the preface that sent him to his grave.

So who wrote and inserted the new preface? The main suspect is Osiander, the clergyman who took on responsibility for publication when Rheticus left Nuremberg for Leipzig. It is likely that he believed that Copernicus would suffer persecution once his ideas became public, and he probably inserted the preface with the best of intentions, hoping that it would assuage critics. Evidence for Osiander's concerns can be found in a letter to Rheticus in which he mentions the Aristotelians, meaning those who believed in the Earth-centred view of the world: 'The Aristotelians and theologians will easily be placated if they are told that . . . the present hypotheses are not proposed because they are in reality true, but because they are the most convenient to calculate the apparent composite motions.'

But in his intended preface, Copernicus had been quite clear that he was willing to adopt a defiant stance against his critics: 'Perhaps there will be babblers who, although completely ignorant of mathematics, nevertheless take it upon themselves to pass judgement on mathematical questions and, badly distorting some passages of Scripture to their purpose, will dare find fault with my undertaking and censure it. I disregard them even to the extent of despising their criticism as unfounded.'

Having finally plucked up the courage to publish the single most important and controversial breakthrough in astronomy since the ancient Greeks, Copernicus tragically died knowing that Osiander had misrepresented his theories as nothing more than artifice. Consequently, *De revolutionibus* was to vanish almost without trace for the first few decades after its publication, as neither the public nor the Church took it seriously. The first edition did not sell out, and the book was reprinted only twice in the next century. In contrast, books promoting the Ptolemaic model were reprinted a hundred times in Germany alone during the same period.

However, Osiander's cowardly and conciliatory preface to *De revolutionibus* was only partly to blame for its lack of impact. Another factor was Copernicus's dreadful writing style, which resulted in four hundred pages of dense, complex text. Worse still, this was his first book on astronomy, and the name Copernicus was not well known in European scholarly circles. This would not have been disastrous, except that Copernicus was now dead and could not promote his own work. The situation could possibly have been rescued by Rheticus, who might have championed *De revolutionibus*, but he had been snubbed and no longer wished to be associated with the Copernican system.

Moreover, just like Aristarchus' original incarnation of the Sun-centred model, *De revolutionibus* was dismissed because the Copernican system was less accurate than Ptolemy's Earth-centred model when it came to predicting future positions of the planets: in this respect the basically correct model was no match for its fundamentally flawed rival. There are two reasons for this strange state of affairs. First, Copernicus's model was missing one vital ingredient, without which its predictions could never be sufficiently accurate to gain its acceptance. Second, Ptolemy's model had achieved its degree of accuracy by tinkering with all the epicycles, deferents, equants and eccentrics, and almost any flawed model can be rescued if such fiddle-factors are introduced.

And, of course, the Copernican model was still plagued with all the problems that had led to the abandonment of Aristarchus'

Sun-centred model (see Table 2, pp. 34–5). In fact, the only attribute of the Sun-centred model that made it clearly better than the Earth-centred model was still its simplicity. Although Copernicus did toy with epicycles, his model essentially employed a simple circular orbit for each planet, whereas Ptolemy's model was inordinately complex, with its finely tuned epicycles, deferents, equants and eccentrics for each and every planet.

Fortunately for Copernicus, simplicity is a prized asset in science, as had been pointed out by William of Occam, a fourteenth-century English Franciscan theologian who became famous during his lifetime for arguing that religious orders should not own property or wealth. He propounded his views with such fervour that he was run out of Oxford University and had to move to Avignon in the south of France, from where he accused Pope John XII of heresy. Not surprisingly, he was excommunicated. After succumbing to the Black Death in 1349, Occam became famous posthumously for his legacy to science, known as Occam's razor, which holds that if there are two competing theories or explanations, then, all other things being equal, the simpler one is more likely to be correct. Occam put it thus: *pluralitas non est ponenda sine necessitate* ('plurality should not be posited without necessity').

Imagine, for instance, that after a stormy night you come across two fallen trees in the middle of a field, and there is no obvious sign of what caused them to fall. The simple hypothesis would be that the trees were blown over by the storm. A more complicated hypothesis might be that two meteorites simultaneously arrived from outer space, each ricocheting off one tree, felling the trees in the process, and then the meteorites collided head on with each other and vaporised, thereby accounting for the lack of any material evidence. Applying Occam's razor, you decide that the storm, rather than the twin meteorites, is the more likely explanation because it is the simpler one. Occam's razor does not guarantee the right answer, but it does usually point us towards the correct one. Doctors often rely on Occam's razor when diagnosing an illness, and medical students are advised: 'When you hear hoof beats, think horses, not zebras.' On the other hand, conspiracy theorists despise Occam's razor, often rejecting a simple explanation in favour of a more convoluted and intriguing line of reasoning.

Occam's razor favoured the Copernican model (one circle per planet) over the Ptolemaic model (one epicycle, deferent, equant and eccentric per planet), but Occam's razor is only decisive if two theories are equally successful, and in the sixteenth century the Ptolemaic model was clearly stronger in several ways; most notably, it made more accurate predictions of planetary positions. So the simplicity of the Sun-centred model was considered irrelevant.

And for many people the Sun-centred model was still too radical even to be contemplated, so much so that Copernicus's work may have resulted in a new meaning for an old word. One etymological theory claims that the word 'revolutionary', referring to an idea that is completely counter to conventional wisdom, was inspired by the title of Copernicus's book, 'On the Revolutions of the Heavenly Spheres'. And as well as revolutionary, the Sun-centred model of the universe also seemed completely impossible. This is why the word *köpperneksch*, based on the German form of Copernicus, has come to be used in northern Bavaria to describe an unbelievable or illogical proposition.

All in all, the Sun-centred model of the universe was an idea ahead of its time, too revolutionary, too unbelievable and still too inaccurate to win any widespread support. *De revolutionibus* sat on a few bookshelves, in a few studies, and was read by just a few astronomers. The idea of a Sun-centred universe had first been suggested by Aristarchus in the fifth century BC, but it was ignored; now it had been reinvented by Copernicus, and it was being ignored again. The model would go into hibernation, waiting for somebody to resuscitate it, examine it, refine it and find the missing ingredient that would prove to the rest of the world that the Copernican model of the universe was the true picture of reality. Indeed, it would be left to the next generation of astronomers to find the evidence that would show that Ptolemy was wrong and that Aristarchus and Copernicus were right.

Castle of the Heavens

Born into the Danish nobility in 1546, Tycho Brahe would earn lasting fame among astronomers for two particular reasons. First, in 1566, Tycho became embroiled in a disagreement with his cousin Manderup Parsberg, possibly because Parsberg had insulted and mocked Tycho over a recent astrological prediction that had fallen flat. Tycho had foretold the death of Suleiman the Great, and even embedded his prophecy within a Latin poem, apparently unaware that the Ottoman leader had already been dead for six months. The dispute culminated in an infamous duel. During the sword fight, a slash from Parsberg cut Tycho's forehead and hacked through the bridge of his nose. An inch deeper and Tycho would have died. Thereafter he glued into place a false metal nose, so cleverly composed of a gold–silver–copper alloy that it blended in with his skin tone.

The second and more important reason for Tycho's fame was that he took observational astronomy to an entirely new level of accuracy. He earned such a high reputation that King Frederick II of Denmark gave him the island of Hven, 10 km off the Danish coast, and paid for him to build an observatory there. Uraniborg (Castle of the Heavens) would grow over the years into a vast ornate citadel that consumed more than 5% of Denmark's gross national product, an all-time world record for research centre funding.

Uraniborg housed a library, a paper mill, a printing press, an alchemist's laboratory, a furnace and a prison for unruly servants.

Figure 11 Uraniborg, on the island of Hven, the best funded and most hedonistic astronomical observatory in history.

The observation turrets contained giant instruments, such as sextants, quadrants and armillary spheres (all naked-eye instruments, as astronomers had not yet learned to exploit the potential of lenses). There were four sets of every instrument for simultaneous and independent measurements, thereby minimising errors in assessing the angular positions of stars and planets. Tycho's observations were generally accurate to $\frac{1}{30}$°, five times better than the best previous measurements. Perhaps Tycho's measurements were aided by his ability to remove his nose and align his eye more perfectly.

Tycho's reputation was such that a stream of VIPs visited his observatory. As well as being interested in his research, these visitors were also attracted by Uraniborg's wild parties, which were famous all over Europe. Tycho provided alcohol in excess and entertainment in the shape of mechanical statues and a story-telling dwarf called Jepp, who was said to be a gifted clairvoyant. To add to the spectacle, Tycho's pet elk was allowed to freely wander the castle, but tragically it died after stumbling down a staircase after drinking too much alcohol. Uraniborg was more like the setting for a Peter Greenaway film than a research institute.

While Tycho had been raised in the traditions of Ptolemaic astronomy, his painstaking observations forced him to reconsider his confidence in the ancient view of the universe. In fact, we know that he had a copy of *De revolutionibus* in his study and that he was sympathetic to Copernicus's ideas, but, instead of adopting them unreservedly, he developed his own model of the universe, which was a faint-hearted halfway house between Ptolemy and Copernicus. In 1588, almost fifty years after Copernicus's death, Tycho published *De mundi ætherei recentioribus phænomenis* ('Concerning the New Phenomena in the Ethereal World'), in which he argued that all the planets orbited the Sun, but that the Sun orbited the Earth, as shown in Figure 12. His liberalism stretched as far as allowing the Sun to be the hub for the planets,

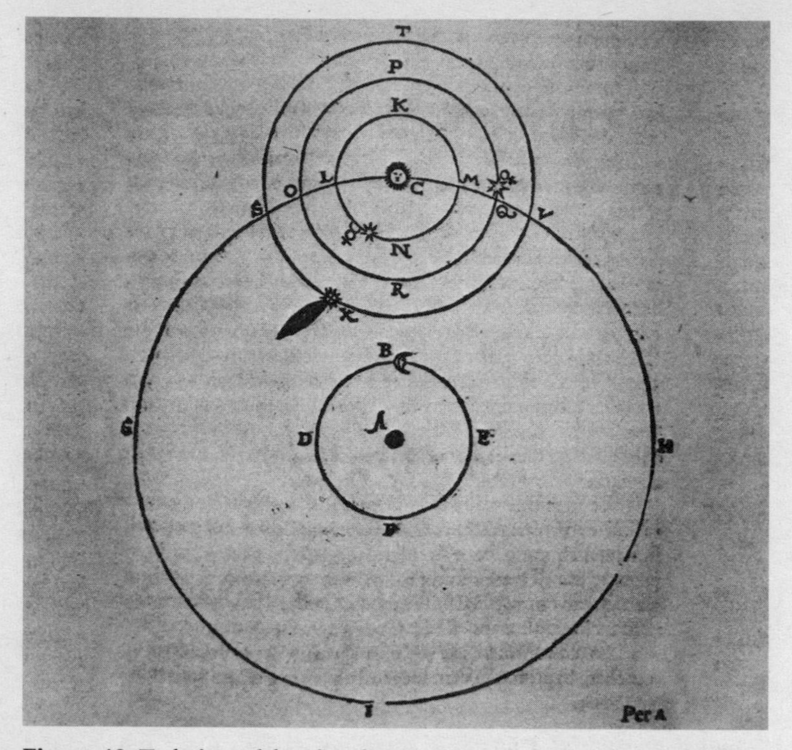

Figure 12 Tycho's model makes the same error as Ptolemy's and places the Earth at the centre of the universe, being orbited by the Moon and the Sun. His main breakthrough was to realise that the planets (and the fiery comet) orbit the Sun. This illustration is from Tycho's *De mundi ætherei.*

---

but his conservatism obliged him to retain the Earth at the centre of the universe. He was reluctant to dislodge the Earth, because its supposed centrality was the only way to explain why objects fall towards the centre of the Earth.

Before Tycho could continue to the next stage of his programme of astronomical observation and theorising, his research suffered a severe blow. His patron, King Frederick, died after a session of binge drinking in the same year that Tycho published *De mundi ætherei*, and the new king, Christian IV, was no longer prepared to fund Tycho's lavish observatory or tolerate his hedonistic lifestyle. Tycho had no option but to abandon Uraniborg and leave Denmark with his family, assistants, Jepp the dwarf and cartloads of astronomical equipment. Fortunately, Tycho's instruments had been designed to be transportable, because he had shrewdly realised: 'An astronomer must be cosmopolitan, because ignorant statesmen cannot be expected to value their services.'

Tycho Brahe migrated to Prague, where Emperor Rudolph II appointed him Imperial Mathematician and allowed him to establish a new observatory in Benatky Castle. The move turned out to have a silver lining, because it was in Prague that Tycho teamed up with a new assistant, Johannes Kepler, who would arrive in the city a few months later. The Lutheran Kepler had been forced to flee his previous home in Graz when the fiercely Catholic Archduke Ferdinand had threatened to execute him, in keeping with his stated declaration that he would rather 'make a desert of the country than rule over heretics'.

Fittingly, Kepler set out on his journey to Prague on 1 January 1600. The start of a new century would mark the start of a new collaboration that would lead to a reinvention of the universe. Together, Tycho and Kepler made the perfect double act. Scientific advance requires both observation and theory. Tycho had accumulated the best collection of observations in the history of astronomy, and Kepler would prove to be an excellent interpreter of those observations. Although Kepler suffered from myopia and multiple vision from birth, he would ultimately see farther than Tycho.

It was a partnership that was formed in the nick of time. Within a few months of Kepler's arrival, Tycho attended a dinner hosted by the Baron of Rosenberg and drank to his usual excess, refusing nonetheless to break etiquette by leaving the table before the Baron. Kepler recorded: 'When he drank more, he felt the tension in his bladder increase, but he put politeness before his health. When he got home, he was scarcely able to urinate.' That night he developed a fever, and from then on he alternated between bouts of unconsciousness and delirium. Ten days later he was dead.

On his deathbed, Tycho repeatedly uttered the phrase: 'May I not have lived in vain.' There was no need to fear, because Kepler would guarantee that Tycho's meticulous observations bore fruit. In fact, it is quite possible that Tycho had to die in order for his work to flourish, because while he was alive he carefully guarded all his notebooks and never shared his observations, always dreaming of publishing a solo masterwork. Tycho certainly never considered embracing Kepler as an equal partner – he was, after all, a Danish aristocrat, whereas Kepler was a mere peasant. However, seeing the deeper meaning of his own observations was beyond Tycho, and required the skills of a trained mathematician such as Kepler.

Kepler was born into a lowly family that struggled to survive the upheavals caused by war, religious strife, a wayward criminal father and a mother who had been exiled after accusations of witchcraft. Not surprisingly, he grew up as an insecure hypochondriac with little self-esteem. In his own self-deprecating horoscope, written in the third person, he described himself as a little dog:

> He likes gnawing bones and dry crusts of bread, and is so greedy that whatever his eyes chance on he grabs; yet, like a dog, he drinks little and is content with the simplest food . . . He continually seeks the goodwill of others, is dependent on others for everything, ministers to their wishes, never gets angry when they berate him and is anxious to get back into their favour . . . He has a dog-like horror of baths, tinctures and lotions. His recklessness knows no limits, which is surely due to Mars in quadrature with Mercury and in trine with the Moon.

His passion for astronomy seems to have been his only respite from self-loathing. At the age of twenty-five he wrote *Mysterium cosmographicum*, the first book to defend Copernicus's *De revolutionibus*. Thereafter, convinced of the veracity of the Sun-centred model, he dedicated himself to identifying just what it was that made it inaccurate. The greatest error was in predicting the exact path of Mars, a problem that had plagued Copernicus's assistant, Rheticus. According to Kepler, Rheticus had been so frustrated with his failure to solve the Mars problem that 'he appealed as a last resort to his guardian angel as an Oracle. The ungracious spirit thereupon seized Rheticus by the hair and alternately banged his head against the ceiling, then let his body down and crashed it against the floor.'

With access at last to Tycho's observations, Kepler was confident that he could solve the problem of Mars and remove the inaccuracies in the Sun-centred model within eight days; in fact, it took him eight years. It is worth stressing the amount of time that Kepler spent perfecting the Sun-centred model – eight years! – because the brief summary that follows could easily underplay his immense achievement. Kepler's eventual solution was the result of arduous and tortuous calculations that filled nine hundred folio pages.

Kepler made his great breakthrough by jettisoning one of the ancient tenets, namely that the planets all move in paths that are circles or combinations of circles. Even Copernicus had clung loyally to this circular dogma, and Kepler pointed out that this was just one of Copernicus's flawed assumptions. In fact, Kepler claimed that his predecessor had wrongly assumed the following three points:

1. the planets move in perfect circles,
2. the planets move at constant speeds,
3. the Sun is at the centre of these orbits.

Although Copernicus was right in stating that the planets orbit the Sun and not the Earth, his belief in these three false assumptions sabotaged his hopes of ever predicting the movements of Mars and the other planets with a high degree of accuracy. However, Kepler would succeed where Copernicus had failed because he discarded these assumptions, believing that the truth emerges only when all ideology, prejudice and dogma are set aside. He opened his eyes and mind, took Tycho's observations as his rock and built his model upon Tycho's data. Gradually an unbiased model of the universe began to emerge. Sure enough, Kepler's new equations for the orbits matched the observations, and the Solar System took shape at last. Kepler exposed Copernicus's errors, and showed that:

1. the planets move in ellipses, not perfect circles,
2. the planets continuously vary their speed,
3. the Sun is not quite at the centre of these orbits.

When he knew he had the solution to the mystery of planetary orbits, Kepler shouted out: 'O, Almighty God, I am thinking Thy thoughts after Thee.'

In fact, the second and third points in Kepler's new model of the Solar System emerge out of the first, which states that planetary orbits are elliptical. A quick guide to ellipses and how they are constructed reveals why this is so. One way to draw an ellipse is to pin a length of string to a board, as shown in Figure 13, and then use a pencil to extend the string. If the pencil is moved around the board, keeping the string taut, it will trace out half an ellipse. Switch to the other side of the string, and make it taut again, and the other half of the ellipse can be traced out. The length of the string is constant and the pins are fixed, so a possible definition of the ellipse is the set of points whose combined distance to the two pins has a specific value.

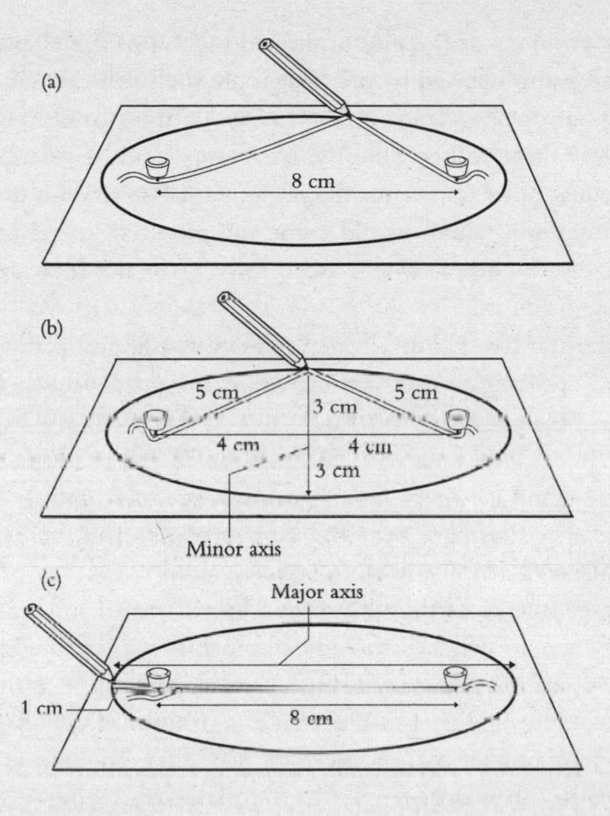

Figure 13 A simple way to draw an ellipse is to use a piece of string attached to two pins, as shown in diagram (a). If the pins are 8 cm apart and the string is 10 cm long, then each point on the ellipse has a combined distance of 10 cm from the two pins. For example, in diagram (b), the 10 cm of string forms two sides of a triangle, both 5 cm long. From Pythagoras' theorem, the distance from the centre of the ellipse to the top must be 3 cm. This means that the total height (or *minor axis*) of the ellipse is 6 cm. In diagram (c), the 10 cm of string is pulled to one side. This indicates that the total width (or *major axis*) of the ellipse is 10 cm, because it is 8 cm from pin to pin plus 1 cm at both ends.

The ellipse is quite squashed, because the minor axis is 6 cm compared with the major axis of 10 cm. As the two pins are brought closer together, the major and minor axes of the ellipse become more equal and the ellipse becomes less squashed. If the pins merge into a single point, then the string would form a constant radius of 5 cm and the resulting shape would be a circle.

The positions of the pins are called the *foci* of the ellipse. The elliptical paths followed by the planets are such that the Sun sits at one of the foci, and not at the centre of the planetary orbits. Therefore there will be times when a planet will be closer to the Sun than at other times, as if the planet has fallen towards the Sun. This process of falling would cause the planet to speed up and, conversely, the planet would slow down as it moved away from the Sun.

Kepler showed that, as a planet follows its elliptical path around the Sun, speeding up and slowing down along the way, an imaginary line joining the planet to the Sun will sweep out equal areas in equal times. This somewhat abstract statement is illustrated in Figure 14, and it is important because it precisely defines how a planet's speed changes over the course of its orbit, contrary to Copernicus's belief in constant planetary speeds.

The geometry of the ellipse had been studied since ancient Greek times, so why had nobody ever before suggested ellipses as the shape of the planetary orbits? One reason, as we have seen, was the enduring belief in the sacred perfection of circles, which seemed to blinker astronomers to all other possibilities. But another reason was that most of the planetary ellipses are only very slightly elliptical, so under all but the closest scrutiny they appear to be circular. For example, the length of the minor axis divided by the length of the major axis (see Figure 13) is a good indication of how close an ellipse is to a circle. The ratio equals 1.0 for a circle, but the Earth's orbit has a ratio of 0.99986. Mars, the planet that had given Rheticus nightmares, was so problematic because its orbit is more squashed, but the ratio of the two axes is still very close to 1, at 0.99566. In short, the Martian orbit was only slightly elliptical, so it duped astronomers into thinking it was circular, but the orbit was elliptical enough to cause real problems for anybody who tried to model it in terms of circles.

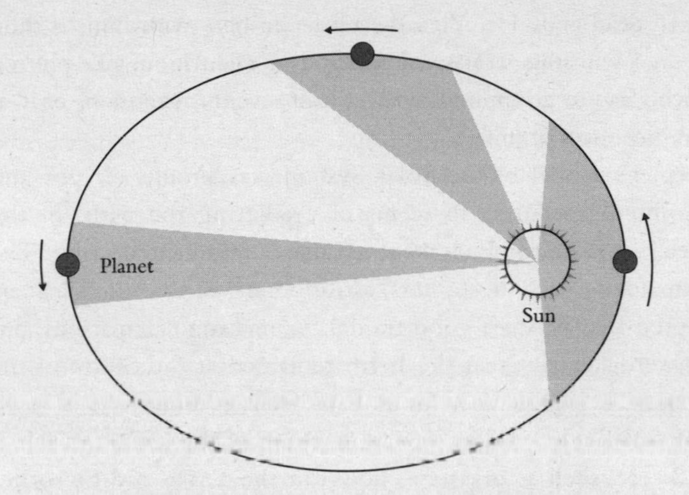

Figure 14 The diagram shows a highly exaggerated planetary orbit. The height of the ellipse is roughly 75% of its width, whereas for most planetary orbits in the Solar System this proportion is typically between 99% and 100%. Similarly, the focus occupied by the Sun is far off-centre, whereas it is only slightly off-centre for actual planetary orbits. The diagram demonstrates Kepler's second law of planetary motion. He explained that the imaginary line joining a planet to the Sun (the radius vector) sweeps out equal areas in equal times, which is a consequence of a planet's increase in speed as it approaches the Sun. The three shaded sectors all have equal areas. When the planet is closer to the Sun the radius vector is short, but this is compensated by its greater speed, which means that it covers more of the ellipse's circumference in a fixed time. When the planet is far from the Sun the radius vector is much longer, but it has a slower speed so it covers a smaller section of the circumference in the same time.

Kepler's ellipses provided a complete and accurate vision of our Solar System. His conclusions were a triumph for science and the scientific method, the result of combining observation, theory and mathematics. He first published his breakthrough in 1609 in a huge treatise entitled *Astronomia nova*, which detailed eight years of meticulous work, including numerous lines of investigation that led only to dead ends. He asked the reader to bear with him: 'If thou art bored with this wearisome method of calculation, take pity on me who had to go through with at least seventy repetitions of it, at a very great loss of time.'

Kepler's model of the Solar System was simple, elegant and undoubtedly accurate in terms of predicting the paths of the planets, yet almost nobody believed that it represented reality. The vast majority of philosophers, astronomers and Church leaders accepted that it was a good model for making calculations, but they were adamant that the Earth remained at the centre of the universe. Their preference for an Earth-centred universe was based largely on Kepler's failure to address some of the issues in Table 2 (pp. 34–5), such as gravity – how can the Earth and the other planets be held in orbit around the Sun, when everything that we see around us is attracted to the Earth?

Also, Kepler's reliance on ellipses, which was contrary to the doctrine of circles, was considered laughable. The Dutch clergyman and astronomer David Fabricius had this to say in a letter to Kepler: 'With your ellipse you abolish the circularity and uniformity of the motions, which appears to me increasingly absurd the more profoundly I think about it . . . If you could only preserve the perfect circular orbit, and justify your elliptic orbit by another little epicycle, it would be much better.' But an ellipse cannot be built from circles and epicycles, so a compromise was impossible.

Disappointed by the poor reception given to *Astronomia nova*, Kepler moved on and began to apply his skills elsewhere. He was forever curious about the world around him, and justified his relentless scientific explorations when he wrote: 'We do not ask for what useful purpose the birds do sing, for song is their pleasure since they were created for singing. Similarly, we ought not to ask why the human mind troubles to fathom the secrets of the heavens . . . The diversity of the phenomena of Nature is so great, and the treasures hidden in the heavens so rich, precisely in order that the human mind shall never be lacking in fresh nourishment.'

Beyond his research into elliptical planetary orbits, Kepler indulged in work of varying quality. He misguidedly revived the Pythagorean theory that the planets resonated with a 'music of the spheres'. According to Kepler, the speed of each planet generated particular notes (e.g. doh, ray, me, fah, soh, lah and te). The Earth emitted the notes fah and me, which gave the Latin word *fames*, meaning 'famine', apparently indicating the true nature of our planet. A better use of his time was his authorship of *Somnium*, one of the precursors of the science fiction genre, recounting how a team of adventurers journey to the Moon. And a couple of years after *Astronomia nova*, Kepler wrote one of his most original research papers, 'On the Six-Cornered Snowflake', in which he pondered the symmetry of snowflakes and put forward an atomistic view of matter.

'On the Six-Cornered Snowflake' was dedicated to Kepler's patron, Johannes Matthaeus Wackher von Wackenfels, who was also responsible for delivering to Kepler the most exciting news that he would ever receive: an account of a technological breakthrough that would transform astronomy in general and the status of the Sun-centred model in particular. The news was so astonishing that Kepler made a special note of Herr Wackher's visit in March 1610: 'I experienced a wonderful emotion while I listened to this curious tale. I felt moved in my deepest being.'

Kepler had just heard for the first time about the telescope, which was being used by Galileo to explore the heavens and reveal completely new features of the night sky. Thanks to this new invention, Galileo would discover the evidence that would prove that Aristarchus, Copernicus and Kepler were all correct.

Seeing Is Believing

Born in Pisa on 15 February 1564, Galileo Galilei has often been referred to as the father of science, and indeed his claim to that title is founded on a staggeringly impressive track record. He may not have been the first to develop a scientific theory, or the first to conduct an experiment, or the first to observe nature, or even the first to prove the power of invention, but he was probably the first to excel at all of these, being a brilliant theorist, a master experimentalist, a meticulous observer and a skilled inventor.

He demonstrated his multiple skills during his student years, when his mind wandered during a cathedral service and he noticed a swinging chandelier. He used his own pulse to measure the time of each swing and observed that the period for the back-and-forth cycle remained constant, even though the wide arc of the swing at the start of the service had faded to just a gentle sway by the end. Once home, he switched from observational to experimental mode and toyed with pendulums of different lengths and weights. He then used his experimental data to develop a theory that explained how the period of swing is independent of the angle of swing and of the weight of the bob, but depends only on the length of the pendulum. After pure research, Galileo switched into invention mode and collaborated on the development of the *pulsilogia*, a simple pendulum whose regular swinging allowed it to act as a timing device.

In particular, the device could be used to measure a patient's pulse rate, thereby reversing the roles in his original observation when he used his pulse to measure the period of the swinging lamp. He was studying to be a doctor at the time, but this was his one and only contribution to medicine. Subsequently he persuaded his father to allow him to abandon medicine and pursue a career in science.

In addition to his undoubted intellect, Galileo's success as a scientist would rely on his tremendous curiosity about the world and everything in it. He was well aware of his inquisitive nature and once exclaimed: 'When shall I cease from wondering?'

This curiosity was coupled with a rebellious streak. He had no respect for authority, inasmuch as he did not accept that anything was true just because it had been stated by teachers, theologians or the ancient Greeks. For example, Aristotle used philosophy to deduce that heavy objects fall faster than light objects, but Galileo conducted an experiment to prove that Aristotle was wrong. He was even courageous enough to say that Aristotle, then the most acclaimed intellect in history, 'wrote the opposite of truth'.

When Kepler first heard about Galileo's use of the telescope to explore the heavens, he probably assumed that Galileo had invented the telescope. Indeed, many people today make the same assumption. In fact, it was Hans Lippershey, a Flemish spectacle-maker, who patented the telescope in October 1608. Within a few months of Lippershey's breakthrough, Galileo noted that 'a rumour came to our ears that a spyglass had been made by a certain Dutchman', and he immediately set about building his own telescopes.

Galileo's great accomplishment was to transform Lippershey's rudimentary design into a truly remarkable instrument. In August 1609, Galileo presented the Doge of Venice with what was then the most powerful telescope in the world. Together they climbed St Mark's bell-tower, set up the telescope and surveyed the lagoon. A week later, in a letter to his brother-in-law, Galileo was able to report that the telescope performed 'to the infinite amazement of all'. Rival instruments had a magnification of about ×10, but Galileo had a better understanding of the optics of the telescope and was able to achieve a magnification of ×60. Not only did the telescope give the Venetians an advantage in warfare, because they could see the enemy before the enemy saw them, but it also enabled the shrewder merchants to spot a distant ship arriving with a new cargo of spices or cloth, which meant that they could sell off their current stock before market prices plummeted.

Galileo profited from his commercialisation of the telescope, but he realised that it also had a scientific value. When he pointed his telescope at the night sky, it enabled him to see farther, clearer and deeper into space than anyone ever before. When Herr Wackher told Kepler about Galileo's telescope, the fellow astronomer immediately recognised its potential and wrote a eulogy: 'O telescope, instrument of much knowledge, more precious than any sceptre! Is not he who holds thee in his hand made king and lord of the works of God?' Galileo would become that king and lord.

First, Galileo studied the Moon and showed it to be 'full of vast protuberances, deep chasms and sinuosities', which was in direct

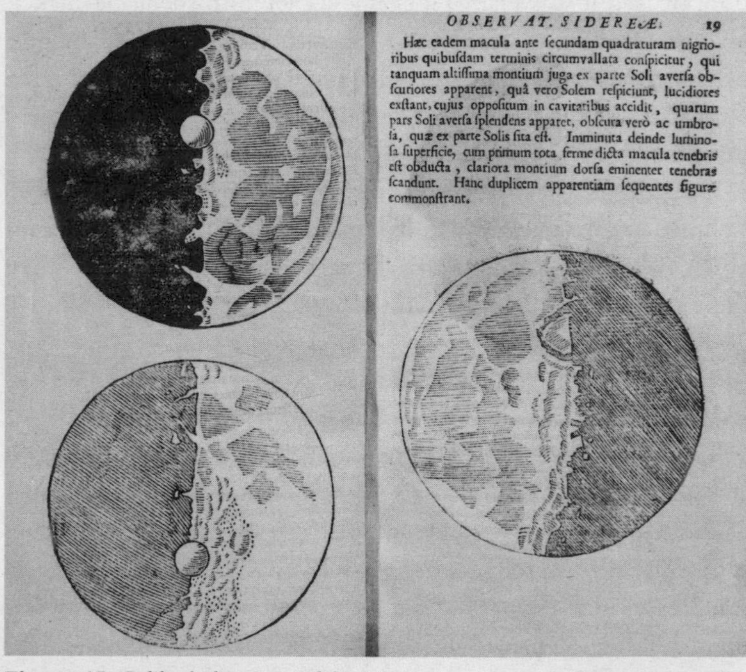

OBSERVAT. SIDEREÆ. 19

Hæc eadem macula ante ſecundam quadraturam nigrioribus quibuſdam terminis circumvallata conſpicitur, qui tanquam altiſſima montium juga ex parte Soli averſa obſcuriores apparent, quâ vero Solem reſpiciunt, lucidiores exſtant, cujus oppoſitum in cavitatibus accidit, quarum pars Soli averſa ſplendens apparet, obſcura verò ac umbroſa, quæ ex parte Solis ſita eſt. Imminuta deinde luminoſa ſuperficie, cum primum tota ferme dicta macula tenebris eſt obducta, clariora montium dorſa eminenter tenebras ſcandunt. Hanc duplicem apparentiam ſequentes figuræ commonſtrant.

Figure 15 Galileo's drawings of the Moon.

contradiction to the Ptolemaic view that the heavenly bodies were flawless spheres. The imperfection of the heavens was later reinforced when Galileo pointed his telescope at the Sun and noticed blotches and blemishes, namely sunspots, which we now know to be cooler patches on the Sun's surface up to 100,000 km across.

Then, during January 1610, Galileo made an even more momentous observation when he spotted what he initially thought were four stars loitering in the vicinity of Jupiter. Soon it became apparent that the objects were not stars, because they moved around Jupiter, which meant that they were Jovian moons. Never before had anybody seen a moon other than our own. Ptolemy had argued that the Earth was the centre of the universe, but here was indisputable evidence that not everything orbited the Earth.

Galileo, who was in correspondence with Kepler, was fully aware of the latest Keplerian version of the Copernican model, and he realised that his discovery of Jupiter's moons was providing further support for the Sun-centred model of the universe. He had no doubt that Copernicus and Kepler were right, yet he continued to search for evidence in favour of this model in the hope of converting the establishment, which still clung to the traditional view of an Earth-centred universe. The only way to break the impasse would be to find a clear-cut prediction that differentiated between the two competing models. If such a prediction could be tested it would confirm one model and refute the other. Good science develops theories that are testable, and it is through testing that science progresses.

In fact, Copernicus had made just such a prediction, one which had been waiting to be tested as soon as the tools were available to make the appropriate observations. In *De revolutionibus*, he had stated that Mercury and Venus should exhibit a series of phases (e.g. full Venus, half Venus, crescent Venus) similar to the phases of the Moon, and the exact pattern of phases would depend on whether the Earth orbited the Sun, or vice versa. In the fifteenth century

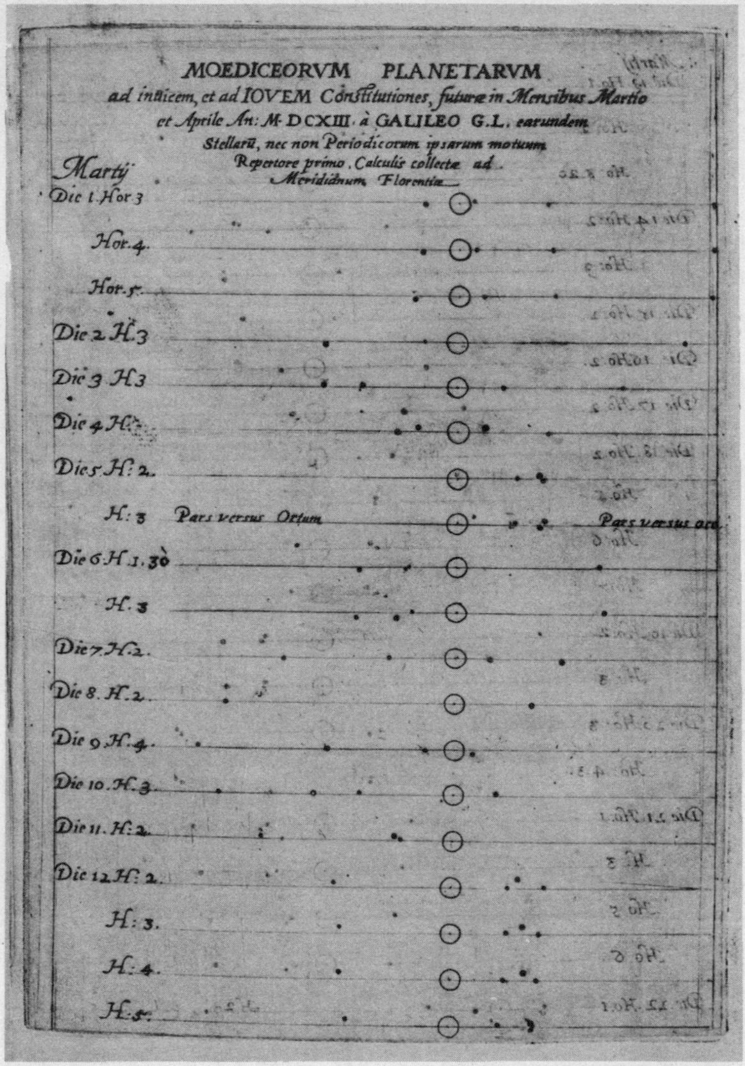
MOEDICEORVM PLANETARVM
ad inuicem, et ad IOVEM Constitutiones, futuræ in Mensibus Martio
et Aprile An: M DCXIII à GALILEO G.L. earundem
Stellarū, nec non Periodicorum ipsarum motuum
Repertore primo. Calculis collectæ ad
Meridianum Florentiæ
Martij
Pars versus Ortum
Pars versus oc

Figure 16 Galileo's sketches of the changing positions of Jupiter's moons. The circles represent Jupiter, and the several dots either side show the changing positions of the moons. Each row represents one observation taken on a particular date and time, with one or more observations per night.

nobody could check the pattern of phases because the telescope had yet to be invented, but Copernicus was confident that it was just a matter of time before he would be proved correct: 'If the sense of sight could ever be made sufficiently powerful, we could see phases in Mercury and Venus.'

Leaving aside Mercury and concentrating on Venus, the significance of the phases is apparent in Figure 17. Venus always has one face illuminated by the Sun, but from our vantage point on the Earth this face is not always towards us, so we see Venus go through a series of phases. In Ptolemy's Earth-centred model, the sequence of phases is determined by Venus's path around the Earth, and its slavish obedience to its epicycle. However, in the Sun-centred model, the sequence of phases is different because it is determined by Venus's path around the Sun without any epicycle. If somebody could identify the actual sequence of Venus's waxing and waning, then it would prove beyond all reasonable doubt which model was correct.

In the autumn of 1610, Galileo became the first person ever to witness and chart the phases of Venus. As he expected, his observations perfectly fitted the predictions of the Sun-centred model, and provided further ammunition to support the Copernican revolution. He reported his results in a cryptic Latin note that read *Haec immatura a me iam frustra leguntur oy* ('These are at present too young to be read by me'). He later revealed that this was a coded anagram that when unravelled read *Cynthiæ figuras æmulatur Mater Amorum* ('Cynthia's figures are imitated by the Mother of Love'). Cynthia was a reference to the Moon, whose phases were already familiar, and Mother of Love was an allusion to Venus, whose phases Galileo had discovered.

The case for a Sun-centred universe was becoming stronger with each new discovery. Table 2 (pp. 34–5) compared the Earth- and Sun-centred models based on pre-Copernican observations, showing

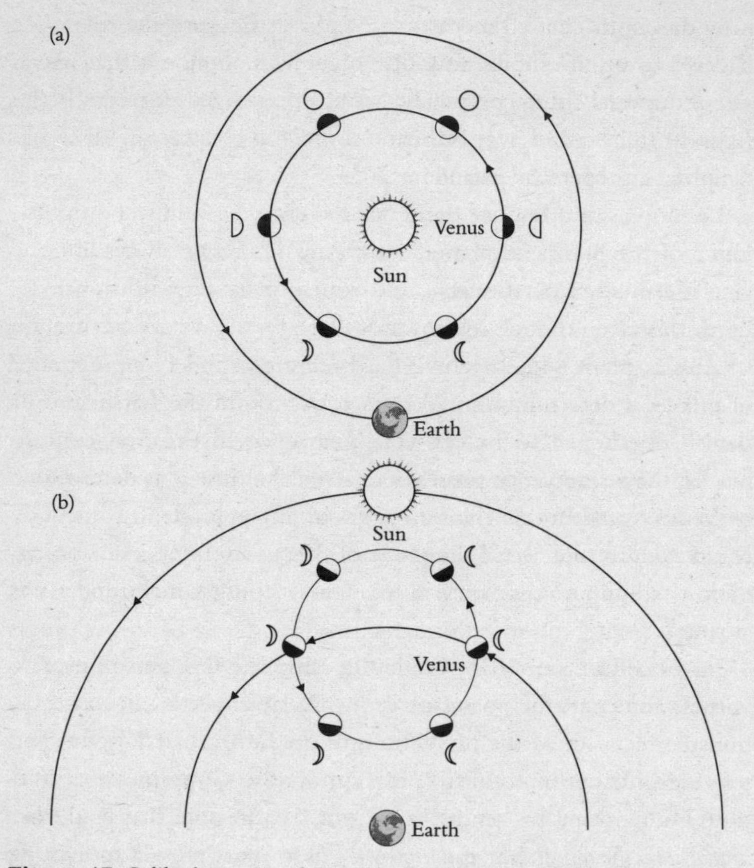

Figure 17 Galileo's precise observations of the phases of Venus proved that Copernicus was right, and Ptolemy wrong. In the Sun-centred model of the universe, shown in diagram (a), both the Earth and Venus orbit the Sun. Although Venus is always half-lit by the Sun, from the Earth's point of view it appears to go through a cycle of phases, turning from a crescent to a disc. The phase is shown next to each position of Venus.

In the Earth-centred model of the universe, both the Sun and Venus orbit the Earth, and in addition Venus moves round its own epicycle. The phases depend on where Venus is on its orbit and on its epicycle. In diagram (b), Venus's orbit is such that it is roughly between the Earth and the Sun, which gives rise to the set of phases shown. By identifying the actual series of phases, Galileo could identify which model was correct.

why the Earth-centred model made more sense in the Middle Ages. Table 3 (overleaf) shows how Galileo's observations made the Sun-centred model more compelling. The remaining weaknesses in the Sun-centred model would be removed later, once scientists had achieved a proper understanding of gravity and were able to appreciate why we do not sense the Earth's motion around the Sun. And although the Sun-centred model did not chime with common sense, one of the criteria in the table, this was not really a weakness because common sense has little to do with science, as discussed earlier.

At this point in history, every astronomer should have switched allegiance to the Sun-centred model, but no such major shift took place. Most astronomers had spent their entire lives convinced that the universe revolved around a static Earth, and they were unable to make the intellectual or emotional leap to a Sun-centred universe. When the astronomer Francesco Sizi heard about Galileo's observation of Jupiter's moons, which seemed to suggest that the Earth was not the hub of everything, he came up with a bizarre counter-argument: 'The moons are invisible to the naked eye and therefore can have no influence on the Earth and therefore would be useless and therefore do not exist.' The philosopher Giulio Libri took a similarly illogical stance and even refused to look through a telescope on a point of principle. When Libri died, Galileo suggested that he might at last see the sunspots, the moons of Jupiter and the phases of Venus on his way to heaven.

The Catholic Church was similarly unwilling to abandon its doctrine that the Earth was fixed at the centre of the universe, even when Jesuit mathematicians confirmed the superior accuracy of the new Sun-centred model. Thereafter, theologians conceded that the Sun-centred model was able to make excellent predictions of planetary orbits, but at the same time they still refused to accept that it was a valid representation of reality. In other words, the Vatican viewed the Sun-centred model in the same way that we regard this

Table 3

This table lists ten important criteria against which the Earth-centred and Sun-centred models could be judged based on what was known in 1610, after Galileo's observations. The ticks and crosses give crude indications of how well each model fared in relation to each criterion, and a question mark

Criterion	Earth-centred model	Success
1. Common sense	It seems obvious that everything revolves around the Earth	✓
2. Awareness of motion	We do not detect any motion, therefore the Earth cannot be moving	✓
3. Falling to the ground	The centrality of the Earth explains why objects appear to fall downwards, i.e. they are being attracted to the centre of the universe	✓
4. Stellar parallax	There is no detection of stellar parallax, absence of which is compatible with a static Earth and a stationary observer	✓
5. Predicting planetary orbits	Very close agreement	✓
6. Retrograde paths of planets	Explained with epicycles and deferents	✓
7. Simplicity	Very complicated – epicycles, deferents, equants and eccentrics for each planet	✗
8. Phases of Venus	Fails to predict the observed phases	✗
9. Blemishes on Sun and Moon	Problematic – this model emerges from an Aristotelian view, which also claims that the heavens are perfect	✗
10. Moons of Jupiter	Problematic – everything is supposed to orbit the Earth!	✗

indicates a lack of data. Compared to the assessment based on the evidence available before Copernicus (Table 2, pp. 34–5), the Sun-centred model now seems more convincing. This is partly down to new observations (points 8, 9 and 10) that were possible only with the advent of the telescope.

Criterion	Sun-centred model	Success
1. Common sense	It still requires a leap of imagination and logic to see that the Earth might circle the Sun	✗
2. Awareness of motion	Galileo was en route to explaining why we do not sense the Earth's motion around the Sun	?
3. Falling to the ground	There is no obvious explanation in a model where the Earth is not centrally located; only later would Newton explain gravity in this context	✗
4. Stellar parallax	The Earth moves, so the apparent lack of parallax must be due to huge stellar distances; parallax should be detected with better telescopes	?
5. Predicting planetary orbits	Perfect agreement, after Kepler's contribution	✓
6. Retrograde paths of planets	A natural consequence of the Earth's motion and our changing vantage point	✓
7. Simplicity	Very simple – everything follows ellipses	✓
8. Phases of Venus	Successfully predicts the observed phases	✓
9. Blemishes on Sun and Moon	No problem – this model makes no claims about the perfection or imperfection of heavenly bodies	✓
10. Moons of Jupiter	No problem – this model tolerates multiple centres	✓

sentence: 'How I need a drink, alcoholic of course, after the heavy lectures involving quantum mechanics.' This phrase is a mnemonic for the number π. By noting the number of letters in each word of the sentence, we obtain 3.141 592 653 589 79, which is the true value of π to fourteen decimal places. The sentence is indeed a highly accurate device for representing the value of π, but at the same time we know that π has nothing to do with alcohol. The Church maintained that the Sun-centred model of the universe had a similar status – accurate and useful, but not reality.

However, the Copernicans continued to argue that the Sun-centred model was good at predicting reality for the very reason that the Sun really was at the centre of the universe. Not surprisingly, this provoked a stern reaction from the Church. In February 1616, a committee of advisors to the Inquisition formally declared that holding the Sun-centred view of the universe was heretical. As a result of this edict, Copernicus's *De revolutionibus* was banned in March 1616, sixty-three years after it had been published.

Galileo was unable to accept the Church's condemnation of his scientific views. Although he was a devout Catholic he was also a fervent rationalist, and had been able to reconcile these two belief systems. He had come to the conclusion that scientists were best qualified to comment on the material world, whereas theologians were best qualified to comment on the spiritual world and how one should live in the material world. Galileo argued: 'Holy Writ was intended to teach men how to go to Heaven, not how the heavens go.'

Had the Church criticised the Sun-centred model by identifying weaknesses in the argument or poor data, then Galileo and his colleagues would have been willing to listen, but their criticisms were purely ideological. Galileo chose to ignore the views of the cardinals, and year after year he continued to press for a new vision of the universe. At last, in 1623, he saw an opportunity to overthrow

Figure 18 Copernicus (top left), Tycho (top right), Kepler (bottom left) and Galileo were responsible for driving the shift from an Earth-centred to a Sun-centred model of the universe. Together their achievements illustrate a key feature of scientific progress, namely how theories and models are developed and refined over time by several scientists building on each other's work.

Copernicus was prepared to make the theoretical leap that relegated the Earth to a mere satellite and promoted the Sun to the central role. Tycho Brahe, despite his brass nose, provided the observational evidence that would later help Johannes Kepler to identify the outstanding flaw in Copernicus's model, namely that the planetary orbits are slightly elliptical, not perfectly circular. Finally, Galileo used a telescope to discover the key evidence that should have convinced doubters. He showed that the Earth is not at the centre of everything, because Jupiter has its own satellites. Also, he showed that the phases of Venus are only compatible with a Sun-centred universe.

the establishment when his friend Cardinal Maffeo Barberini was elected to the papal throne as Urban VIII.

Galileo and the new pope had known each other ever since they had attended the same university in Pisa, and soon after his election Urban VIII granted Galileo six lengthy audiences. During one audience, Galileo mentioned the idea of writing a book that compared the two rival views of the universe, and when he departed the Vatican he was left with the firm impression that he had received the Pope's blessing. He returned to his study and made a start on what would turn out to be one of the most controversial books ever published in the history of science.

In his *Dialogue Concerning the Two Chief World Systems*, Galileo used three characters to explore the merits of the Sun-centred and Earth-centred world-views. Salviati presented Galileo's preferred Sun-centred view and was clearly an intelligent, well-read and eloquent man. Simplicio, the buffoon, attempted to defend the Earth-centred position. And Sagredo acted as a mediator, guiding the conversation between the other two characters, although his bias sometimes emerged when he scolded and mocked Simplicio along the way. This was a scholarly text, but the device of using characters to explain the arguments and counter-arguments made it accessible to a wider readership. Also, it was written in Italian, not Latin, so clearly Galileo's objective was to win widespread popular backing for a Sun-centred universe.

The *Dialogue* was eventually published in 1632, almost a decade after Galileo had apparently won the Pope's approval. That huge delay between inception and publication turned out to have severe consequences, because the ongoing Thirty Years' War had changed the political and religious landscape, and Pope Urban VIII was now ready to quash Galileo and his argument. The Thirty Years' War had begun in 1618, when a group of Protestants marched into the Royal Palace in Prague and threw two of the town's officials out of an upper window, an event known as the Defenestration of Prague. The local people had been angered because of the continual persecution of Protestants, and by taking this action they sparked a violent uprising by Protestant communities in Hungary, Transylvania, Bohemia and other parts of Europe.

By the time the *Dialogue* was published, the war had been raging for fourteen years, and the Catholic Church felt increasingly alarmed by the growing Protestant threat. The Pope had to be seen to be a strong champion of the Catholic faith, and he decided that part of his new hard-hitting populist strategy would be to make a deft U-turn and condemn the blasphemous writings of any heretical scientists who dared question the traditional Earth-centred view of the universe.

A more personal explanation for the Pope's dramatic change of heart is that astronomers jealous of Galileo's fame, together with the more conservative cardinals, had stirred up trouble by highlighting parallels between some of the Pope's earlier and more naive pronouncements on astronomy and statements uttered by the *Dialogue*'s buffoon, Simplicio. For example, Urban had argued, much as Simplicio does, that an omnipotent God created a universe with no regard to the laws of physics, so the Pope must have been humiliated by Salviati's sarcastic response to Simplicio in the *Dialogue*: 'Surely, God could have caused birds to fly with their bones made of solid gold, with their veins full of quicksilver, with their flesh heavier than lead, and with their wings exceedingly small. He did not, and that ought to show something. It is only in order to shield your ignorance that you put the Lord at every turn.'

Soon after the *Dialogue*'s publication, the Inquisition ordered Galileo to appear before them on a charge of 'vehement suspicion of heresy'. When Galileo protested that he was too ill to travel, the Inquisition threatened to arrest him and drag him to Rome in chains, whereupon he acquiesced and prepared for the journey.

While waiting for Galileo's arrival, the Pope attempted to impound the *Dialogue* and ordered the printer to send all copies to Rome, but it was too late – every single copy had been sold.

The trial began in April 1633. The accusation of heresy centred on the conflict between Galileo's views and the Biblical statement that 'God fixed the Earth upon its foundation, not to be moved for ever.' Most members of the Inquisition took the view expressed by Cardinal Bellarmine: 'To assert that the Earth revolves around the Sun is as erroneous as to claim that Jesus was not born of a virgin.' However, among the ten cardinals presiding over the trial, there was a sympathetic rationalist faction led by Francesco Barberini, the nephew of Pope Urban VIII. For two weeks, the evidence mounted against Galileo and there were even threats of torture, but Barberini continually called for leniency and tolerance. To some extent he was successful. After being found guilty, Galileo was neither executed nor thrown into a dungeon, but sentenced instead to indefinite house arrest, and the *Dialogue* was added to the list of banned books, the *Index librorum prohibitorum*. Barberini was one of three judges who did not sign the sentence.

Galileo's trial and subsequent punishment was one of the darkest episodes in the history of science, a triumph for irrationality over logic. At the end of the trial, Galileo was forced to recant, to deny the truth of his argument. However, he did manage to salvage some small pride in the name of science. After sentencing, as he rose from his knees, he reputedly muttered the words '*Eppur si muove!*' ('And yet it moves!'). In other words, the truth is dictated by reality, not by the Inquisition. Regardless of what the Church might have claimed, the universe still operated according to its own immutable scientific laws, and the Earth did indeed orbit the Sun.

Galileo slipped into isolation. Confined to his house, he continued to think about the laws that governed the universe, but his research was severely limited when he became blind in 1637, perhaps through glaucoma caused by staring at the Sun through his telescope. The great observer could no longer observe. Galileo died on 8 January 1642. As a final act of punishment, the Church refused to let him be buried in consecrated ground.

The Ultimate Question

The Sun-centred model gradually became widely accepted by astronomers over the course of the next century, partly because there was more observational evidence being gathered with the aid of better telescopes, and partly because there were theoretical breakthroughs to explain the physics behind the model. Another important factor was that a generation of astronomers had passed away. Death is an essential element in the progress of science, since it takes care of conservative scientists of a previous generation reluctant to let go of an old, fallacious theory and embrace a new and accurate one. Their recalcitrance is understandable, because they had framed their entire life's work around one model and were faced with the possibility of having to abandon it in favour of a new model. As Max Planck, one of the greatest physicists of the twentieth century, commented: 'An important scientific innovation rarely makes its way by gradually winning over and converting its opponents: it rarely happens that Saul becomes Paul. What does happen is that its opponents gradually die out, and the growing generation is familiarised with the ideas from the beginning.'

In parallel with the acceptance of the Sun-centred view of the universe by the astronomical establishment, there was also a shift in the attitude of the Church. Theologians came to realise that they would look foolish if they continued to deny what men of learning regarded as reality. The Church softened its stance towards astronomy and many other areas of science, which gave rise to a new period of intellectual freedom. Throughout the eighteenth century, scientists would apply their skills to a wide variety of questions about the world around them, replacing supernatural myths, philosophical blunders and religious dogmas with accurate, logical, verifiable, natural explanations and answers. Scientists studied everything from the nature of light to the process of reproduction, from the constituents of matter to the mechanics of volcanoes.

However, one particular question was conspicuously ignored, because scientists agreed that it was beyond their remit, indeed inaccessible to rational endeavour of any kind. Nobody, it seemed, was keen to tackle the ultimate question of how the universe was created. Scientists restricted themselves to explaining natural phenomena, and the creation of the universe was acknowledged to be a supernatural event. Also, addressing such a question would have jeopardised the mutual respect that had developed between science and religion. Modern notions of a godless Big Bang would have seemed heretical to eighteenth-century theologians, much as the Sun-centred universe had offended the Inquisition back in the seventeenth century. In Europe, the Bible continued to be the indisputable authority on the creation of the universe, and the overwhelming majority of scholars accepted that God had created the Heavens and the Earth.

It seemed that the only issue open to discussion was *when* God had created the universe. Scholars trawled through the lists of Biblical begats from Genesis onwards, adding up the years between each birth, taking into account Adam, the prophets, the reigns of the kings, and so on, keeping a careful running total as they went along. There were sufficient uncertainties for the estimated date of creation to vary by up to three thousand years, depending on who was doing the reckoning. Alfonso X of Castile and León, for instance, the king responsible for the *Alphonsine Tables*, quoted the oldest date for creation, 6904 BC, while Johannes Kepler preferred a date at the lower end of the range, 3992 BC.

The most fastidious calculation was by James Ussher, who became the Archbishop of Armagh in 1624. He employed an agent in the Middle East to seek out the oldest known Biblical texts, to make his estimate less susceptible to errors in transcription and translation. He also put an enormous effort into anchoring the Old Testament chronology to an event in recorded history. In the end, he spotted that Nebuchadnezzar's death was indirectly mentioned in the Second Book of Kings, so it could be dated in terms of Biblical history; the death and its date also appeared in a list of the Babylonian kings compiled by the astronomer Ptolemy, so it could be linked to the modern historical record. Consequently, after much tallying and historical research, Ussher was able to pronounce that the date of creation was Saturday 22 October, 4004 BC. To be even more precise, Ussher announced that time began at 6 p.m. on that day, based on a passage from the Book of Genesis which proclaimed: 'And the evening and the morning were the first day.'

While this may seem an absurdly literal interpretation of the Bible, it made perfect sense in a society that judged Scripture to be the definitive authority on the great question of creation. Indeed, Bishop Ussher's date was recognised by the Church of England in 1701, and was thereafter published in the opening margin of the King James Bible right the way through to the twentieth century. Even scientists and philosophers were happy to accept Ussher's date well into the nineteenth century.

However, the scientific pressure to question 4004 BC as the year of creation emerged strongly when Charles Darwin published his theory of evolution by natural selection. While Darwin and his supporters found natural selection compelling, they had to admit that it was a painfully slow mechanism for evolution, wholly incompatible with Ussher's statement that the world was just six thousand years old. Consequently, there was a coordinated effort to date the age of the Earth by scientific means, with the hope of establishing an age of millions or even billions of years.

Victorian geologists analysed the rate of sedimentary rock deposition and estimated that the Earth was at least several million years old. In 1897 Lord Kelvin used a different technique: assuming that the world was molten hot when it was formed, he worked out that it must have taken at least 20 million years to cool to its current temperature. A couple of years later, John Joly used a different assumption, namely that the oceans started off pure, and estimated how long it would have taken for the salt to have been dissolved to give the current salinity, which seemed to imply an age of roughly 100 million years. In the early years of the twentieth century, physicists showed that radioactivity could be used to date the Earth, which led to an estimate of 500 million years in 1905. Technical refinements of this technique raised the age to over a billion years in 1907. The dating game was proving to be an enormous scientific challenge, but it was becoming clear that each new measurement was making the Earth appear increasingly ancient.

As scientists witnessed this huge change in their perception of the Earth's age, there was a parallel shift in how they viewed the universe. Before the nineteenth century, scientists generally subscribed to the *catastrophist* view, believing that catastrophes could explain the history of the universe. In other words, our world had been created and shaped by a series of sudden cataclysmic events, such as a massive upheaval of rock to create mountains, or the Biblical flood to sculpt the geological formations that we see today. Such catastrophes were essential for the Earth to have been shaped over the course of just a few thousand years. But by the end of the nineteenth century, after studying the Earth in more detail and in light of the latest results from dating rock samples, scientists moved towards a *uniformitarian* view of the world, believing in gradual and uniform change to explain the history of the universe.

Uniformitarians were convinced that mountains did not appear overnight, but were uplifted at a rate of a few millimetres per year over the course of millions of years.

The growing uniformitarian movement came to the consensus that the Earth is more than a billion years old, and that the universe must therefore be even older, perhaps even infinitely old. An eternal universe seemed to strike a chord with the scientific community, because the theory had a certain elegance, simplicity and completeness. If the universe has existed for eternity, then there was no need to explain how it was created, when it was created, why it was created or Who created it. Scientists were particularly proud that they had developed a theory of the universe that no longer relied on invoking God.

Charles Lyell, the most prominent uniformitarian, stated that the start of time was 'beyond the reach of mortal ken'. This view was reinforced by the Scottish geologist James Hutton: 'The result therefore of our present enquiry is, that we find no vestige of a beginning, no prospect of an end.'

Uniformitarianism would have met with the approval of some of the early Greek cosmologists, such as Anaximander, who argued that planets and stars 'are born and perish within an eternal and ageless infinity'. A few decades later, in around 500 BC, Heraclitus of Ephesus reiterated the eternal nature of the universe: 'This cosmos, the same for all, was made by neither god nor man, but was, is and always will be: an ever-living fire, kindling and extinguishing according to measure.'

So, by the start of the twentieth century, scientists were content to live in an eternal universe. This theory, however, was based on quite flimsy evidence. Although there was dating evidence that pointed towards a truly ancient universe, at least billions of years old, the idea that the universe was eternal was largely based on a leap of faith. There was simply no scientific justification for extrapolating from an Earthly age of at least billions of years to a universe that was eternal. Sure enough, an infinitely old universe constituted a coherent and consistent cosmological view, but this was nothing more than wishful thinking unless somebody could find some scientific evidence to back it up. In fact, the eternal universe model was built upon such fragile foundations that it probably deserved the title of myth rather than scientific theory. The eternal universe model of 1900 was almost as flimsy as the explanation that it was the giant blue god Wulbari who separated the sky from the land.

Eventually, cosmologists would confront this embarrassing state of affairs. Indeed, they would spend the rest of the twentieth century struggling to replace the last great myth with a respectable and rigorous scientific explanation. They strove to develop a detailed theory and sought the concrete evidence to back it up, so that they could confidently address the ultimate question: is the universe eternal, or was it created?

The battle over the history of the universe, finite or infinite, would be fought by obsessive theorists, heroic astronomers and brilliant experimenters. A rebel alliance would attempt to overthrow an implacable establishment, employing the latest in technology, from giant telescopes to space satellites. Answering the ultimate question would result in one of the greatest, most controversial, most daring adventures in the history of science.

CHAPTER 1 - IN THE BEGINNING
SUMMARY NOTES

INITIALLY SOCIETIES EXPLAINED EVERYTHING IN TERMS OF MYTHS, GODS AND MONSTERS.

① IN 6TH CENTURY BC GREECE:
PHILOSOPHERS BEGAN TO DESCRIBE THE UNIVERSE IN TERMS OF NATURAL (NOT SUPERNATURAL) PHENOMENA.

GREEK PROTO-SCIENTISTS SOUGHT THEORIES AND MODELS THAT WERE:
- SIMPLE
- ACCURATE
- NATURAL
- VIABLE

THEY WERE ABLE TO MEASURE THE SIZE OF THE EARTH, SUN AND MOON AND THE DISTANCES BETWEEN THEM USING:
- EXPERIMENT/OBSERVATIONS
- LOGIC/THEORY
(+ MATHEMATICS)

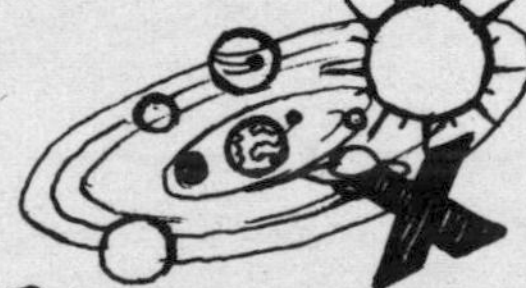

GREEK ASTRONOMERS ESTABLISHED A FALSE EARTH-CENTRED MODEL OF THE UNIVERSE WITH THE SUN, STARS AND PLANETS ORBITING A FIXED EARTH.

② WHEN THE EARTH-CENTRED MODEL WAS FOUND WANTING, ASTRONOMERS RESPONDED WITH AD-HOC FIXES.
(eg PTOLEMY'S EPICYCLES EXPLAINED THE RETROGRADE MOTION OF PLANETS)

THEOLOGIANS ENCOURAGED ASTRONOMERS TO STAY LOYAL TO THE EARTH-CENTRED MODEL AS IT WAS CONSISTENT WITH THE BIBLE.

③ IN 16TH CENTURY:
COPERNICUS CONSTRUCTED A SUN-CENTRED MODEL OF THE UNIVERSE IN WHICH THE EARTH AND OTHER PLANETS ORBITED THE SUN. IT WAS SIMPLE AND REASONABLY ACCURATE.

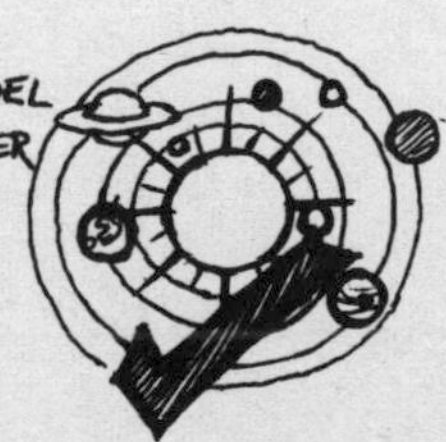

UNFORTUNATELY COPERNICUS'S SUN-CENTRED MODEL WAS IGNORED BECAUSE:

- HE WAS VIRTUALLY UNKNOWN
- HIS MODEL DEFIED COMMON-SENSE
- HIS MODEL WAS LESS ACCURATE THAN PTOLEMY'S
- RELIGIOUS (AND SCIENTIFIC) ORTHODOXY QUASHED ORIGINAL THOUGHT.

(4) COPERNICUS'S MODEL WAS IMPROVED BY KEPLER USING TYCHO'S OBSERVATIONS. HE SHOWED THAT PLANETS FOLLOW (SLIGHTLY) ELLIPTICAL, NOT CIRCULAR, ORBITS. THE SUN-CENTRED MODEL WAS NOW SIMPLER AND MORE ACCURATE THAN THE EARTH-CENTRED MODEL.

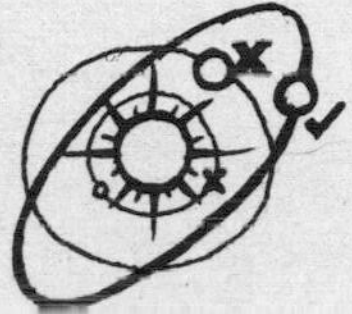

(5) GALILEO CHAMPIONED THE SUN-CENTRED MODEL. HE USED THE TELESCOPE TO SHOW THAT JUPITER HAS MOONS, THE SUN HAS SPOTS AND VENUS HAS PHASES, WHICH CONTRADICTED THE ANCIENT THEORY AND SUPPORTED THE NEW ONE.

GALILEO WROTE A BOOK EXPLAINING WHY THE SUN-CENTRED MODEL WAS CORRECT. UNFORTUNATELY THE CHURCH BULLIED GALILEO AND FORCED HIM TO RECANT IN 1633.

IN LATER CENTURIES, THE CHURCH BECAME MORE TOLERANT. ASTRONOMERS ADOPTED THE SUN-CENTRED MODEL AND SCIENCE FLOURISHED.

(6) BY 1900 COSMOLOGISTS CONCLUDED THAT THE UNIVERSE WAS NOT CREATED BUT HAD EXISTED FOR ETERNITY. BUT THERE WAS NO EVIDENCE TO BACK THIS THEORY. THE ETERNAL UNIVERSE HYPOTHESIS WAS NOT MUCH MORE THAN A MYTH.

(7) 20TH CENTURY COSMOLOGISTS WOULD RETURN TO THE BIG QUESTION AND ADDRESS IT SCIENTIFICALLY

WAS THE UNIVERSE CREATED?

OR

HAS IT EXISTED FOR ALL ETERNITY?

Chapter 2

THEORIES OF THE UNIVERSE

[Einstein's theory of relativity] is probably the greatest synthetic achievement of the human intellect up to the present time.

BERTRAND RUSSELL

It is as if a wall which separated us from the Truth has collapsed. Wider expanses and greater depths are now exposed to the searching eye of knowledge, regions of which we had not even a presentiment. It has brought us much nearer to grasping that plan that underlies all physical happening.

HERMANN WEYL

But the years of anxious searching in the dark for a truth that one feels but cannot express, the intense desire and the alternations of confidence and misgiving, and the final emergence into light – only those who have experienced it can appreciate it.

ALBERT EINSTEIN

It is impossible to travel faster than the speed of light, and certainly not desirable, as one's hat keeps blowing off.

WOODY ALLEN

During the course of the early twentieth century, cosmologists would develop and test a whole variety of models of the universe. These candidate models emerged as physicists gained a clearer understanding of the universe and the scientific laws that underpin it. What were the substances that made up the universe and how did they behave? What caused the force of gravity and how did gravity govern the interactions between the stars and planets? And the universe was made up of space and evolved in time, so what exactly did physicists mean by space and time? Crucially, answering all these fundamental questions would be possible only after physicists had addressed one seemingly simple and innocent question: what is the speed of light?

When we see a flash of lightning, it is because the lightning is emitting light, which might have to travel several kilometres towards us before reaching our eyes. Ancient philosophers wondered how the speed of light affected the act of seeing. If light travels at a finite speed, then it would take some time to reach us, so by the time we see the lightning it may no longer actually exist. Alternatively, if light travels infinitely fast then the light would reach our eyes instantaneously, and we would see the lightning strike as it is happening. Deciding which scenario was correct seemed to be beyond the wit of the ancients.

The same question could be asked about sound, but this time the answer was more obvious. Thunder and lightning are generated simultaneously, but we hear the thunder after we see the lightning. For the ancient philosophers, it was reasonable to assume that sound has a finite speed and certainly travels much slower than light. They thus established a theory of light and sound based on the following incomplete chain of reasoning:

1. A lightning strike creates light and sound.
2. Light travels either very fast or infinitely fast towards us.
3. We see lightning very soon after the event, or instantaneously.
4. Sound travels at a slower speed (roughly 1,000 km/h).
5. Therefore we hear the thunder some time later, depending on the distance to the lightning strike.

But still the fundamental question relating to the speed of light – whether it was finite or infinite – continued to exercise the world's greatest minds for centuries. In the fourth century BC, Aristotle argued that light travelled with infinite speed, so the event and the observation of the event would be simultaneous. In the eleventh century AD, the Islamic scientists Ibn Sina and al-Haytham both took the opposite view, believing that the speed of light, though exceedingly high, was finite, so any event could be observed only some time after it had happened.

There was clearly a difference of opinion, but either way the debate remained merely philosophical until 1638, when Galileo proposed a method for measuring the speed of light. Two observers with lamps and shutters would stand some distance apart. The first observer would flash a signal to the second, who would then immediately flash a signal back. The first observer could then estimate the speed of light by measuring the time between sending and receiving signals. Unfortunately Galileo was already blind and under house arrest when he came up with this idea, so he was never able to conduct his experiment.

In 1667, twenty-five years after Galileo's death, Florence's illustrious Accademia del Cimento decided to put Galileo's idea to the test. Initially, two observers stood relatively close together. One flashed a lantern at the other, and the other would see the signal and flash back. The first man estimated the time between sending the original flash and seeing the response flash, and the result was an interval of a fraction of a second. This, however, could be attributed to their reaction times. The experiment was repeated over and over again, with the two men moving farther apart, measuring the time of the return flash over increasing distances. Had the return time increased with distance, it would have indicated a relatively low and finite speed of light, but in fact the return time remained constant. This implied that the speed of light was either infinite, or so fast that the time taken by the light to travel between the two observers was insignificant compared with their reaction times. The experimenters could draw only the limited conclusion that the speed of light was somewhere between 10,000 km/h and infinity. Had it been any slower, they would have detected a steadily increasing delay as the men moved apart.

Whether the speed of light was finite or infinite remained an open question until a Danish astronomer named Ole Römer addressed the issue a few years later. As a young man, he had worked at Tycho Brahe's former observatory at Uraniborg, measuring the observatory's exact location so that Tycho's observations could be correlated with others made elsewhere in Europe. In 1672, having earned a reputation as an excellent surveyor of the heavens, he was offered a post at the prestigious Academy of Sciences in Paris, which had been set up so that scientists could pursue independent research, free from having to pander to the whims of kings, queens or popes. It was in Paris that fellow

Academician Giovanni Domenico Cassini encouraged Römer to study a strange anomaly associated with Jupiter's moons, in particular Io. Each moon should orbit Jupiter in a perfectly regular manner, just as our Moon orbits the Earth regularly, so astronomers were shocked to discover that Io's timings were slightly irregular. Sometimes Io appeared from behind Jupiter ahead of schedule by a few minutes, while at other times it was a few minutes late. A moon should not behave in this way, and everybody was baffled by Io's lackadaisical attitude.

In order to investigate the mystery, Römer studied in minute detail a table of Io's positions and timings that had been logged by Cassini. Nothing made sense, until it gradually dawned on Römer that he could explain everything if light had a finite speed, as shown in Figure 19. Sometimes the Earth and Jupiter were on the same side of the Sun, whereas at other times they were on opposite sides of the Sun and farther apart. When the Earth and Jupiter were farthest apart, then the light from Io had to travel 300,000,000 km farther before reaching the Earth compared with when the two planets were closest together. If light had a finite speed, then it would take longer for the light to cover this extra distance and it would seem as if Io was running behind schedule. In short, Römer argued that Io was perfectly regular, and its apparent irregularity was an illusion caused by the different times required for the light from Io to cover different distances to the Earth.

To help understand what is going on, imagine that you are near a cannon that is fired exactly on the hour. You hear the cannon, start your stopwatch and then start driving away in a straight line at 100 km/h, so that you are 100 km away by the time the cannon is fired again. You stop the car and hear a very faint cannon blast. Given that sound travels at roughly 1,000 km/h, you will perceive that it was 66 minutes, not 60 minutes, between the first and second cannon blasts. The 66 minutes consists of 60 minutes for the

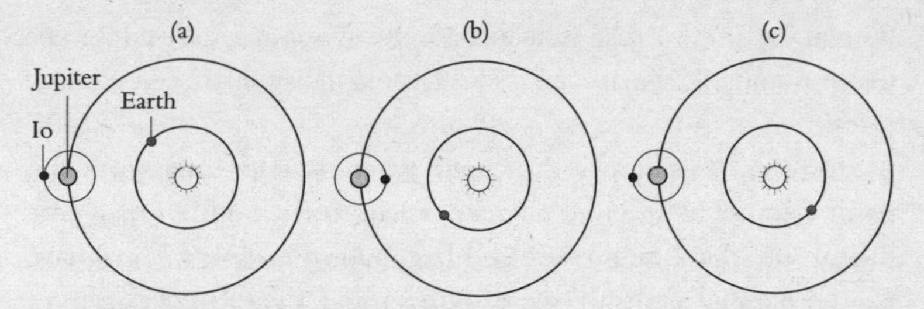

Figure 19 Ole Römer measured the speed of light by studying the movements of Jupiter's moon Io. These diagrams represent a slight variation on his actual method. In diagram (a), Io is about to disappear behind Jupiter; in diagram (b) Io has completed half a revolution so that it is in front of Jupiter. Meanwhile, Jupiter has hardly moved and the Earth has moved significantly, because the Earth orbits the Sun twelve times more quickly than Jupiter. An astronomer on the Earth measures the time that has elapsed between (a) and (b), namely the time taken for Io to complete half a revolution.

In diagram (c), Io has completed another half-revolution back to where it started, while the Earth has moved on to a position that is farther from Jupiter. The astronomer measures the time between (b) and (c), which should be the same as the time between (a) and (b), but in fact it turns out to be significantly longer. The reason for the extra time is that it takes the light from Io a little longer to cover the extra distance to the Earth in diagram (c), because the Earth is now farther away from Jupiter. The time delay and the distance between Earth and Jupiter can be used to estimate the speed of light. (The distances moved by the Earth in these diagrams are exaggerated, because Io orbits Jupiter in less than two days. Also, Jupiter's position would change and complicate matters.)

---

actual interval between firings and 6 minutes for the time taken for the sound of the second blast to cover the 100 km and reach you. The cannon is perfectly regular in its firings, but you will experience a delay of 6 minutes because of the finite speed of sound and your new position.

Having spent three years analysing the observed timings of Io and the relative positions of the Earth and Jupiter, Römer was able to estimate the speed of light to be 190,000 km/s. In fact, the true value is almost 300,000 km/s, but the important point was that

Römer had shown that light had a finite speed and derived a value that was not wildly inaccurate. The age-old debate had been resolved at last.

However, Cassini was distraught when Römer announced his result, because he received no acknowledgement from Römer, even though the calculation was based largely on his observational data. Cassini became a harsh critic of Römer and a vocal spokesman for the majority who still favoured the theory that the speed of light was infinite. Römer did not relent, and used his finite light speed to predict that an eclipse of Io on 9 November 1676 would occur 10 minutes later than predicted by his opponents. In a classic case of 'I told you so', Io's eclipse was indeed several minutes behind schedule. Römer was proved right, and he published another paper confirming his measurement of the speed of light.

This eclipse prediction should have settled the argument once and for all. Yet, as we have already seen in the case of the Sun-centred versus Earth-centred debate, factors beyond pure logic and reason sometimes influence the scientific consensus. Cassini was senior to Römer and also outlived him, so by political clout and simply by being alive he was able to sway opinion against Römer's argument that light had a finite speed. A few decades later, however, Cassini and his colleagues gave way to a new generation of scientists who would take an unbiased look at Römer's conclusion, test it for themselves and accept it.

Once scientists had established that the speed of light was finite, they set about trying to solve yet another mystery concerning its propagation: what was the medium responsible for carrying light? Scientists knew that sound could travel in a variety of media – talkative humans send sound waves through the medium of gaseous air, whales sing to each other through the medium of liquid water, and we can hear the chattering of our teeth through the medium of the solid bones between teeth and ears. Light can also travel through gases, liquids and solids, such as air, water and glass, but there was a fundamental difference between light and sound, as demonstrated by Otto von Guericke, the Burgomeister of Magdeburg, Germany, who conducted a whole series of famous experiments in 1657.

Von Guericke had invented the first vacuum pump and was keen to explore the strange properties of the vacuum. In one experiment he placed two large brass hemispheres face to face and evacuated the air from inside them so that they behaved like two exceedingly powerful suction cups. Then, in a marvellous display of scientific showmanship, he demonstrated that it was impossible for two teams of eight horses to pull the hemispheres apart.

Although this equine tug-of-war showed the power of the vacuum, it said nothing about the nature of light. This question was addressed in a somewhat daintier experiment, which required von Guericke to evacuate a glass jar containing a ringing bell. As the air was sucked out of the jar, the audience could no longer hear the ringing, but they could still see the clapper hitting the bell. It was clear, therefore, that sound could not travel through a vacuum. At the same time, the experiment showed that light could travel through a vacuum because the bell did not vanish and the jar did not darken. Bizarrely, if light could travel through a vacuum, then something could travel through nothing.

Confronted with this apparent paradox, scientists began to wonder if a vacuum was really empty. The jar had been evacuated of air, but perhaps there was something remaining inside, something that provided the medium for conveying light. By the nineteenth century, physicists had proposed that the entire universe was permeated by a substance they termed the *luminiferous ether*, which somehow acted as a medium for carrying light. This hypothetical substance had to possess some remarkable properties, as pointed out by the great Victorian scientist Lord Kelvin:

> Now what is the luminiferous ether? It is matter prodigiously less dense than air – millions and millions and millions of times less dense than air. We can form some sort of idea of its limitations. We believe it is a real thing, with great rigidity in comparison with its density: it may be made to vibrate 400 million million times per second; and yet be of such density as not to produce the slightest resistance to any body going through it.

In other words, the ether was incredibly strong, yet strangely insubstantial. It was also transparent, frictionless and chemically inert. It was all around us, yet it was clearly hard to identify because nobody had ever seen it, grabbed it or bumped into it. Nevertheless, Albert Michelson, America's first Nobel Laureate in physics, believed that he could prove its existence.

Michelson's Jewish parents had fled persecution in Prussia in 1854, when he was just two years old. He grew up and studied in San Francisco before going on to join the US Naval Academy, where he graduated a lowly twenty-fifth in seamanship, but top in optics. This prompted the Academy's superintendent to remark: 'If in the future you'd give less attention to those scientific things and more to your naval gunnery, there might come a time when you would know enough to be of some service to your country.' Michelson sensibly moved into full-time optics research, and in 1878, aged just twenty-five, he determined the speed of light to be 299,910 ± 50 km/s, which was twenty times more accurate than any previous estimation.

Then, in 1880, Michelson devised the experiment that he hoped would prove the existence of the light-bearing ether. His equipment split a single light beam into two separate perpendicular beams. One beam travelled in the same direction as the Earth's movement through space, while the other beam moved in a direction at a right angle to the first beam. Both beams travelled an equal distance, were reflected off mirrors, and then returned to combine into a single beam. Upon combining they underwent a process known as interference, which allowed Michelson to compare the two beams and identify any discrepancy in travel times.

Michelson knew that the Earth travels at roughly 100,000 km/h around the Sun, which presumably meant that it also passed through the ether at this speed. Since the ether was supposed to be a steady medium that permeated the universe, the Earth's passage through the universe would create a sort of *ether wind*. This would be similar to the sort of pseudo-wind you would feel if you were speeding along in an open-top car on a still day – there is no actual wind, but there seems to be one due to your own motion. Therefore, if light is carried in and by the ether, its speed should be affected by the ether wind. More specifically, in Michelson's experiment one light beam would be travelling into and against the ether wind and should thus have its speed significantly affected, while the other beam would be travelling across the ether wind and its speed should be less affected. If the travel times for the two beams were different, then Michelson would be able to use this discrepancy as strong evidence in favour of the ether's existence.

This experiment to detect the ether wind was complicated, so Michelson explained the underlying premise in terms of a puzzle:

> Suppose we have a river of width 100 feet, and two swimmers who both swim at the same speed, say 5 feet per second. The river flows at a steady rate of 3 feet per second. The swimmers race in the following way: they both start at the same point on one bank. One swims directly across the river to the closest point on the opposite bank, then turns around and swims back. The other stays on one side of the river, swimming downstream a distance (measured along the bank) exactly equal to the width of the river, then swims back to the start. Who wins? [See Figure 20 for the solution.]

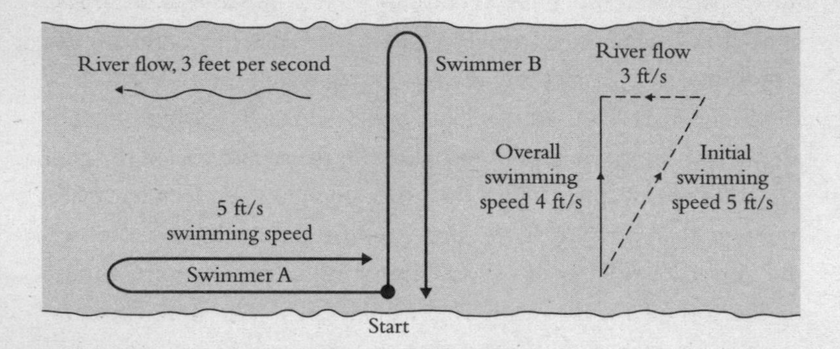

Figure 20 Albert Michelson used this swimming puzzle to explain his ether experiment. The two swimmers play the same role as the two beams of light heading in perpendicular directions, then both returning to the same starting point. One swims first with and then against the current, while the other swims across the current – just as one light beam travels with and against the ether wind, and the other across it. The puzzle is to work out the winner of a race over a distance of 200 feet between two swimmers who both can swim at 5 feet per second in still water. Swimmer A goes downstream 100 feet and back upstream 100 feet, whereas swimmer B goes across the river and back, also covering two legs of 100 feet. The river has a 3 ft/s current.

The time of swimmer A, going downstream and then upstream, is easy to analyse. With the current, the swimmer has an overall speed of 8 ft/s (5 + 3 ft/s), so the 100 feet takes just 12.5 seconds. Coming back against the current means that he is swimming at only 2 ft/s (5 – 3 ft/s), so swimming this 100 feet takes him 50 seconds. Therefore his total time is 62.5 seconds to swim 200 feet.

Swimmer B, going across the river, has to swim at an angle in order to compensate for the current. Pythagoras' theorem tells us that if he swims at 5 ft/s at the correct angle, he will have an upstream component of 3 ft/s, which cancels the effect of the current, and a cross-stream component of 4 ft/s. Therefore he swims the first width of 100 feet in just 25 seconds, and then takes another 25 seconds to return, giving a total time of 50 seconds to swim 200 feet.

Although both swimmers would swim at the same speed in still water, the swimmer crossing the current wins the race against the swimmer who goes with and against the current. Hence, Michelson suspected that a light beam travelling across the ether wind would have a shorter travel time than a beam travelling with and then against the ether wind. He designed an experiment to see if this was really the case.

Michelson invested in the best possible light sources and mirrors for his experiment and took every conceivable precaution in assembling the apparatus. Everything was carefully aligned, levelled and polished. To increase the sensitivity of his equipment and minimise errors, he even floated the main assembly in a vast bath of mercury, thereby isolating it from external influences such as the tremors caused by distant footsteps. The whole point of this experiment was to prove the existence of the ether, and Michelson had done everything possible to maximise the chance of its detection – which is why he was so astonished by his complete and utter failure to detect any difference in the arrival times of the two perpendicular beams of light. There was no sign of the ether whatsoever. It was a shocking result.

Desperate to find out what had gone wrong, Michelson recruited the chemist Edward Morley. Together they rebuilt the apparatus, improving each piece of equipment to make the experiment even more sensitive, and then they carried out the measurements over and over again. Eventually, in 1887, after seven years of repeating their experiment, they published their definitive results. There was still no sign of the ether. Therefore they were forced to conclude that the ether did not exist.

Bearing in mind its ridiculous set of properties – it was supposed to be the least dense yet the most rigid substance in the universe – it should have come as no surprise that the ether was a fiction. Nevertheless, scientists discarded it with great reluctance because it had been the only conceivable way to explain how light was transmitted. Even Michelson had problems coming to terms with his own conclusion. He once nostalgically referred to the 'beloved old ether, which is now abandoned, though I personally still cling a little to it'.

The crisis of the non-existent ether was magnified because it was supposed to have been responsible for carrying both the electric and magnetic fields as well as light. The dire situation was nicely summarised by the science writer Banesh Hoffmann:

> First we had the luminiferous ether,
> Then we had the electromagnetic ether,
> And now we haven't e(i)ther.

So, by the end of the nineteenth century Michelson had proved that the ether did not exist. Ironically, he had built his career on a whole series of successful experiments relating to optics, but his greatest triumph was the result of a failed experiment. His goal all along had been to prove the existence of the ether, not its absence. Physicists now had to accept that light could somehow travel through a vacuum – through space devoid of any medium.

Michelson's achievement had required expensive, specialist experimental apparatus and years of dedicated effort. At roughly the same time, a lone teenager, unaware of Michelson's experimental breakthrough, had also concluded that the ether did not exist, but on the basis of theoretical arguments alone. His name was Albert Einstein.

Einstein's Thought Experiments

Einstein's youthful prowess and his later full-blown genius sprang largely from his immense inquisitiveness about the world around him. Throughout his prolific, revolutionary and visionary career he never stopped wondering about the underlying laws that governed the universe. Even at the age of five, he became engrossed in the mysterious workings of a compass given to him by his father. What was the invisible force that tugged at the needle, and why did it always point to the north? The nature of magnetism became a lifelong fascination, typical of Einstein's insatiable appetite for exploring apparently trivial phenomena.

As Einstein told his biographer Carl Selig: 'I have no special talents. I am only passionately curious.' He also noted: 'The important thing is not to stop questioning. Curiosity has its own reason for existing. One cannot help but be in awe when one contemplates the mysteries of eternity, of life, of the marvellous structure of reality. It is enough if one tries to comprehend only a little of this mystery every day.' The Nobel Laureate Isidor Isaac Rabi reinforced this point: 'I think physicists are the Peter Pans of the human race. They never grow up and they keep their curiosity.'

In this respect, Einstein had much in common with Galileo. Einstein once wrote: 'We are in the position of a little child entering a huge library, whose walls are covered to the ceiling with books in many different languages.' Galileo made a similar analogy, but he condensed the entire library of nature into a single grand book and a single language, which his curiosity compelled him to decipher: 'It is written in the language of mathematics, and its characters are triangles, circles and other geometrical figures, without which it is humanly impossible to understand a single word of it; without these one is wandering about in a dark labyrinth.'

Also linking Galileo and Einstein was a common interest in the principle of relativity. Galileo had discovered the principle of relativity, but it was Einstein who would fully exploit it. Simply stated, Galilean relativity argues that all motion is relative, which means that it is impossible to detect whether or not you are moving without referring to an external reference frame. Galileo stated vividly what he meant by relativity in the *Dialogue*:

> Shut yourself up with a friend in the main cabin below deck on some large ship, and have with you there some flies, butterflies and other small flying animals. Have a large bowl of water with some fish in it; hang up a bottle that empties drop by drop into a wide vessel beneath it. With the ship standing still, observe carefully how all the little animals fly with equal speed to all sides of the cabin;

> how the fish swim indifferently in all directions; how the drops fall into the vessel beneath. And, in throwing something to your friend, you need to throw it no more strongly in one direction than another, the distances being equal; and jumping with your feet together, you pass equal spaces in every direction.
>
> When you have observed all these things carefully . . . have the ship proceed with any speed you like, so long as the motion is uniform and not fluctuating this way and that. You will discover not the least change in all the effects named, nor could you tell from any of them whether the ship moves or stands still.

In other words, as long as you are moving at constant speed in a straight line, there is nothing you can do to measure how fast you are travelling, or indeed to tell whether you are moving at all. This is because everything around you is moving at the same velocity, and all phenomena (e.g. dripping bottles, flying butterflies) happen the same regardless of whether you are moving or stationary. Also, Galileo's scenario takes place 'in the main cabin below deck', so you are isolated, which removes any hope of detecting any relative motion by referring to an external frame of reference. If you isolate yourself in a similar way by sitting with your ears plugged and your eyes shut inside a train on a smooth track, then it is very difficult to tell if the train is racing along at 100 km/h or whether it is still stuck at the station, which is another demonstration of Galilean relativity.

This was one of Galileo's greatest discoveries, because it helped to convince sceptical astronomers that the Earth does indeed go round the Sun. Anti-Copernican critics had argued that the Earth could not go around the Sun because we would feel this motion as a constant wind or as the ground being pulled from under our feet, and clearly this does not happen. However, Galileo's principle of relativity explained that we would not sense the Earth's tremendous velocity through space because everything from the ground to the atmosphere is moving through space at the same speed as we are. A moving Earth is effectively the same environment as the one we would experience if the Earth were static.

In general, Galileo's theory of relativity stated that you can never tell if you are moving quickly or moving slowly or moving at all. This holds true whether you are isolated on the Earth, or ear-plugged and blinkered on a train, or tucked away below deck on a ship, or cut off from an external reference frame in some other way.

Unaware that Michelson and Morley had disproved the existence of the ether, Einstein used Galileo's principle of relativity as his bedrock for exploring whether or not the ether existed. In particular, he invoked Galilean relativity in the context of a *thought experiment*, also known as a *gedanken experiment* (from the German word for 'thought'). This is a purely imaginary experiment that is conducted only in the physicist's head, usually because it involves a procedure that is not in practice achievable in the real world. Although a purely theoretical construct, a thought experiment can often lead to a deep understanding of the real world.

In a thought experiment he conducted in 1896, when just sixteen years old, Einstein wondered what would happen if he could travel at the speed of light while holding out a mirror in front of him. In particular, he wondered whether he would be able to see his own reflection. The Victorian theory of the ether pictured it as a static substance that permeated the entire universe. Light was supposedly carried by the ether, so this implied that it travelled at the speed of light (300,000 km/s) relative to the ether. In Einstein's thought experiment, he, his face and his mirror were also travelling through the ether at the speed of light. Therefore light would try to leave Einstein's face and try to travel towards the mirror in his hand, but it would never actually leave his face, let alone reach the mirror because everything is moving at the speed of light. If light could not reach the mirror, then it could not be reflected back, and consequently Einstein would not be able to see his own reflection.

This imaginary scenario was shocking because it completely defied Galileo's principle of relativity, according to which someone travelling at constant velocity should not be able to ascertain whether they are moving quickly, slowly, forwards, backwards – or indeed whether they are moving at all. Einstein's thought experiment implied that he would know when he was moving at the speed of light because his reflection would vanish.

The boy wonder had conducted a thought experiment based on an ether-filled universe, and the result was paradoxical because it contradicted Galileo's principle of relativity. Einstein's thought experiment can be recast in terms of Galileo's below-deck scenario: the sailor would know if the ship was moving at the speed of light because his reflection would vanish. However, Galileo had firmly declared that the sailor should be unable to tell whether his ship was moving.

Something had to give. Either Galilean relativity was wrong, or Einstein's thought experiment was fundamentally flawed. In the end, Einstein realised that his thought experiment was at fault because it was based on an ether-filled universe. To resolve the paradox, he concluded that light did not travel at some fixed speed relative to the ether, that light was not carried by the ether, and that the ether did not even exist. Unbeknown to Einstein, this is exactly what Michelson and Morley had already discovered.

You might feel wary of Einstein's slightly tortuous thought experiment, especially if you view physics as a discipline reliant on real experiments with real equipment and real measurements. Indeed, thought experiments are at the fringe of physics and are not wholly reliable, which is why Michelson and Morley's real experiment was so important. Nevertheless, Einstein's thought experiment demonstrated the brilliance of his young mind and, even more importantly, it set him on the road to addressing the implications for a universe devoid of ether and what this meant in terms of the speed of light.

The Victorian notion of the ether had been very comforting, because it provided an adequate enough context for what scientists meant when they talked about the speed of light. Everybody accepted that light travelled at a constant speed, 300,000 km/s, and everybody had assumed that this meant 300,000 km/s relative to the medium in which it travelled, which was thought to be the ether. Everything made sense in the Victorian ether-filled universe. But Michelson, Morley and Einstein had shown that there was no ether. So, if light did not require a medium in which to travel, what did it mean when scientists talked about the speed of light? The speed of light was 300,000 km/s, but relative to what?

Einstein thought about the question intermittently over the next few years. He eventually came up with a solution to the problem, but one that depended heavily on intuition. At first sight his solution seemed nonsensical, yet later he would be proved to be absolutely right. According to Einstein, light travels at a constant velocity of 300,000 km/s *relative to the observer*. In other words, no matter what our circumstances or how the light is being emitted, each one of us personally measures the same speed of light, which is 300,000 km/s, or 300,000,000 m/s (more accurately, 299,792,458 m/s). This seems absurd because it runs counter to our everyday experience of the velocities of ordinary objects.

Imagine a schoolboy with a peashooter which always fires peas at 40 m/s. You are leaning against a wall some way down the street from the schoolboy. He fires his peashooter at you, so the pea leaves the peashooter at 40 m/s, it crosses the intervening space at 40 m/s, and when it hits your forehead it certainly feels as if it was moving at 40 m/s. If the schoolboy gets on his bike and cycles towards you at

10 m/s and fires the peashooter again, then the pea still leaves the peashooter at 40 m/s, but it covers the ground at 50 m/s and feels like 50 m/s when it hits you. The extra speed is down to the pea being launched from a moving bicycle. And if you march towards the schoolboy at 4 m/s then the situation gets even worse, because the pea now feels like it is moving at 54 m/s. In summary, you (the observer) perceive a different pea speed depending on a variety of factors.

Einstein believed that light behaved differently. When the boy is not riding his bicycle, then the light from his bicycle lamp strikes you at a speed of 299,792,458 m/s. When the bike is ridden towards you at 10 m/s, then the light from the lamp still strikes you at a speed of 299,792,458 m/s. And even when you start moving towards the bike while it is moving towards you, then the light still strikes you at 299,792,458 m/s. Light, insisted Einstein, travels at a constant velocity relative to the observer. Whoever is measuring the speed of light always comes up with the same answer, whatever the situation. Experiments would later demonstrate that Einstein was correct. The distinction between the behaviour of light and other things, such as peas, is laid out below.

	Your perception of the speed of peas	Your perception of the speed of light
Nobody is moving	40 m/s	299,792,458 m/s
Schoolboy cycles towards you at 10 m/s	50 m/s	299,792,458 m/s
. . . and you walk towards the boy at 4 m/s	54 m/s	299,792,458 m/s

Einstein was convinced that the speed of light must be constant for the observer because it seemed to be the only way to make sense of his mirror-based thought experiment. We can re-examine the thought experiment according to this new rule for the speed of light. If Einstein, who was the observer in his thought experiment, were to travel at the speed of light, he would nonetheless see the light leaving his face at the speed of light, because it travels relative to the observer. So the light would leave Einstein at the speed of light, and would be reflected back at the speed of light, which means that he would now be able to see his reflection. Exactly the same thing would happen if he were to stand still in front of his bathroom mirror – the light would leave his face at the speed of light and be reflected back at the speed of light, and he would see his reflection. In other words, by assuming that the speed of light was constant relative to the observer, then Einstein would not be able to tell whether he was moving at the speed of light or standing still in his bathroom. This is exactly what Galileo's principle of relativity required, namely that you have the same experience whether or not you are moving.

The constancy of the speed of light relative to the observer was a striking conclusion, and it continued to dominate Einstein's thoughts. He was still only a teenager, so it was with the ambition and naivety of youth that he explored the implications of his ideas. Eventually, he would go public and shake the world with his revolutionary ideas, but for the time being he worked in private and continued with his mainstream education.

Crucially, throughout this period of contemplation, Einstein maintained his natural verve, creativity and curiosity, despite the authoritarian nature of his college. He once said: 'The only thing that interferes with my learning is my education.' He paid little attention to his lecturers, including the distinguished Hermann Minkowski, who responded by dismissing him as 'a lazy dog'.

Another lecturer, Heinrich Weber, told him: 'You are a smart boy, Einstein, a very smart boy. But you have one great fault: you do not let yourself be told anything.' Einstein's attitude was partly due to Weber's refusal to teach the latest ideas in physics, which is also the reason why Einstein addressed him as plain Herr Weber, rather than Herr Professor Weber.

As a result of this battle of wills, Weber did not write the letter of recommendation that Einstein required to pursue an academic career. Consequently, Einstein spent the next seven years after graduation as a clerk in the patent office at Berne, Switzerland. As it turned out, this was not such a terrible predicament. Instead of being constrained by the mainstream theories promulgated at the great universities, Einstein could now sit in his office and think about the implications of his teenage thought experiment – exactly the sort of speculative deliberations that Herr Professor Weber would have pooh-poohed. Also, Einstein's prosaic office job, initially 'probationary technical expert, third class', allowed him to squeeze all of his patenting responsibilities into just a few hours each day, leaving him plenty of time to conduct his personal research. Had he been a university academic, he would have wasted day after day dealing with institutional politics, endless administrative chores and burdensome teaching responsibilities. In a letter to a friend, he described his office as 'that secular cloister, where I hatched my most beautiful ideas'.

These years as a patent clerk would prove to be one of the most fruitful periods of his intellectual life. At the same time, it was a highly emotional time for the maturing genius. In 1902, Einstein experienced the deepest shock of his entire life when his father fell fatally ill. On his deathbed, Hermann Einstein gave Albert his blessing to marry Mileva Marić, unaware that the couple already had a daughter, Lieserl. In fact, historians were also unaware of Albert and Mileva's daughter until they were given access to Einstein's personal correspondence in the late 1980s. It emerged that Mileva had returned to her native Serbia to give birth, and as soon as Einstein heard the news of their daughter's arrival he wrote to Mileva: 'Is she healthy and does she already cry properly? What kind of little eyes does she have? Who of us two does she resemble more? Who is giving her milk? Is she hungry? And is she completely bald? I love her so much and I do not even know her yet! . . . She certainly can cry already, but will learn to laugh only much later. Therein lies a deep truth.' Albert would never hear his daughter cry or watch her laugh. The couple could not risk the social disgrace of having an illegitimate daughter, and Lieserl was put up for adoption in Serbia.

Albert and Mileva were married in 1903, and their first son, Hans Albert, was born the next year. In 1905, while juggling the responsibilities of fatherhood and his obligations as a patent clerk, Einstein finally managed to crystallise his thoughts about the universe. His theoretical research climaxed in a burst of scientific papers which appeared in the journal *Annalen der Physik*. In one paper, he analysed a phenomenon known as Brownian motion and thereby presented a brilliant argument to support the theory that matter is composed of atoms and molecules. In another paper, he showed that a well-established phenomenon called the photoelectric effect could be fully explained using the newly developed theory of quantum physics. Not surprisingly, this paper went on to win Einstein a Nobel prize.

The third paper, however, was even more brilliant. It summarised Einstein's thoughts over the previous decade on the speed of light and its constancy relative to the observer. The paper created an entirely new foundation for physics and would ultimately lay the ground rules for studying the universe. It was not so much the constancy of the speed of light itself that was so important, but the consequences that Einstein predicted. The

Figure 21 Albert Einstein pictured in 1905, the year he published his special theory of relativity and established his reputation.

repercussions were mind-boggling, even to Einstein himself. He was still a young man, barely twenty-six years old when he published his research, and he had experienced periods of enormous self-doubt as he worked towards what has become known as his *special theory of relativity*: 'I must confess that at the very beginning when the special theory of relativity began to germinate in me, I was visited by all sorts of nervous conflicts. When young I used to go away for weeks in a state of confusion, as one who at the time had yet to overcome the state of stupefaction in his first encounter with such questions.'

One of the most amazing outcomes of Einstein's special theory of relativity is that our familiar notion of time is fundamentally wrong. Scientists and non-scientists had always pictured time as the progression of some kind of universal clock that ticked relentlessly, a cosmic heartbeat, a benchmark against which all other clocks could be set. Time would therefore be the same for everybody, because we would all live by the same universal clock: the same pendulum would swing at the same rate today and tomorrow, in London or in Sydney, for you and for me. Time was assumed to be absolute, regular and universal. No, said Einstein: time is flexible, stretchable and personal, so your time may be different from my time. In particular, a clock moving relative to you ticks more slowly than a static clock alongside you. So if you were on a moving train and I was standing on a station platform looking at your watch as you whizzed by, then I would perceive your watch to be running more slowly than my own watch.

This seems impossible, but for Einstein it was logically unavoidable. What follows in the next few paragraphs is a brief explanation of why time is personal to the observer and depends on the travelling speed of the clock being observed. Although there is a small amount of mathematics, the formulas are quite simple, and if you can follow the logic then you will understand exactly why special relativity forces us to change our view of the world. However, if you do skip the mathematics or get stuck, then don't worry, because the most important points will be summarised when the mathematics is complete.

To understand the impact of the special theory of relativity on the concept of time, let us consider an inventor, Alice, and her very unusual clock. All clocks require a ticker, something with a regular beat that can be used to count time, such as a swinging pendulum in a grandfather clock or a constant dripping in a water-clock. In Alice's clock, the ticker is a pulse of light that is reflected between two parallel mirrors 1.8 metres apart, as shown in Figure 22(a). The reflections are ideal for keeping time, because the speed of light is constant and so the clock will be highly accurate. The speed of light is 300,000,000 m/s (which can be written as 3×10^8 m/s), so if one tick is defined as the time for the light pulse to travel from one mirror to the other and back again, then Alice sees that the time between ticks is

$$\text{Time}_{\text{Alice}} = \frac{\text{distance}}{\text{speed}} = \frac{3.6\,\text{m}}{3 \times 10^8\,\text{m/s}} = 1.2 \times 10^{-8}\,\text{s}$$

Alice takes her clock inside a train carriage, which moves at a constant velocity down a straight track. She sees that the duration for each tick remains the same – remember, everything should remain the same because Galileo's principle of relativity says that it should be impossible for her to tell whether she is stationary or moving by studying objects that are travelling with her.

Meanwhile, Alice's friend Bob is standing on a station platform as her train whizzes past at 80% of the speed of light, which is 2.4×10^8 m/s (this is an express train in the most extreme sense of the word). Bob can see Alice and her clock through a large window in her carriage, and from his point of view the light pulse traces out an angled path, as shown in Figure 22(b). He sees the light pulse as following its usual up-and-down motion, but for him it is also moving sideways, along with the train.

In other words, in between leaving the lower mirror and arriving at the upper mirror, the clock has moved forward, so the light has to follow a longer diagonal path. In fact, from Bob's perspective, the train has moved forward 2.4 metres by the time the pulse has reached the upper mirror, which leads to a diagonal path length of 3.0 metres, so the light pulse has to cover 6.0 metres (up and down) between ticks. Because, according to Einstein, the speed of light is constant for any observer, for Bob the time between ticks must be

(a)

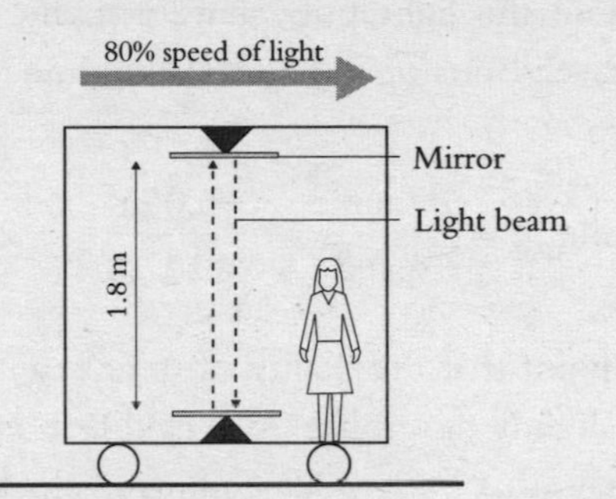

(b)

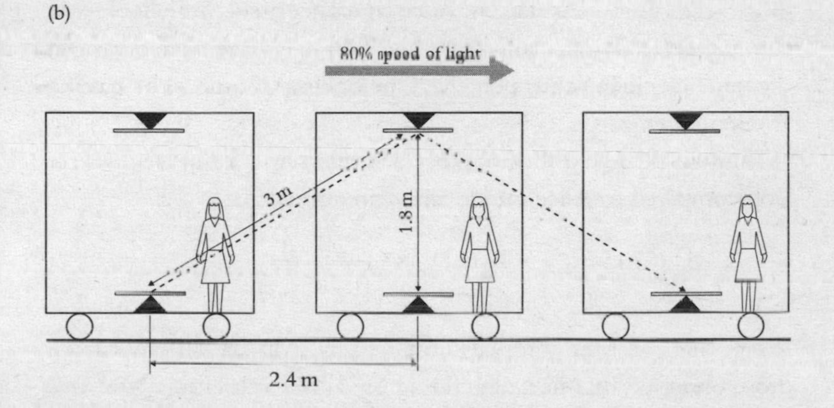

Figure 22 The following scenario demonstrates one of the main consequences of Einstein's special theory of relativity. Alice is inside her railway carriage with her mirror-clock, which 'ticks' regularly as the light pulse is reflected between the two mirrors. Diagram (a) shows the situation from Alice's perspective. The carriage is moving at 80% of the speed of light, but the clock is not moving relative to Alice, so she sees it behaving quite normally and ticking at the same rate as it always has.

Diagram (b) shows the same situation (Alice and her clock) from Bob's perspective. The carriage is moving at 80% of the speed of light, so Bob sees the light pulse follow a diagonal path. Because the speed of light is constant for any observer, Bob perceives that it takes longer for the light pulse to follow the longer diagonal path, so he thinks that Alice's clock is ticking more slowly than Alice herself perceives the ticking.

longer because the light pulse travels at the same speed but has farther to travel. Bob's perception of the time between ticks is easy to calculate:

$$\text{Time}_{\text{Bob}} = \frac{\text{distance}}{\text{speed}} = \frac{6.0\text{ m}}{3\times10^{8}\text{ m/s}} = 2.0\times10^{-8}\text{ s}$$

It is at this point that the reality of time begins to look extremely bizarre and slightly disturbing. Alice and Bob meet up and compare notes. Bob says that he saw Alice's mirror-clock ticking once every 2.0×10^{-8} s, whereas Alice maintains that her clock was ticking once every 1.2×10^{-8} s. As far as Alice is concerned, her clock was running perfectly normally. Alice and Bob may have been staring at the same clock, but they perceived the ticking of time to be passing at different rates.

Einstein devised a formula that described how time changes for Bob compared to Alice under every circumstance:

$$\text{Time}_{\text{Bob}} = \text{Time}_{\text{Alice}} \times \frac{1}{\sqrt{\left(1 - v_{\text{A}}^2/c^2\right)}}$$

It says that the time intervals observed by Bob are different from those observed by Alice, depending on Alice's velocity (v_{A}) relative to Bob and the speed of light (c). If we insert the numbers appropriate to the case described above, then we can see how the formula works:

$$\text{Time}_{\text{Bob}} = 1.2\times10^{-8}\text{ s} \times \frac{1}{\sqrt{\left(1 - (0.8c)^2/c^2\right)}}$$

$$\text{Time}_{\text{Bob}} = 1.2\times10^{-8}\text{ s} \times \frac{1}{\sqrt{(1 - 0.64)}}$$

$$\text{Time}_{\text{Bob}} = 2.0\times10^{-8}\text{ s}$$

Einstein once quipped: 'Put your hand on a hot stove for a minute, and it seems like an hour. Sit with a pretty girl for an hour, and it seems like a minute. *That's* relativity.' But the theory of special relativity was no joke. Einstein's mathematical formula described exactly how any observer would genuinely perceive time to slow down when looking at a moving clock, a phenomenon known as *time dilation*. This seems so utterly perverse that it raises four immediate questions:

1. *Why don't we ever notice this peculiar effect?*
The extent of the time dilation depends on the speed of the clock or object in question compared with the speed of light. In the above example the time dilation is significant because Alice's carriage is travelling at 80% of the speed of light, which is 240,000,000 m/s. However, if the carriage were travelling at a more reasonable speed of 100 m/s (360 km/h), then Bob's perception of Alice's clock would be almost the same as her own. Plugging the appropriate numbers into Einstein's equation would show that the difference in their perception of time would be just one part in a trillion. In other words, it is impossible for humans to detect the everyday effects of time dilation.

2. *Is this difference in time real?*
Yes, it is very real. There are numerous pieces of sophisticated hi-tech gadgetry that have to take into account time dilation in order to work properly. The Global Positioning System (GPS), which relies on satellites to pinpoint locations for devices such as car navigation systems, can function accurately only because it takes into account the effects of special relativity. These effects are significant because the GPS satellites travel at very high speeds and they make use of high-precision timings.

3. *Does Einstein's special theory of relativity apply only to clocks relying on light pulses?*
The theory applies to all clocks and, indeed, to all phenomena. This is because light actually determines the interactions that take place at the atomic level. Therefore all the atomic interactions taking place in the carriage slow down from Bob's point of view. He cannot view these individual atomic interactions, but he can view the combined effect of this atomic slowing-down. As well as seeing Alice's mirror-clock ticking more slowly, Bob would see her waving to him more slowly as she passed by; she would blink and think more slowly, and even her heartbeat would slow down. Everything would be similarly affected by the same degree of time dilation.

4. *Why can't Alice use the slowing of her clock and her own movements to prove that she is moving?*
All the peculiar effects described above are as observed by Bob from outside the moving train. As far as Alice is concerned, everything inside the train is perfectly normal, because neither her clock nor anything else in her carriage is moving relative to herself. Zero relative motion means zero time dilation. We should not be surprised that there is no time dilation, because if Alice noticed any change in her immediate surroundings as a result of her carriage's motion, it would contravene Galileo's principle of relativity. However, if Alice looked at Bob as she whizzed past him, it would appear to her that it was Bob and his environment that was undergoing time dilation, because he is moving relative to her.

The special theory of relativity impacts on other aspects of physics in equally staggering ways. Einstein showed that as Alice approaches, Bob perceives that she contracts along her direction of motion. In other words, if Alice is 2 m tall and 25 cm from front to back, and she is facing the front of the train as it approaches Bob, then he will see her as still 2 m tall but only 15 cm from front to back. She appears to be thinner. This is nothing as trivial as a perspective-based illusion, but is in fact a reality in Bob's view of distance and space. It is a consequence of the same sort of reasoning that showed that Bob observes Alice's clock ticking more slowly.

So, as well as assaulting traditional notions of time, special relativity was forcing physicists to reconsider their rock-solid notion of space. Instead of time and space being constant and universal, they were flexible and personal. It is not surprising that Einstein himself, as he developed his theory, sometimes found it difficult to trust his own logic and conclusions. 'The argument is amusing and seductive,' he said, 'but for all I know, the Lord might be laughing over it and leading me around by the nose.'

Nevertheless, Einstein overcame his doubts and continued to pursue the logic of his equations. After his research was published, scholars were forced to acknowledge that a lone patent clerk had made one of the most important discoveries in the history of physics. Max Planck, the father of quantum theory, said of Einstein: 'If [relativity] should prove to be correct, as I expect it will, he will be considered the Copernicus of the twentieth century.'

Einstein's predictions of time dilation and length contraction were all confirmed by experiments in due course. His special theory of relativity alone would have been enough to make him one of the most brilliant physicists of the twentieth century, providing as it did a radical overhaul of Victorian physics, but Einstein's stature was set to reach even greater heights.

Soon after publishing his 1905 papers, he set to work on a programme of research that was even more ambitious. To put it into context, Einstein once called his special theory of relativity 'child's play' compared with what came after it. The rewards, however, would be well worth the effort. His next great discovery would reveal how the universe behaved on the grandest scale and provide cosmologists with the tools they needed to address the most fundamental questions imaginable.

The Gravity Battle: Newton v. Einstein

Einstein's ideas were so iconoclastic that it took time for mainstream scientists to welcome this deskbound civil servant into their community. Although he published his special theory of relativity in 1905, it was not until 1908 that he received his first junior academic post at Berne University. Between 1905 and 1908, Einstein continued to work at the patent office in Berne, where he was promoted to 'technical expert, second class' and given the time to push ahead with his effort to extend the power and remit of his theory of relativity.

The special theory of relativity is labelled *special* because it applies only to special situations, namely those in which objects are moving at constant velocity. In other words, it could deal with Bob observing Alice's train travelling at a fixed speed on a straight track, but not with a train that was speeding up or slowing down. Consequently, Einstein attempted to reformulate his theory so that it would cope with situations involving acceleration and deceleration. This grand extension of special relativity would soon become known as *general relativity*, because it would apply to more general situations.

When Einstein made his first breakthrough in building general relativity in 1907, he called it 'the happiest thought of my life'. What followed, however, was eight years of torment. He told a friend how his obsession with general relativity was forcing him to neglect every other aspect of his life: 'I cannot find the time to write because I am occupied with truly great things. Day and night I rack my brain in an effort to penetrate more deeply into the things that I gradually discovered in the past two years and that represent an unprecedented advance in the fundamental problems of physics.'

In speaking of 'truly great things' and 'fundamental problems', Einstein was referring to the fact that the general theory of relativity seemed to be leading him towards an entirely new theory of gravity. If Einstein was right, then physicists would be forced to question the work of Isaac Newton, one of the icons of physics.

Newton was born in tragic circumstances on Christmas Day 1642, his father having died just three months earlier. While Isaac was still an infant, his mother married a sixty-three-year-old rector, Barnabas Smith, who refused to accept Isaac into his home. It fell to Isaac's grandparents to bring him up, and as each year passed he developed a growing hatred towards the mother and stepfather who had abandoned him. Indeed, as an undergraduate, he compiled a list of childhood sins that included the admission of 'threatening my father and mother Smith to burne them and the house over them'.

Not surprisingly, Newton grew into an embittered, isolated and sometimes cruel man. For example, when he was appointed Warden of the Royal Mint in 1696, he implemented a harsh regime of capturing counterfeiters, making sure that those convicted were hung, drawn and quartered. Forgery had brought Britain to the brink of economic collapse, and Newton judged that his punishments were necessary. In addition to brutality, Newton also used his brains to save the nation's currency. One of his most important innovations at the Mint was to introduce milled edges on coins to combat the practice of clipping, whereby counterfeiters would shave off the edges of coins and use the clippings to make new coins.

In recognition of Newton's contribution, the British £2 coin issued in 1997 had the phrase STANDING ON THE SHOULDERS OF GIANTS around its milled edge. These words are taken from a letter that Newton sent to fellow scientist Robert Hooke, in which he wrote: 'If I have seen further it is by standing on the shoulders of giants.' This appears to be a statement of modesty, an admission that

Newton's own ideas were built upon those of illustrious predecessors such as Galileo and Pythagoras. In fact, the phrase was a veiled and spiteful reference to Hooke's crooked back and severe stoop. In other words, Newton was pointing out that Hooke was neither a physical giant, nor, by implication, an intellectual giant.

Whatever his personal failings, Newton made an unparalleled contribution to seventeenth-century science. He laid the foundations for a new scientific era with a research blitz that lasted barely eighteen months, culminating in 1666, which is today known as Newton's *annus mirabilis*. The term was originally the title of a John Dryden poem about other more sensational events that took place in 1666, namely London's survival after the Great Fire and the victory of the British fleet over the Dutch. Scientists, however, judge Newton's discoveries to be the true miracles of 1666. His *annus mirabilis* included major breakthroughs in calculus, optics and, most famously, gravity.

In essence, Newton's law of gravity states that every object in the universe attracts every other object. More exactly, Newton defined the force of attraction between any two objects as

$$F = \frac{G \times m_1 \times m_2}{r^2}$$

The force (F) between the two objects depends on the masses of the objects (m_1 and m_2) – the bigger the masses, the bigger the force. Also, the force is inversely proportional to the square of the distance between the objects (r^2), which means that the force gets smaller as the objects move farther apart. The gravitational constant (G) is always equal to $6.67 \times 10^{-11}\ \mathrm{N\,m^2\,kg^{-2}}$, and reflects the strength of gravity compared with other forces such as magnetism.

The power of this formula is that it encapsulates everything that Copernicus, Kepler and Galileo had been trying to explain about the Solar System. For example, the fact that an apple falls towards the ground is not because it wants to get to the centre of the universe, but simply because the Earth and the apple both have mass, and so are naturally attracted towards each other by the force of gravity. The apple accelerates towards the Earth, and at the same time the Earth even accelerates up towards the apple, although the effect on the Earth is imperceptible because it is much more massive than the apple. Similarly, Newton's gravity equation can be used to explain how the Earth orbits the Sun because both bodies have a mass and therefore there is a mutual attraction between them. Again, the Earth orbits the Sun and not vice versa because the Earth is much less massive than the Sun. In fact, Newton's gravity formula can even be used to predict that moons and planets will follow elliptical paths, which is exactly what Kepler demonstrated after analysing Tycho Brahe's observations.

For centuries after his death, Newton's law of gravity ruled the cosmos. Scientists assumed that the problem of gravity had been solved and used Newton's formula to explain everything from the flight of an arrow to the trajectory of a comet. Newton himself, however, suspected that his understanding of the universe was incomplete: 'I do not know what I may appear to the world, but to myself I seem to have been only a little boy playing on the seashore, and diverting myself now and then in finding a smoother pebble or a prettier shell than ordinary, whilst the great ocean of truth lay undiscovered before me.'

And it was Albert Einstein who first realised that there might be more to gravity than Newton had imagined. After his own *annus mirabilis* in 1905, when Einstein published several historic papers, he concentrated on expanding his special theory of relativity into a general theory. This involved a radically different interpretation of gravity based on a fundamentally different vision of how planets, moons and apples attract one another.

At the heart of Einstein's new approach was his discovery that both distance and time are flexible, which was a consequence of his special theory of relativity. Remember, Bob sees a clock slowing down and Alice getting thinner as they move towards him. So time is flexible, as are the three dimensions of space (width, height, depth). Furthermore, the flexibility of both space and time are inextricably linked, which led Einstein to consider a single flexible entity known as *spacetime*. And it turned out that this flexible spacetime was the underlying cause of gravity. This cavalcade of weird flexibility is undoubtedly mind-bending, but the following paragraph provides a reasonably easy way to visualise Einstein's philosophy of gravity.

Spacetime consists of four dimensions, three of space and one of time, which is unimaginable for most mortals, so it is generally easier to consider just two dimensions of space, as shown in Figure 23. Fortunately, this rudimentary spacetime illustrates many of the key features of authentic spacetime, so this is a convenient simplification. Figure 23(a) shows that space (and indeed spacetime) is rather like a piece of stretchy fabric; the gridlines help to show that if nothing is occupying space, then its 'fabric' is flat and undisturbed. Figure 23(b) shows how two-dimensional space changes severely if an object is placed upon it. This second diagram could represent space being warped by the massive Sun, rather like a trampoline curving under the weight of a bowling ball.

In fact, the trampoline analogy can be extended. If the bowling ball represents the Sun, then a tennis ball representing the Earth could be launched into orbit around it, as shown in Figure 23(c). The tennis ball actually creates its own tiny dimple in the trampoline and it carries this dimple with it around the trampoline. If we wanted to model the Moon, then we could try to roll a marble in the tennis ball dimple and make it race around the tennis ball, while the tennis ball and its dimple raced around the hollow caused by the bowling ball.

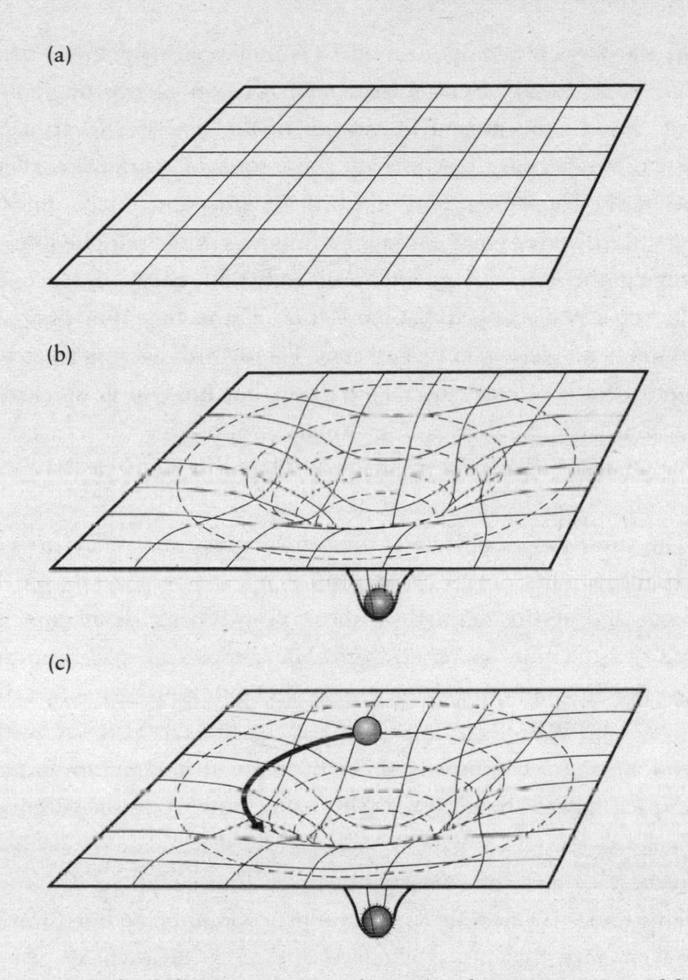

Figure 23 These diagrams are two-dimensional representations of four-dimensional spacetime, ignoring time and one space dimension. Diagram (a) shows a flat, smooth, undisturbed grid, representing empty space. If a planet were to pass through this space, then it would follow a straight line.

Diagram (b) shows space warped by an object such as the Sun. The depth of the depression depends on the mass of the Sun.

Diagram (c) shows a planet orbiting the depression caused by the Sun. The planet causes its own little depression in space, but it is too small to be represented in this diagram because the planet is relatively light.

In practice, any attempt to model a complicated system on a trampoline soon breaks down, because the friction of the trampoline fabric disturbs the natural movement of the objects. Nevertheless, Einstein was arguing that exactly these sorts of trampoline effects were really happening in the fabric of spacetime. According to Einstein, whenever physicists and astronomers witnessed phenomena involving the force of gravitational attraction, they were actually seeing objects reacting to the curvature of spacetime. For example, Newton would have said that an apple fell to Earth because there was a mutual force of gravitational attraction, but Einstein now felt that he had a deeper understanding of what was driving this attraction: the apple fell to Earth because it was falling into the deep hollow in spacetime caused by the mass of the Earth.

The presence of objects in spacetime gives rise to a two-way relationship. The shape of spacetime influences the motion of objects, and at the same time those very objects determine the shape of spacetime. In other words, the dimples in spacetime that guide the Sun and the planets are caused by those selfsame Sun and planets. John Wheeler, one of the leading general relativists of the twentieth century, summed up the theory with the dictum 'Matter tells space how to bend; space tells matter how to move.' Although Wheeler sacrificed accuracy for snappiness ('space' should have been 'spacetime'), this is still a neat summary of Einstein's theory.

This notion of flexible spacetime may sound crazy, but Einstein was convinced that it was right. According to his own set of aesthetic criteria, the link between flexible spacetime and gravity had to be true, or as Einstein put it: 'When I am judging a theory, I ask myself whether, if I were God, I would have arranged the world in such a way.' However, if Einstein was to convince the rest of the world that he was right, he had to develop a formula that encapsulated his theory. His greatest challenge was to transform the rather vague notion of spacetime and gravity described above into a formal theory of general relativity, set in a rigorous mathematical framework.

It would take Einstein eight years of arduous theoretical research before he could back up his intuition with a detailed, reasoned mathematical argument, during which time he suffered major setbacks and had to endure periods when his calculations seemed to fall apart. The intellectual effort would push Einstein to the brink of a nervous breakdown. His state of mind and level of frustration are revealed in brief comments he made in letters to friends during these years. He begged Marcel Grossman: 'You must help me or else I'll go crazy!' He told Paul Ehrenfest that working on relativity was like enduring 'a rain of fire and brimstone'. And in another letter, he worried that he had 'again perpetrated something about gravitation theory which somewhat exposes me to the danger of being confined to a madhouse'.

The courage required to venture into uncharted intellectual territory cannot be underestimated. In 1913 Max Planck even warned Einstein against working on general relativity: 'As an older friend I must advise you against it for in the first place you will not succeed, and even if you succeed no one will believe you.'

Einstein persevered, endured his ordeal and finally completed his theory of general relativity in 1915. Like Newton, Einstein had finally developed a mathematical formula to explain and calculate the force of gravity in every conceivable situation, but Einstein's formula was very different and was built on a completely separate premise – the existence of a flexible spacetime.

Newton's theory of gravity had been sufficient for the previous two centuries of physics, so why should physicists suddenly abandon it for Einstein's newfangled theory? Newton's theory could successfully predict the behaviour of everything from apples to planets, from cannonballs to raindrops, so just what was the point of Einstein's theory?

The answer lies in the nature of scientific progress. Scientists attempt to create theories to explain and predict natural phenomena as accurately as possible. A theory could work satisfactorily for a few years, decades or centuries, but eventually scientists might develop and adopt a better theory, one that is more accurate, one that works in a wider range of situations, one that accounts for previously unexplained phenomena. This is exactly what happened with early astronomers and their understanding of the position of the Earth in the cosmos. Initially, astronomers believed that the Sun orbited a stationary Earth and, thanks to Ptolemy's epicycles and deferents, this was a fairly successful theory. Indeed, astronomers used it to predict the motions of the planets with reasonable accuracy. However, the Earth-centred theory was eventually replaced by the Sun-centred theory of the universe because this new theory, based on Kepler's elliptical orbits, was more accurate and could explain new telescopic observations such as the phases of Venus. It was a long and painful transition from one theory to the other, but once the Sun-centred theory had proved itself, there was no turning back.

In much the same way, Einstein believed that he was providing physics with an improved theory of gravity, one that was more accurate and closer to reality. In particular, Einstein suspected that Newton's theory of gravity might fail in certain circumstances, whereas his own theory would be successful in every situation. According to Einstein, Newton's theory would give incorrect results when predicting phenomena in situations where the gravitational force was extreme. Therefore, in order to prove that he was right, Einstein merely had to find one of these extreme scenarios and put both his and Newton's theories of gravity to the test. Whichever theory could mimic reality most accurately would win the contest and reveal itself to be the true theory of gravity.

The problem for Einstein was that every single scenario on Earth involved the same level of mediocre gravity, and in these conditions both theories of gravity were equally successful and matched each other. Consequently, he realised that he would have to look beyond the Earth and into space to find an extreme gravity environment that might expose the flaws in Newton's theory. In particular, he knew that the Sun has a tremendous gravitational field and that the planet closest to the Sun, Mercury, would feel a high gravitational attraction. He wondered if the Sun's attraction was strong enough to make Mercury behave in a way that was inconsistent with Newton's theory of gravity and perfectly in keeping with his own theory. On 18 November 1915, Einstein came across the test case that he needed – a piece of planetary behaviour which had been bothering astronomers for decades.

Back in 1859, the French astronomer Urbain Le Verrier had analysed an anomaly in the orbit of Mercury. The planet has an elliptical orbit, but instead of being fixed the ellipse itself rotates around the Sun, as shown in Figure 24. The elliptical orbit twists around the Sun, tracing out a classic Spirograph pattern. The twisting is very slight, amounting to just 574 arcseconds per century, and it takes a million orbits and over 200,000 years for Mercury to cycle its way around the Sun and return to its original orbital orientation.

Astronomers had assumed that Mercury's peculiar behaviour was caused by the gravitational tug of the other planets in the Solar System pulling at its orbit, but when Le Verrier used Newton's formula for gravity he found that the combined effect of the other planets could account for only 531 out of the 574 arcseconds of twisting that took place each century. This meant that 43 arcseconds of the twisting was unexplained. According to some, there had to be an extra, unseen influence on Mercury's orbit that was causing the 43 arcseconds of twisting, such as an inner asteroid belt or an unknown moon of Mercury. Some even suggested the existence of a hitherto undiscovered planet, dubbed Vulcan, within Mercury's orbit. In other words, astronomers assumed that

Newton's gravity formula was correct, and that the problem must lie with a failure on their part to input all the necessary factors. Once they found the new asteroid belt, moon or planet, they expected that redoing the calculation would yield the right answer, 574 arcseconds.

Einstein, however, was sure that there was no undiscovered asteroid belt, moon or planet, and that the problem lay with Newton's gravity formula. Newton's theory worked fine in terms of describing what happened in the lesser gravity of the Earth, but Einstein was confident that the extreme gravity found close to the Sun was outside Newton's comfort zone. This was a perfect arena for the contest between the two rival theories of gravitation, and Einstein fully expected that his own theory would accurately account for Mercury's twisting orbit.

He sat down, performed the necessary calculations using his own formula, and the result was 574 arcseconds, in exact agreement with observation. 'For a few days', wrote Einstein, 'I was beside myself with joyous excitement.'

Unfortunately, the physics community was not entirely convinced by Einstein's calculation. The scientific establishment is inherently conservative, as we already know, partly for practical reasons and partly for emotional reasons. If a new theory overturns an old one, the old theory has to be abandoned and what remains of the scientific framework has to be reconciled with the new theory. Such an upheaval is justified only if the establishment is utterly convinced that the new idea really works. In other words, the burden of proof always falls on the advocates of any new theory. The emotional barrier to acceptance is equally high. Senior scientists who had spent their entire lives believing in Newton were naturally reluctant to discard what they understood and trusted in favour of some upstart theory. Mark Twain also made a perceptive point: 'A scientist will never show any kindness for a theory which he did not start himself.'

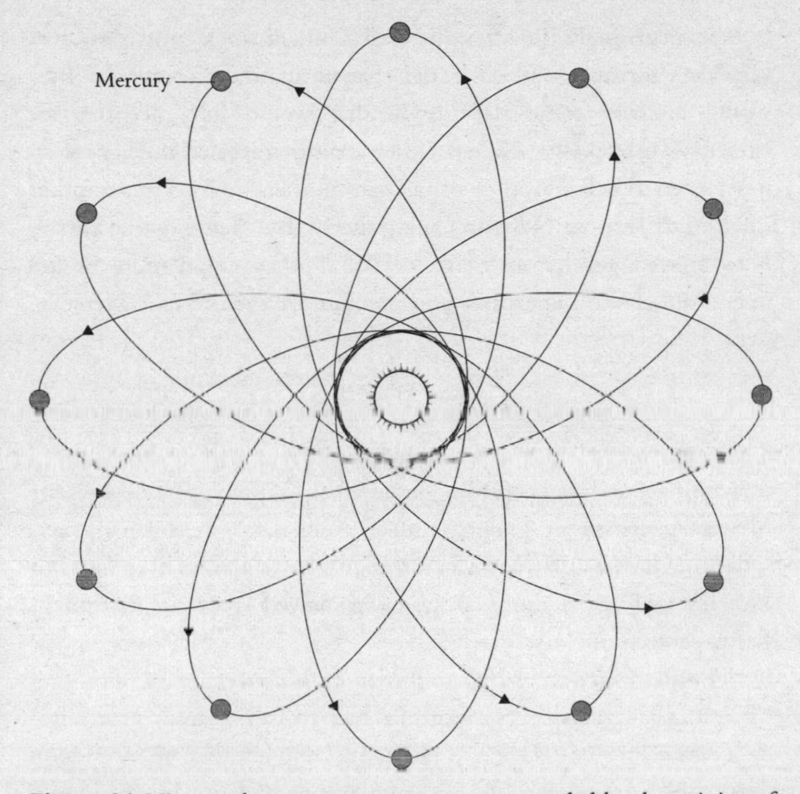

Figure 24 Nineteenth-century astronomers were puzzled by the twisting of Mercury's orbit. This is an exaggerated diagram, inasmuch as Mercury's orbit is less elliptical (i.e. more circular) and the Sun is closer to the centre of that orbit. More importantly, the twisting of the orbit is highly exaggerated. In reality, each orbit advances by just 0.00038° with respect to the previous orbit. When dealing with such small angles, scientists tend to use arcminutes and arcseconds rather than degrees:

$$1 \text{ arcminute} = \tfrac{1}{60}{}^{\circ}$$
$$1 \text{ arcsecond} = \tfrac{1}{60} \text{ arcminute} = \tfrac{1}{3,600}{}^{\circ}$$

So each orbit of Mercury advances by roughly 0.00038°, or 0.023 arcminutes, or 1.383 arcseconds with respect to the previous orbit. It takes Mercury 88 Earth days to orbit the Sun, so after one Earth century Mercury completes 415 orbits, and its orbit has advanced by 415 × 1.383 = 574 arcseconds.

Not surprisingly, the scientific establishment stuck to its view that Newton's formula was right and that astronomers sooner or later would discover some new body that would fully account for Mercury's orbital twist. When closer scrutiny revealed no sign of an inner asteroid belt, moon or planet, astronomers then offered another solution to prop up Newton's ailing theory. By changing one part of Newton's equation from r^2 to $r^{2.00000016}$, they could more or less rescue the classical approach and account for the orbit of Mercury:

$$F = \frac{G \times m_1 \times m_2}{r^{2.00000016}}$$

This, however, was just a mathematical trick. It had no justification in physics, but was merely a desperate last-ditch effort to rescue Newton's theory of gravity. Indeed, such ad hoc tinkering was indicative of the sort of blinkered logic that had earlier resulted in Ptolemy adding yet more circles to his flawed epicyclic view of an Earth-centred universe.

If Einstein was going to overcome such conservatism, win over his critics and depose Newton, he had to gather even more evidence in favour of his theory. He had to find another phenomenon that could be explained by his own theory and not by Newton's, something so extraordinary that it would provide overwhelming, incontrovertible proof in favour of Einsteinian gravity, general relativity and spacetime.

The Ultimate Partnership: Theory and Experiment

If a new scientific theory wants to be taken seriously, then it should pass two critical tests. First, it needs to be able to produce theoretical results that match all the existing observations of reality. Einstein's theory of gravity had passed this test, because among other things it had indicated exactly the right amount of twisting for Mercury's orbit. The second test, which is even more exacting, is that the theory should predict results for observations that have not yet been made. Once scientists are able to make those observations, and if they match the theoretical predictions, then this is compelling evidence that the theory is correct. When Kepler and Galileo argued that the Earth orbited the Sun, they were rapidly able to pass the first test, which was to produce theoretical results that matched the known movements of the planets. However, the second test was passed only when Galileo's observation of the phases of Venus matched a theoretical prediction that had been made by Copernicus decades earlier.

The reason why the first type of test alone is not sufficient to convince doubters is the fear that the theory might have been tinkered with to generate the right result. However, it is impossible to adjust a theory to make it agree with the result of an observation that has not yet been made. Imagine that you are thinking of investing money with either Alice or Bob, who both claim to have their own perfect systems or theories for playing the stock market. Bob tries to convince you that his theory is better by showing you yesterday's stock market figures and then reveals how his theory would have predicted them perfectly. Alice, on the other hand, shows you her predictions for the next day's trading. Sure enough, twenty-four hours later, she is proved right. Who do you invest with, Bob or Alice? Clearly, there is a suspicion that Bob may have adjusted his theory to fit the previous day's data after trading had finished, so his theory is not wholly convincing. But Alice's theory on playing the stock market genuinely seems to work.

Similarly, if Einstein was going to prove that he was right and Newton was wrong, he would have to use his theory to make a robust prediction about an as yet unobserved phenomenon. Of course, this phenomenon would have to take place in an environment of extreme gravity, otherwise the Newtonian and Einsteinian predictions would coincide and there would be no winner.

In the end, the make-or-break test was to be a phenomenon involving the behaviour of light. Even before he had applied his theory to Mercury – in fact, before he had even finished developing his theory of general relativity – Einstein had begun to explore the interaction between light and gravity. According to his space-time formulation of gravity, any beam of light that passed by a star or massive planet would be attracted by the force of gravity towards the star or planet, and the light would be slightly deflected from its original path. Newton's theory of gravity also predicted that heavy objects would bend light, but to a lesser extent. Consequently, if somebody could measure the bending of light by a massive celestial body, then whether it was slight or very slight would determine who was right, Einstein or Newton.

As early as 1912, Einstein began collaborating with Erwin Freundlich on how to make the crucial measurement. Whereas Einstein was a theoretical physicist, Freundlich was an accomplished astronomer and therefore in a better position to say how one might go about making the observations that would discern the optical warping predicted by general relativity. Initially, they wondered whether Jupiter, the most massive planet in the Solar System, might be big enough to bend the light from a distant star, as shown in Figure 25. But when Einstein performed the relevant calculation using his formula, it was clear that the amount of bending caused by Jupiter would be too feeble to be detected, even though the planet has 300 times the mass of the Earth. Einstein wrote to Freundlich: 'If only nature had given us a planet bigger than Jupiter!'

Next, they focused on the Sun, which is a thousand times as massive as Jupiter. This time Einstein's calculations showed that the Sun's gravitational attraction would have a significant influence on

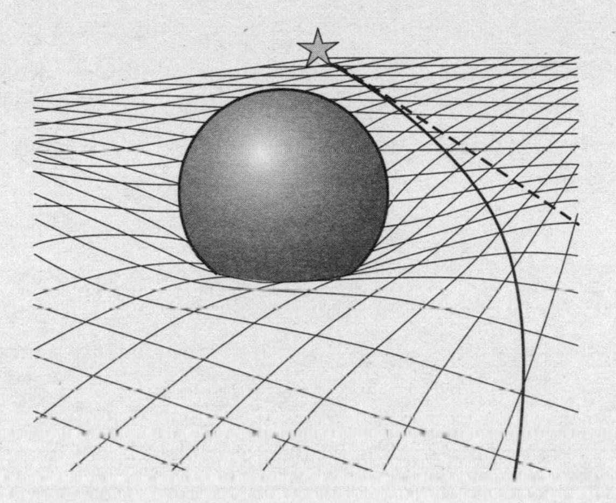

Figure 25 Einstein was interested in the possible bending of starlight by Jupiter, a planet massive enough to make a deep hollow in the fabric of spacetime. The diagram shows a distant star emitting a ray of light, which crosses space. The straight path shows how the light would have travelled across flat space had Jupiter not been present. The curved path shows how the light is deflected by Jupiter's warping of space. Unfortunately for Einstein, Jupiter's bending of starlight was too small to be detected.

---

a ray of light from a distant star, and that the bending of the light should be detectable. For example, if a star was behind the rim of the Sun, thus not in our line of sight, we would not expect to see it from the Earth, as shown in Figure 26. However, the immense gravitational force of the Sun and warping of spacetime should deflect the star's light towards the Earth, making it just visible. The star, which is still behind the Sun, should appear to be slightly to the side of the Sun. The amount of movement from actual to apparent position would be very slight, but it would indicate who was right, because Newton's formula predicted an even smaller shift than did Einstein's formula.

But there was a problem: a star whose light was deflected by the Sun so that its position was apparently shifted just to the side of the

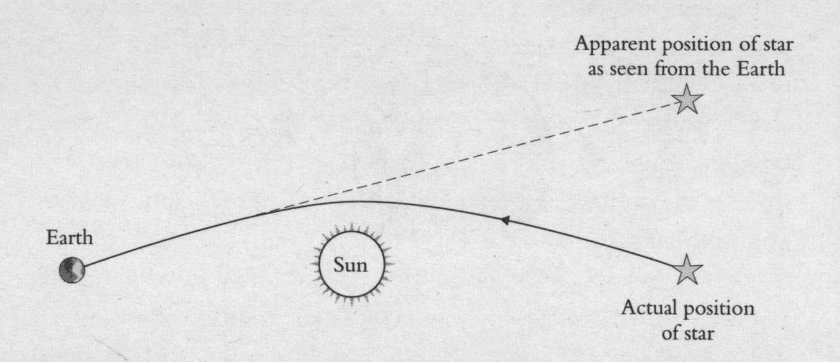

Figure 26 Einstein hoped that the bending of starlight by the Sun could be used to prove his general theory of relativity. The line of sight between the Earth and the distant star is blocked by the Sun, but the mass of the Sun distorts spacetime, and the starlight is deflected to follow a curved path towards the Earth. Our instinct tells us that light travels in straight lines, so from Earth we project the path of the light back along the straight line on which it appears to have arrived, and it seems that the star has shifted. Einstein's theory of gravity predicted a greater apparent stellar shift than did Newton's theory of gravity, so measuring the shift would indicate which theory of gravity was correct.

Sun would still be impossible to see because of the overwhelming brilliance of the Sun. In fact, the region around the Sun is always sprinkled with stars, but they all remain invisible because their brightness is negligible in comparison with the Sun's. There is, however, one circumstance when the stars beyond the Sun do reveal themselves. In 1913, Einstein wrote to Freundlich suggesting that they look for stellar shifts during a total solar eclipse.

When the Moon obliterates the Sun during an eclipse, day temporarily becomes night and the stars emerge. The Moon's disc fits over the Sun's so perfectly that it ought to be possible to identify a star just a fraction of a degree from the rim of the Sun – or rather a star whose light has been warped so that it *appears* to be a fraction of a degree outside the solar disc.

Einstein hoped that Freundlich could examine photographs of past eclipses to find the changes in position that he needed in order to prove that his gravity formula was correct, but it soon became clear that second-hand data would not suffice. The exposure and framing of photographs would have to be perfect to detect slight shifts in the positions of stars, and past eclipse photographs were just not up to scratch.

There was only one option. Freundlich would have to mount a special expedition to photograph the next solar eclipse, which would be observable from the Crimea on 21 August 1914. Einstein's reputation depended on this observation, so he was prepared to fund the mission if necessary. He became so obsessed that he would visit Freundlich for dinner, rush through the meal and start scribbling on the tablecloth, checking over calculations with his partner to make sure that there was no room for error. Later, Freundlich's widow would regret washing the tablecloths, as they would have been worth a fortune with their Einstein jottings intact.

Freundlich left Berlin for the Crimea on 19 July. In hindsight it was a foolish trip to undertake, because Archduke Franz Ferdinand had been assassinated in Sarajevo the previous month, and the events that would set off the First World War were well under way. Freundlich arrived in Russia in plenty of time to set up his telescope in readiness for the day of the eclipse, seemingly oblivious to the fact that Germany had declared war on Russia during his travels. German nationals carrying telescopes and photographic equipment around Russia at this time were asking for trouble, and not surprisingly Freundlich and his party were arrested on suspicion of spying. Worse still, they were detained before the eclipse took place, so the expedition was a complete failure. Fortunately for Freundlich, Germany had arrested a group of Russian officers at roughly the same time, so a prisoner exchange was arranged and Freundlich was safely back in Berlin by 2 September.

This ill-fated enterprise was symbolic of how warfare would freeze progress in physics and astronomy for the next four years. Pure science ground to a halt as all research was focused on winning the war, and many of Europe's most brilliant young minds volunteered to fight for their country. For example, Harry Moseley, who had already made his name as an atomic physicist at Oxford, volunteered to join one of Kitchener's New Army divisions. He was shipped off to Gallipoli in the summer of 1915 to join the Allied forces that were attacking Turkish territory. He described the conditions at Gallipoli in a letter to his mother: 'The one real interest in life is the flies. No mosquitoes, but flies by day and flies by night, flies in the water, flies in the food.' At dawn on 10 August, 30,000 Turkish soldiers launched an assault, resulting in some of the fiercest hand-to-hand combat of the entire war. By the time that the assault was over, Moseley had lost his life. Even the German press mourned his death, calling it 'a severe loss' (*ein schwerer Verlust*) for science.

Similarly, Karl Schwarzschild, the director of Germany's Potsdam Observatory, volunteered to fight for his country. He continued to write papers while stuck in the trenches, including one on Einstein's general theory of relativity which later led to an understanding of black holes. On 24 February 1916, Einstein presented the paper to the Prussian Academy. Just four months later, Schwarzschild was dead. He had contracted a fatal disease on the Eastern front.

While Schwarzschild volunteered to fight, his counterpart at the Cambridge Observatory, Arthur Eddington, refused to enlist on principle. Raised as a devout Quaker, Eddington made his position clear: 'My objection to war is based on religious grounds . . . Even if the abstention of conscientious objectors were to make the difference between victory and defeat, we cannot truly benefit the nation by wilful disobedience to the divine will.' Eddington's colleagues pressed for him to be exempted from military service on the grounds that he was of more value to the country as a scientist, but the Home Office rejected the petition. It seemed inevitable that Eddington's stance as a conscientious objector would land him in a detention camp.

Then Frank Dyson, the Astronomer Royal, came to the rescue. Dyson knew there would be a total eclipse of the Sun on 29 May 1919, which would take place against a rich cluster of stars known as the Hyades – an excellent scenario for measuring any gravitational deflection of starlight. The path of the eclipse crossed South America and Central Africa, so making observations would require mounting a major expedition to the tropics. Dyson suggested to the Admiralty that Eddington could serve his country by organising and leading such an eclipse expedition, and in the meantime he should remain in Cambridge in order to prepare for it. He threw in a jingoistic justification, suggesting that it was the duty of an Englishman to defend Newtonian gravity against the German theory of general relativity. In his heart and mind Dyson was pro-Einstein, but he hoped that this subterfuge would convince the authorities. His lobbying paid off. The threat of the detention camp was duly lifted, and Eddington was allowed to continue working at the observatory in preparation for the 1919 eclipse.

As it happened, Eddington was the perfect man to attempt a verification of Einstein's theory. He had a lifelong fascination with mathematics and astronomy, dating back to the age of four when he attempted to count all the stars in the sky. He went on to become a brilliant pupil, winning a scholarship to Cambridge University, where he came top of his year, earning the title Senior Wrangler. He maintained his reputation by graduating a year ahead of fellow students. As a researcher, he became well known as an advocate of general relativity, and in due course he would write *The Mathematical Theory of Relativity*, which Einstein praised as 'the finest presentation of the subject in any language'. Eddington became so closely associated with the theory that the physicist Ludwig

Silberstein, who also considered himself an authority on general relativity, once said to Eddington, 'You must be one of three persons in the world who understands general relativity.' Eddington stared back in silence, until Silberstein told him not to be so modest. 'On the contrary,' replied Eddington, 'I am trying to think who the third person is.'

As well as being intellectually gifted and having the confidence required to lead an expedition, Eddington was also strong enough to survive the rigours of a tropical adventure. This was important because astronomical expeditions had a reputation for being arduous journeys that pushed scientists to the limit. In the late eighteenth century, for instance, the French scientist Jean d'Auteroche made two expeditions to observe the planet Venus passing across the face of the Sun. First, in 1761, he went to Siberia where he had to be guarded by Cossacks, because the locals believed that the strange equipment he had aimed at the Sun was responsible for the severe spring floods they had recently suffered. Then, eight years later, he repeated his observations of the transit of Venus, this time from the Baja peninsula in Mexico, but fever killed d'Auteroche and two of his party soon afterwards, leaving only one man to carry the precious measurements back to Paris.

Other expeditions were less hazardous to the body but more gruelling for the mind. Guillaume le Gentil, one of d'Auteroche's colleagues, also planned to observe the 1761 transit of Venus, but he journeyed to Pondicherry in French India for the event. By the time he arrived, the British were at war with the French, Pondicherry was under siege, and le Gentil could not land in India. Instead he decided to sit tight in Mauritius and earn a living by trading while he waited eight years for the 1769 transit. This time he was able to reach Pondicherry and enjoyed weeks of glorious sunshine in the run-up to the transit, only for clouds to appear at the crucial moment, completely obscuring his view. 'I was more than two weeks in a singular dejection', he wrote, 'and almost did not have the courage to take up my pen to continue my journal; and several times it fell from my hands, when the moment came to report to France the fate of my operations.' After an absence of 11 years, 6 months and 13 days, he eventually returned home to France, only to find his house looted. He managed to rebuild his life by writing his memoirs, which became a great commercial success.

On 8 March 1919, Eddington and his team left Liverpool on board HMS *Anselm* and headed for the island of Madeira, where the scientists split into two groups. One group remained on board the *Anselm* and voyaged to Brazil to observe the eclipse from Sobral, in the Brazilian jungle, while Eddington and a second group boarded the cargo vessel *Portugal* and headed to the island of Principe, just off the coast of Equatorial Guinea in West Africa. The hope was that if cloudy weather obscured the eclipse in the Amazon, then maybe the African team would strike lucky, or vice versa. Weather would make or break the expeditions, so both teams began scouting for the ideal observation site as soon as they arrived at their respective locations. Eddington used one of the earliest four-wheel-drive vehicles to explore Principe, and eventually decided to set up his equipment at Roca Sundy, an elevated site in the north-west of the island, which seemed less prone to cloudy skies. His team proceeded to take test plates and check the equipment, making sure that everything was perfect for the big day.

The eclipse observations could lead to three possible results. Perhaps the starlight would be very slightly deflected, as predicted by Newton's theory of gravity. Or, as Einstein hoped, there would be a more significant deflection in keeping with general relativity. Or maybe the result would disagree with both theories of gravity, which would imply that Newton and Einstein were both wrong. Einstein predicted that a star appearing at the edge of the Sun should be deflected by 1.74 arcseconds (0.0005°), which was just about within the tolerances of Eddington's equipment and twice the deflection predicted by Newton. Such an angular deflection is equivalent to a candle at a distance of 1 km being moved to the left by just 1 cm.

As the day of the eclipse approached, ominous clouds gathered over both Sobral and Principe, followed by a flurry of thunderstorms. The storms relented at Eddington's observation site just an hour before the Moon's disc first touched the edge of the Sun, but the sky still looked gloomy and viewing conditions were still far from ideal. The mission was in jeopardy. Eddington recorded what happened next in his notebook: 'The rain stopped about noon and about 1.30, when the partial phase was well advanced, we began to get a glimpse of the Sun. We had to carry out our programme of photographs in faith. I did not see the eclipse, being too busy changing plates, except for one glance to make sure it had begun and another half-way through to see how much cloud there was . . .'

The team of observers operated with military precision. The plates were mounted, exposed and then removed with split-second timing. Eddington noted: 'We are conscious only of the weird half-light of the landscape and the hush of nature, broken by the calls of the observers, and the beat of the metronome ticking out the 302 seconds of totality.'

Of the sixteen photographs taken by the Principe team, the majority were spoilt by wisps of cloud obscuring the stars. In fact, during the brief precious moments of clear sky, it was possible to take only one photograph of scientific significance. In his book *Space, Time and Gravitation*, Eddington described what happened to this precious photograph:

> This one was measured . . . a few days after the eclipse in a micrometric measuring machine. The problem was to determine how the

> apparent positions of the stars were affected by the Sun's gravitational field, compared with the normal position on a photograph taken when the Sun was out of the way. Normal photographs for comparison had been taken with the same telescope in England in January. The eclipse photograph and a comparison photograph were placed film to film in a measuring machine so that the corresponding images fell close together, and the small distances were measured in two rectangular directions. From these the relative displacements of the stars could be ascertained ... The results from this plate gave a definite displacement in good accordance with Einstein's theory and disagreeing with the Newtonian prediction.

The stars immediately around the eclipse had been obliterated by the Sun's corona, which appeared as a bright halo as soon as the body of the Sun was completely covered by the Moon. However, those stars a little further from the Sun were visible, and they had been deflected by roughly 1 arcsecond from their usual positions. Eddington then extrapolated the extent of the shift to those imperceptible stars that would have been at the edge of the Sun, and estimated that the maximum deflection would have been 1.61 arcseconds. After allowing for misalignments and other possible inaccuracies, Eddington calculated that the error on the maximum deflection was anything up to 0.3 arcseconds, so his final result was that the gravitational deflection caused by the Sun was 1.61 ± 0.3 arcseconds. Einstein had predicted a deflection of 1.74 arcseconds. This meant that Einstein's prediction was in agreement with the actual measurement, whereas the Newtonian prediction, which was just 0.87 arcseconds, was far too low. Eddington despatched a guardedly optimistic telegram to his colleagues back home: 'Through clouds, hopeful. Eddington.'

As Eddington headed back to Britain, the Brazil team was also homeward bound. The storms at Sobral had abated several hours before the eclipse, clearing the air of dust and blessing the observers with ideal viewing conditions. The Brazil plates could not be examined until they were returned to Europe, because they were of a type that would not tolerate being developed in the hot, moist Amazonian climate. The result from Brazil, based on measurements of the positions of several stars, implied a maximum deflection of 1.98 arcseconds, which was higher than Einstein's prediction but still in agreement, given the margins of error. This corroborated the conclusion from the Principe team.

Even before they were formally announced, Eddington's results were the subject of rumours that spread rapidly across Europe. One such leak reached the Dutch physicist Hendrik Lorentz, who then told Einstein that Eddington had found strong evidence for the general theory of relativity and his gravity formula. In turn, Einstein sent a brief postcard to his mother: 'Joyful news today. H.A. Lorentz has telegraphed me that the English expedition has really proved the deflection of light by the Sun.'

On 6 November 1919, Eddington's results were officially presented at a joint meeting of the Royal Astronomical Society and the Royal Society. The event was witnessed by the mathematician and philosopher Alfred North Whitehead: 'The whole atmosphere of tense interest was exactly that of the Greek drama: we were the chorus commenting on the decree of destiny as disclosed in the development of a supreme incident. There was a dramatic quality in the very staging – the traditional ceremonial, and in the background the picture of Newton to remind us that the greatest of scientific generalisations was now, after more than two centuries, to receive its first modification.'

Eddington took the stage and described with clarity and passion the observations he had made, concluding with an explanation of their astounding implications. It was a bravura performance, delivered by a man who was convinced that the photographic plates taken in

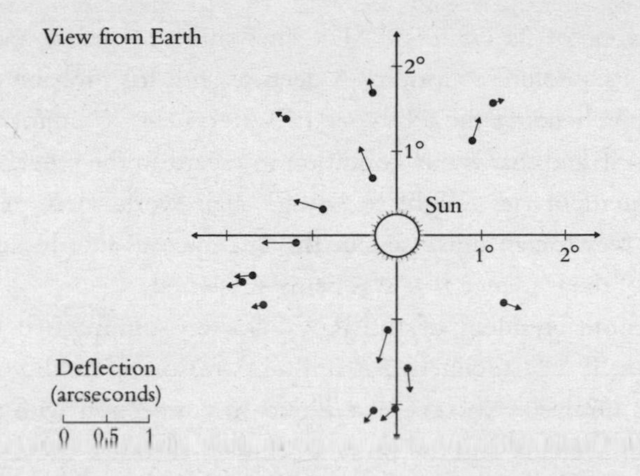

Figure 27 Eddington's results from the 1919 eclipse expedition were confirmed in 1922 by a team of astronomers who observed a solar eclipse from Australia. This chart shows the actual positions of fifteen stars around the Sun (the dots) and the arrows point to the observed positions, which all show an outward deflection. Figure 26 explains why starlight that has been bent towards the Sun makes the star appear to move away from the Sun.

On a technical point, astronomers who want to compare observed results with predictions based on Newtonian or Einsteinian theories often extrapolate their data and estimate the deflection of a hypothetical star right on the edge of the Sun's disc. Also, the actual positions of the stars are marked in degrees relative to the Sun, but the shifts are indicated according to a separate arcsecond scale – otherwise they would be too small to see on this diagram.

Principe and Brazil were indisputable proof that Einstein's view of the universe was right. Cecilia Payne, who would go on to become a celebrated astronomer, was just a nineteen-year-old student when she watched Eddington's lecture: 'The result was a complete transformation of my world picture. My world had been so shaken that I experienced something very like a nervous breakdown.'

However, there were voices of dissent, most notably from the radio pioneer Oliver Lodge. Born in 1851, Lodge was very much a Victorian scientist, grounded in the teachings of Newton. In fact, he was still a devout believer in the ether and would continue to argue in favour of its existence: 'The first thing to realise about the ether is its absolute continuity. A deep sea fish has probably no means of apprehending the existence of water; it is too uniformly immersed in it: and that is our condition in regard to the ether.' He and his contemporaries fought to salvage their world-view of an ether-filled Newtonian universe, but the attempt was utterly futile in the face of the evidence that was being presented.

J.J. Thomson, president of the Royal Society, summarised the meeting thus: 'If it is sustained that Einstein's reasoning holds good – and it has survived two very severe tests in connection with the perihelion of Mercury and the present eclipse – then it is the result of one of the highest achievements of human thought.'

The next day, *The Times* broke the story with the headline REVOLUTION IN SCIENCE – NEW THEORY OF THE UNIVERSE – NEWTONIAN IDEAS OVERTHROWN. A few days later the *New York Times* announced: LIGHT ALL ASKEW IN THE HEAVENS, EINSTEIN'S THEORY TRIUMPHS. Suddenly Albert Einstein had become the world's first science superstar. He had demonstrated an unrivalled understanding of the forces that guided the universe and at the same time was charismatic, witty and philosophical. He was a journalist's dream. Although Einstein initially enjoyed the attention, he soon began to tire of the media frenzy, expressing his concern in a letter to the physicist Max Born: 'Your excellent article in the *Frankfurter Zeitung* gave me much pleasure. But now you, as well as I, will be persecuted by the press and other rabble, although you to a lesser extent. It is so bad that I can hardly come up for air, let alone work properly.'

In 1921 Einstein made the first of several trips to the United States, and on each occasion he was surrounded by huge crowds and addressed packed lecture theatres. No physicist before or since Einstein has achieved such worldwide fame or attracted such admiration and adulation. Perhaps Einstein's impact on the general public was best summarised by a slightly hysterical journalist, describing the consequences of a lecture that Einstein gave at the American Museum of Natural History in New York:

> The crowd, which had gathered in the main auditorium among the big meteorites, resented the fact that the uniformed attendants were trying to exclude those who did not have tickets. Fearful of being excluded from the lecture altogether, a group of young men suddenly charged the four or five attendants who were guarding the door which leads into the Hall of the North American Indians . . . After the attendants had once been butted aside, the men, women and children in the meteorite hall surged through. The less agile were knocked down and stepped on. Women screamed. The man-handled attendants, as soon as they could find an opening, ran for help. The doorman telephoned for the police, and in a few minutes uniformed men were rushing into the great scientific institution on a mission that was new to Police Department history – quelling a science riot.

Although the theory of general relativity was entirely Einstein's work, he was well aware that Eddington's observations had been crucial to the acceptance of this revolution in physics. Einstein had developed the theory; Eddington had checked it against reality. Observation and experiment are the ultimate arbiters of truth, and general relativity had passed the test.

Nevertheless, Einstein once made a tongue-in-cheek comment when asked by a student how he would have reacted if God's universe had turned out to behave differently from the way the general theory of relativity had predicted. In a wonderful demonstration of mock hubris, Einstein answered: 'Then I would feel sorry for the Good Lord. The theory is correct anyway.'

Figure 28 Albert Einstein, who developed the theoretical framework of general relativity, and Sir Arthur Eddington, who proved it by observing the 1919 eclipse. This photograph was taken in 1930, when Einstein visited Cambridge to collect an honorary degree.

Einstein's Universe

Newton's theory of gravity is still widely used today to calculate everything from the flight of a tennis ball to the forces on a suspension bridge, from the swinging of a pendulum to the trajectory of a missile. Newton's formula remains highly accurate when applied to phenomena that take place within the realm of low terrestrial gravity, where the forces are comparatively weak. However, Einstein's theory of gravity was ultimately better because it could be applied equally to the weak gravity environment of Earth and to the intense gravity environments that surround stars. Although Einstein's theory was superior to Newton's, the creator of general relativity was quick to praise the seventeenth-century giant upon whose shoulders he had stood: 'You found the only way which, in your age, was just about possible for a man of highest thought and creative power.'

It has been a somewhat tortuous journey that has brought us to Einstein's theory of gravity, involving the measurement of the speed of light, the rejection of the ether, Galilean relativity, special relativity and, finally, general relativity. After all the twists and turns in the story so far, the only truly important point to remember is that astronomers now had a new and improved theory of gravity, one which was accurate and reliable.

Understanding gravity is critical to astronomy and cosmology, because gravity is the force that guides the movements and interactions of all the celestial bodies. Gravity dictates whether an asteroid will collide with the Earth or swing harmlessly by; it determines how two stars orbit each other in a binary star system; and it explains why an especially massive star might eventually collapse under its own weight to form a black hole.

Einstein was anxious to see how his new theory of gravity would affect our understanding of the universe, so in February 1917 he wrote a scientific paper entitled 'Cosmological Considerations of the General Theory of Relativity'. The key word in the title was 'cosmological'. Einstein was no longer interested in the twisting orbit of our fellow planet Mercury or the way in which our own local Sun tugged at starlight, but instead he focused on the role of gravity on the grand cosmic scale.

Einstein wanted to understand the properties and interactions of the entire universe. When Copernicus, Kepler and Galileo formulated their vision of the universe, they effectively focused their attention on the Solar System, but Einstein was truly interested in the whole universe, as far as any telescope could see and beyond. Soon after publishing this paper, Einstein commented: 'The state of mind which enables a man to do work of this kind . . . is akin to that of the religious worshipper or the lover; the daily effort comes from no deliberate intention or programme, but straight from the heart.'

Using a gravity formula to predict the behaviour of Mercury's orbit entails little more than plugging in a few masses and distances and making a straightforward calculation. To do the same for the whole universe would require taking into consideration all the stars and planets, known and unknown. That seems an absurd ambition – surely such a calculation is impossible? But Einstein reduced his task to a manageable level by making a single simplifying assumption about the universe.

Einstein's assumption is known as the *cosmological principle*, which states that the universe is more or less the same everywhere. More specifically, the principle assumes that the universe is *isotropic*, which means that it looks the same in every direction – which certainly seems to be the case when astronomers stare into deep space. The cosmological principle also assumes that the universe is *homogeneous*, which means that the universe looks the same wherever you happen to be, which is another way of saying that the Earth does not occupy a special place in the universe.

When Einstein applied general relativity and his gravity formula to the universe at large, he was a little surprised and disappointed by the theory's prediction of how the universe operates. What he found implied that the universe was ominously unstable. Einstein's gravity formula showed that every object in the universe was pulled towards every other object on a cosmic scale. This would cause every object to move closer to every other object. The attraction might start as a steady creep, but it would gradually turn into an avalanche which would end in an almighty crunch – the universe was apparently destined to destroy itself. Returning to our trampoline analogy for the fabric of spacetime, we can imagine a giant elastic sheet occupied by several bowling balls, each creating its own hollow. Sooner or later, two of the balls will roll towards each other's hollows, forming an even deeper hollow, which would in turn attract the other balls, until they all crashed together into a single, very deep well.

This was a preposterous result. As discussed in Chapter 1, the scientific establishment at the start of the twentieth century was confident that the universe was static and eternal, not contracting and temporary. Not surprisingly, Einstein disliked the notion of a collapsing universe: 'To admit such a possibility seems senseless.'

Although Isaac Newton's theory of gravity was different, it also gave rise to a collapsing universe, and Newton had also been troubled by this implication of his theory. One of his solutions was to envisage an infinite, symmetric universe, in which every object would therefore be pulled equally in all directions, and there would be no overall movement and no collapse. Unfortunately, he soon realised that this carefully balanced universe would be unstable. An infinite universe could theoretically exist in a state of equilibrium, but in practice the tiniest disturbance in the gravitational equilibrium would upset this balance and end in catastrophe. For example, a comet passing through the Solar System would momentarily increase the mass density of each part of space through which it passed, attracting more material towards those regions and thus initiating the process of total collapse. Even turning a page in a book would alter the balance of the universe and, given enough time, this too would trigger a cataclysmic collapse. To solve the problem, Newton suggested that God intervened from time to time to keep the stars and other celestial objects apart.

Einstein was not prepared to acknowledge a role for God in holding the universe apart, but at the same time he was anxious to find a way to maintain an eternal and static universe in keeping with the scientific consensus. After re-examining his theory of general relativity, he discovered a mathematical trick that would rescue the universe from collapse. He saw that his formula for gravity could be adapted to include a new feature known as the *cosmological constant*. This imbued empty space with an inherent pressure that pushed the universe apart. In other words, the cosmological constant gave rise to a new repulsive force throughout the universe which effectively worked against the gravitational attraction of all the stars. This was a sort of anti-gravity, whose strength depended on the value given to the constant (which in theory could adopt any arbitrary value). Einstein realised that by carefully selecting the value of the cosmological constant he could exactly counteract conventional gravitational attraction and stop the universe from collapsing.

Crucially, this anti-gravity was significant over huge cosmic distances, but negligible over shorter distances. Therefore it did not disrupt general relativity's proven ability to successfully model gravity on the relatively intimate terrestrial or stellar scales. In short, Einstein's revised formula for general relativity could claim three distinct successes in terms of describing gravity. It could:

1. explain a static, eternal universe,
2. mimic all Newton's successes in low gravity (e.g. Earth),
3. succeed where Newton failed in high gravity (e.g. Mercury).

Many cosmologists were happy with Einstein's cosmological constant, because it seemed to do the trick of making general relativity compatible with a static eternal universe. But no one had much of a clue about what the cosmological constant actually represented. In some ways it was on a par with Ptolemy's epicycles, inasmuch as it was an ad hoc tweak that allowed Einstein to get the right result. Even Einstein sheepishly admitted that this was the case when he confessed that the cosmological constant was 'necessary only for the purpose of making a quasi-static distribution of matter'. In other words, it was a fudge that Einstein used to get the result that was expected, namely a stable and eternal universe.

Einstein also admitted that he found the cosmological constant ugly. Talking of its role in general relativity, he once said that it was 'gravely detrimental to the formal beauty of the theory'. This was a problem, because physicists are often motivated in their theorising by a desire for beauty. There is a consensus that the laws of physics should be elegant, simple and harmonious, and these factors often act as excellent guides for pointing physicists towards laws that might be valid and away from those that are false. Beauty in any context is hard to define, but we all know it when we see it, and when Einstein looked at his cosmological constant he had to admit that it was not very pretty. Nevertheless, he was prepared to sacrifice a degree of beauty in his formula because it allowed the theory of general relativity to accommodate an eternal universe, which is what scientific orthodoxy demanded.

Meanwhile, another scientist would take the opposing view and set beauty above orthodoxy in a radically different vision of the universe. Having read Einstein's cosmological paper with relish, Alexander Friedmann would question the role of the cosmological constant and defy the scientific establishment.

Born in St Petersburg in 1888, Friedmann grew up amid great political turmoil, and learned to challenge the establishment from an early age. He was a teenage activist who led school strikes as part of a national protest against the repressive Tsarist government. The 1905 Revolution that followed the protests resulted in a reformed constitution and a period of relative calm, although Tsar Nicholas II remained in power.

When Friedmann entered the University of St Petersburg in 1906 to study mathematics, he became a protégé of Professor Vladimir Steklov, himself an anti-Tsarist, who encouraged Friedmann to tackle problems that would have intimidated many

Figure 29 Alexander Friedmann, the Russian mathematician whose cosmological model indicated an evolving and expanding universe.

other students. Steklov kept fastidious records and noted what happened when he set Friedmann a formidable mathematical problem related to the Laplace equation: 'I touched on this problem in my doctoral thesis, but did not treat it in detail. I suggested that Mr Friedmann should try to solve this problem, in view of his outstanding working capacity and knowledge compared with other persons of his age. In January of this year, Mr Friedmann submitted to me an extensive study of about 130 pages, in which he gave a quite satisfactory solution of the problem.'

Although Friedmann clearly had a passion and talent for mathematics, which can be a highly abstract discipline, he also had a penchant for science and technology, and he was prepared to engage in military research during the First World War. He even volunteered to fly on bombing missions and applied his mathematical skills to the practical problem of dropping the bombs with better accuracy. He wrote to Steklov: 'I have recently had a chance to verify my ideas during a flight over Przemysl; the bombs turned out to be falling almost the way the theory predicts. To have conclusive proof of the theory I'm going to fly again in a few days.'

As well as the First World War, Friedmann also endured the 1917 Revolution and the ensuing civil war. When he eventually returned to his academic life, he was confronted by the delayed arrival of Einstein's theory of general relativity, which had spent several years maturing in Western Europe before being properly noticed in Russian academic circles. Indeed, perhaps it was Russia's very isolation from the Western scientific community that allowed Friedmann to ignore Einstein's approach to cosmology and forge his own model of the universe.

While Einstein had started with the assumption of an eternal universe and then added the cosmological constant to make his theory fit expectation, Friedmann adopted the opposite stance. He started with the theory of general relativity in its simplest and most aesthetically appealing form – without the cosmological constant – which gave him the freedom to see what sort of universe logically emerged from the theory. This was a typically mathematical approach, for Friedmann was a mathematician at heart. Obviously he hoped that his purer approach would lead to an accurate description of the universe, but for Friedmann it was the beauty of the equation and the majesty of the theory that took precedence over reality – or, indeed, over expectation.

Friedmann's research came to a climax in 1922, when he published an article in the journal *Zeitschrift für Physik*. Whereas Einstein had argued for a finely tuned cosmological constant and a finely balanced universe, Friedmann now described how different models of the universe could be created with various values of the cosmological constant. Most importantly, he outlined a model of the universe in which the cosmological constant was set to zero. Such a model was effectively based on Einstein's original formula for gravity, without any cosmological constant. With no cosmological constant to counteract gravitational attraction, Friedmann's model was vulnerable to gravity's relentless pull. This gave rise to a dynamic and evolving model of the universe.

For Einstein and his colleagues, such dynamism was associated with a universe that would be doomed to cataclysmic collapse. Therefore the majority of cosmologists found it unthinkable. For Friedmann, however, such dynamism was associated with a universe that might have been kick-started with an initial expansion, so it would have an impetus with which to fight against the pull of gravity. This was a radically new vision of the universe.

Friedmann explained how his model of the universe could react to gravity in three possible ways, depending on how quickly the universe started expanding and how much matter it contained. The first possibility assumed that the average density of the universe was high, with lots of stars in a given volume. Lots of stars would mean a strong gravitational attraction, which would eventually pull all the stars back, halting the expansion and gradually causing a contraction of the universe until it collapsed completely. The second variation of Friedmann's model assumed that the average density of stars was low, in which case the pull of gravity would never overcome the expansion of the universe, which would therefore continue to expand for ever. The third variation considered a density between the two extremes, leading to a universe in which gravity would slow but never quite halt the expansion. Thus the universe would neither collapse to a point nor expand to infinity.

A useful analogy is to think of firing a cannonball out of a cannon and into the air at a fixed launch speed. Imagine that this takes place on three different-sized planets, as shown in Figure 30. If the planet is massive, then the cannonball will fly a few hundred metres through the air before the strong gravity will make it fall down to the ground. This scenario is akin to Friedmann's first model of a very dense universe that expands and then collapses. If the planet is very small, then it has weak gravity and the cannonball flies off into space, never to be seen again, which is akin to Friedmann's second scenario of a universe that expands for ever. However, if the planet is just the right middling size with the right gravity, then the cannonball travels in a straight line and then goes into orbit, moving neither farther away from nor closer to the planet, which is akin to Friedmann's third scenario.

Something that was common to all three of Friedmann's world-views was the notion of a changing universe. He believed in a universe that was different yesterday and would be different again tomorrow. This was Friedmann's revolutionary contribution to cosmology: the prospect of a universe that evolves on a cosmic scale rather than remaining static throughout eternity.

As the hypotheticals proliferate, perhaps it is time to take stock. Einstein had offered two versions of general relativity, one with the cosmological constant and one without. He then created a static model of the universe based on his theory with the cosmological constant, whereas Friedmann had created a model (with three variations) based on a theory without the cosmological constant. Of course, there might be many models, but there is only one reality. The question was this – which model fitted reality?

As far as Einstein was concerned, the answer was obvious: he was right and Friedmann was wrong. He even thought that the Russian's work was mathematically flawed, and wrote a letter of complaint to the journal that published Friedmann's paper: 'The results concerning the non-stationary world, contained in [Friedmann's] work, appear to me suspicious. In reality it turns out that the solution given in it does not satisfy the [general relativity] equations.' In fact, Friedmann's calculations were correct, so

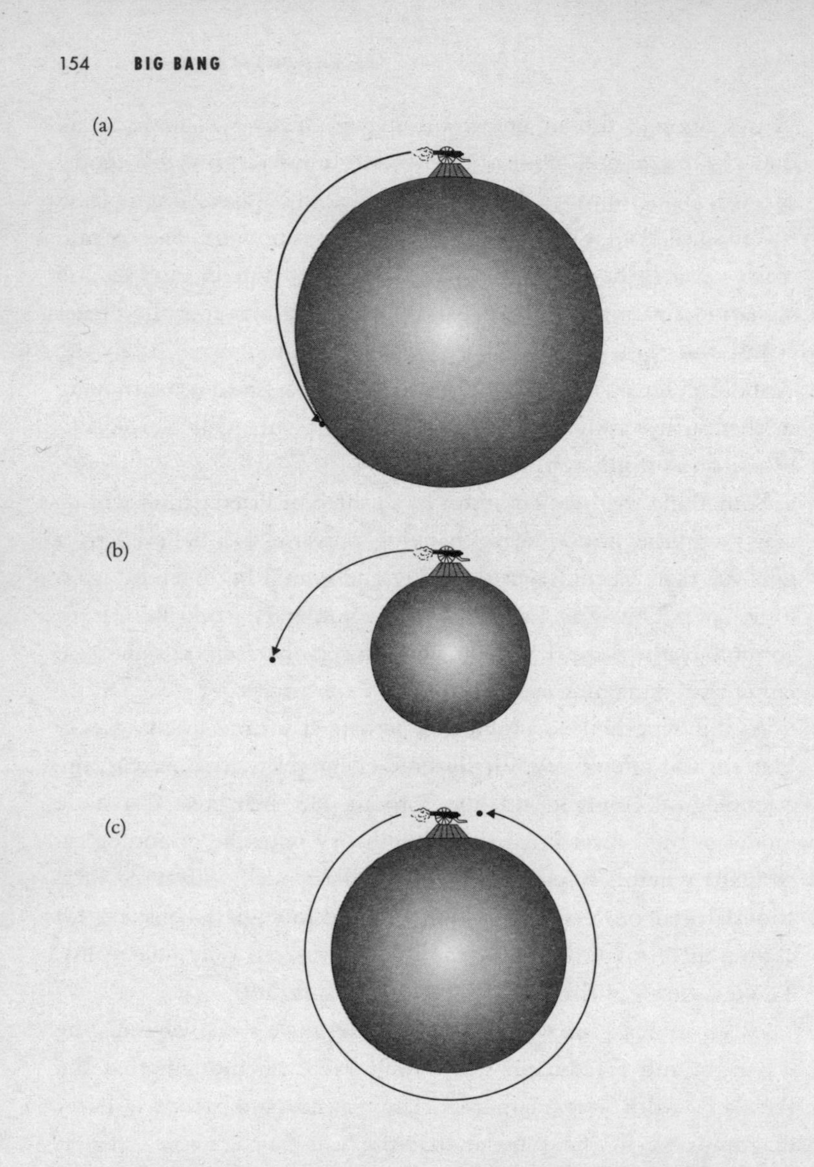

Figure 30 A cannonball is fired from a cannon at the same speed on three different-sized planets. Planet (a) is so massive and its gravitational attraction so strong that the cannonball falls to the ground. Planet (b) is so light and its gravitational attraction so weak that the cannonball flies off into space. Planet (c) has the perfect mass for the cannonball to enter orbit.

his models were mathematically valid even if their resemblance to reality was debatable. Perhaps Einstein had given the paper only a cursory glance and assumed that it must be flawed because it disagreed with his belief in a static universe.

When Friedmann lobbied for a retraction, Einstein found himself humbled into admission: 'I am convinced that Mr Friedmann's results are both correct and clarifying. They show that in addition to the static solutions to the [general relativity] equations there are time varying solutions with a spatially symmetric structure.' Although he now agreed that Friedmann's dynamic solutions were mathematically correct, Einstein still persisted in considering them to be scientifically irrelevant. Significantly, in the original draft of Einstein's retraction he had belittled Friedmann's solutions by claiming that 'a physical significance can hardly be ascribed', but then he crossed out the criticism, probably remembering that this letter was supposed to be an apology.

Despite Einstein's objections, Friedmann continued to promote his own ideas. However, before he could mount any serious assault on the scientific establishment, fate intervened. In 1925, Friedmann's wife was about to give birth to their first child, so he had everything to live for. While working away from home, he wrote a letter to her: 'Now everybody is gone from the Observatory, and I am alone among the statues and portraits of my predecessors, my soul after the day's bustle is becoming calmer and calmer, and it gives me joy to think that thousands of miles away the beloved heart is beating, the gentle soul is living, the new life is growing . . . the life whose future is a mystery, and which has no past.' But Friedmann would not live to witness the birth of his child. He contracted a serious illness, probably typhoid fever, and died in a state of delirium. One of the Leningrad newspapers reported that he had tried to carry out calculations on his deathbed, while muttering about his students and lecturing to an imaginary audience.

Friedmann had developed a new vision of the universe, yet he had died virtually unknown. His ideas had been published, but in his lifetime they were largely unread and completely ignored. Part of the problem was that Friedmann was simply too radical. It seems that Friedmann had much in common with Copernicus.

To make matters worse, Friedmann had been condemned by Einstein, the world's most prominent cosmologist. And although Einstein had issued a grudging apology, the fact that it was not widely circulated meant that Friedmann's reputation remained tarnished. Also, Friedmann had a background in mathematics rather than astronomy, so he was considered an outsider by the cosmological community. To cap it all, Friedmann was simply ahead of his time. Astronomers were not yet capable of making the sort of detailed observations that might support a model that described an expanding universe. Friedmann openly acknowledged that there was no evidence in favour of his models: 'All this should at present be considered as curious facts which cannot be reliably supported by the inadequate astronomical experimental material.'

Fortunately, the notion of an expanding and evolving universe did not disappear completely. The idea resurfaced just a few years after Friedmann's death, but again the Russian received virtually no credit. This was because the expanding universe model would be independently reinvented from scratch by Georges Lemaître, a Belgian cleric and cosmologist whose education had also been severely disrupted by the First World War.

Lemaître, who was born in Charleroi in 1894, took a degree in engineering at the University of Louvain, but had to abandon his studies when German forces invaded Belgium. He spent the next four years in the army, witnessing the first German poison gas attacks and winning the Croix de Guerre for his bravery. After the war he resumed his studies at Louvain, but this time he switched from engineering to theoretical physics, and in 1920 he also enrolled in a seminary at Maline. He was ordained in 1923, and for the rest of his life would maintain parallel careers as a physicist and a priest. 'There were two ways of arriving at the truth', he said. 'I decided to follow them both.'

After ordination, Lemaître spent a year in Cambridge with Arthur Eddington, who described him as 'a very brilliant student, wonderfully quick and clear-sighted, and of great mathematical ability'. The following year he went to America, spending time making astronomical measurements at the Harvard Observatory and starting his Ph.D. at the Massachusetts Institute of Technology. Lemaître was embedding himself within the community of cosmologists and astronomers, and familiarising himself with the observational side of the subject in a bid to complement his preference for theory.

In 1925 he returned to the University of Louvain, took up an academic post and began to develop his own cosmological models

Figure 31 Georges Lemaître, the Belgian priest and cosmologist who unwittingly resurrected Friedmann's model of an evolving and expanding universe. His theory that the universe started with an exploding primeval atom was a forerunner of the Big Bang model.

based on Einstein's equations of general relativity, but largely ignoring the role of the cosmological constant. Over the next two years he rediscovered the models that described an expanding universe, oblivious to the fact that Friedmann had been through the same thought processes earlier in the decade.

Lemaître, however, went beyond his Russian predecessor by relentlessly pursuing the implications of an expanding universe. While Friedmann was a mathematician, Lemaître was a cosmologist who wanted to understand the reality behind the equations. In particular, Lemaître was interested in the physical history of the cosmos. If the universe really is expanding, then yesterday it must have been smaller than it is today. Similarly, last year it must have been smaller still. And logically, if we go back far enough, then the entirety of space must have been compacted into a tiny region. In other words, Lemaître was prepared to run the clock backwards until he reached an apparent start of the universe.

Lemaître's great insight was that general relativity implied a moment of creation. Although his pursuit for scientific truth was not coloured by his search for theological truth, such a realisation must have resonated with the young priest. He concluded that the universe began in a small compact region from which it exploded outwards and evolved over time to become the universe in which we find ourselves today. Indeed, he believed that the universe would continue to evolve into the future.

Having developed this model of the universe, Lemaître started searching for the physics that could corroborate or explain his theory of cosmic creation and evolution. He alighted on an area of growing interest among astronomers, namely cosmic-ray physics. Back in 1912, the Austrian scientist Viktor Hess had reached an altitude of almost 6 km in a balloon and detected evidence of highly energetic particles coming from outer space. Lemaître was also familiar with the process of *radioactive decay*, in which large atoms such as uranium break down into smaller atoms, emitting particles, radiation and energy. Lemaître began to speculate that a similar process, albeit on a vastly greater scale, might have given birth to the universe. By extrapolating backwards in time, Lemaître envisaged all the stars squeezed into a super-compact universe, which he dubbed the *primeval atom*. He then viewed the moment of creation as the moment when this single, all-encompassing atom suddenly decayed, generating all the matter in the universe.

Lemaître speculated that the cosmic rays observable today might be remnants of this initial decay, and that the bulk of the ejected matter would have condensed over time to form today's stars and planets. He later summarised his theory thus: 'The primeval atom hypothesis is a cosmogenic hypothesis which pictures the present universe as the result of the radioactive disintegration of an atom.' Furthermore, the energy released in this mother of all radioactive decays could have powered the expansion that was central to his model of the universe.

To summarise, Lemaître was the first scientist to give a reasonably confident and detailed description of what we now refer to as the Big Bang model of the universe. Indeed, he maintained that this was not just *a* model of the universe, but *the* model of the universe. He had started with Einstein's general theory of relativity, developed a theoretical model of cosmological creation and expansion, and then integrated it with known observations of phenomena such as cosmic rays and radioactive decay.

A moment of creation was at the core of Lemaître's model, but he was also interested in the processes that had transformed a shapeless explosion into the stars and planets we see today. He was developing a theory of the creation, evolution and history of the universe. Although his research was rational and logical, he wrote about it in poetic terms: 'The evolution of the universe can be likened to a display of fireworks that has just ended: some few wisps, ashes and smoke. Standing on a well-cooled cinder, we see the fading of the suns, and try to recall the vanished brilliance of the origins of the worlds.'

By coupling theory with observation and setting his Big Bang within a framework of physics and observational astronomy, Lemaître had moved far beyond Friedmann's earlier work. Nevertheless, when the Belgian cleric announced his theory of creation in 1927, he was met by the same damning silence that had greeted Friedmann's models. It did not help that Lemaître chose to publish his ideas in a little-known Belgian journal, the *Annales de la Société Scientifique de Bruxelles*.

The situation was made worse by an encounter with Einstein soon after Lemaître published his *Hypothèse de l'atome primitif*. Lemaître was attending the 1927 Solvay Conference in Brussels, a gathering of the world's greatest physicists, where he quickly established his presence thanks to his eye-catching dog collar. He managed to corner Einstein and explained his vision of a created and expanding universe. Einstein responded by mentioning that he had already heard about the idea from Friedmann, introducing the Belgian to the work of his deceased Russian counterpart for the first time. Then Einstein rebuffed Lemaître: 'Your calculations are correct, but your physics is abominable.'

Einstein had now been offered two chances to accept or at least consider the expanding Big Bang scenario, but he had rejected the idea twice over. And rejection by Einstein meant rejection by the establishment. In the absence of hard evidence, Einstein's blessing or criticism had the power to make or break a nascent theory. Einstein, who had once been the epitome of rebellion, had become an unwitting dictator. He eventually came to appreciate the irony of his position, and once lamented: 'To punish me for my contempt for authority, Fate made me an authority myself.'

Lemaître was devastated by the events at Solvay and decided not to promote his ideas any further. He still believed in his expanding universe model, but he had no influence in the scientific establishment and could see no point in advocating a Big Bang model that everybody else considered foolish. Meanwhile, the world focused on Einstein's static universe – which was also a perfectly legitimate model, although the finely tuned cosmological constant was somewhat contrived. In any case, the static universe was consistent with the prevailing belief in an eternal universe, so any scientific blemishes were overlooked.

In hindsight, we can see that both models had similar strengths and weaknesses, and were very much on a par with each other. After all, both models were mathematically consistent and scientifically valid: they both emerged out of the general relativity formula and neither conflicted with any known physical laws. However, both theories suffered from a complete lack of any supporting observational or experimental data to back them up. It was this absence of evidence that allowed the scientific establishment to be swayed by prejudice, favouring Einstein's eternal static model over Friedmann and Lemaître's expanding Big Bang model.

In truth, cosmologists were still in that uncomfortable no-man's land between myth and science. If they were going to make progress, it would be necessary to find some concrete evidence. The theorists turned to the observational astronomers in the hope that they could peer deep into space and distinguish between the competing models, proving one of them and disproving the other. Astronomers would indeed spend the rest of the twentieth century building bigger, better and more powerful telescopes, ultimately making the key observation that would transform our view of the universe.

CHAPTER 2 - THEORIES OF THE UNIVERSE
SUMMARY NOTES

① 1670s CASSINI PROVED THAT LIGHT HAS A FINITE SPEED BY OBSERVING ONE OF JUPITER'S MOONS.

THE SPEED OF LIGHT TURNED OUT TO BE 300,000 km/s

② THE VICTORIANS BELIEVED THAT THE UNIVERSE IS FILLED WITH ETHER: A MEDIUM WHICH CARRIES LIGHT.

THE MEASURED SPEED OF LIGHT WAS THOUGHT TO BE ITS SPEED RELATIVE TO THE ETHER

THEREFORE, AS THE EARTH MOVED THROUGH SPACE, IT SHOULD MOVE THROUGH THE ETHER, GIVING RISE TO AN 'ETHER WIND.'

SO THE SPEED OF LIGHT AGAINST THE ETHER WIND SHOULD BE DIFFERENT FROM ITS SPEED ACROSS THE ETHER WIND.

1880s - MICHELSON AND MORLEY TESTED THIS - THEY FOUND NO EVIDENCE OF A DIFFERENCE IN SPEED. THUS THEY DISPROVED THE EXISTENCE OF THE ETHER.

③ IF LIGHT DOES NOT TRAVEL RELATIVE TO THE NON-EXISTENT ETHER, THEN ALBERT EINSTEIN ARGUED THAT:

THE SPEED OF LIGHT IS CONSTANT RELATIVE TO THE OBSERVER.

- WHICH CONTRADICTED OUR EXPERIENCE WITH ALL OTHER FORMS OF MOVEMENT.

HE WAS RIGHT.

FROM THIS ASSUMPTION (+ GALILEAN RELATIVITY) EINSTEIN DEVELOPED HIS:

SPECIAL THEORY OF RELATIVITY (1905)

THIS SAID THAT BOTH SPACE AND TIME ARE FLEXIBLE.

THEY FORM A SINGLE UNIFIED ENTITY - SPACETIME.

1915 - EINSTEIN DEVELOPED HIS GENERAL THEORY OF RELATIVITY

THIS GAVE A NEW THEORY OF GRAVITY WHICH WAS BETTER THAN NEWTON'S THEORY OF GRAVITY BECAUSE IT ALSO WORKED IN HIGH-GRAVITY ENVIRONMENTS (eg STARS)

④ EINSTEIN'S AND NEWTON'S THEORIES OF GRAVITY WERE TESTED BY STUDYING THE ORBIT OF MERCURY AND THE BENDING OF LIGHT AROUND THE SUN (1919). IN BOTH CASES EINSTEIN WAS RIGHT, AND NEWTON WAS WRONG.

⑤ WITH HIS NEW THEORY OF GRAVITY, EINSTEIN STUDIED THE ENTIRE UNIVERSE:

PROBLEM - GRAVITATIONAL ATTRACTION WOULD CAUSE THE UNIVERSE TO COLLAPSE.

SOLUTION - EINSTEIN ADDED THE COSMOLOGICAL CONSTANT TO GENERAL RELATIVITY.

- THIS GAVE RISE TO AN ANTI-GRAVITATIONAL EFFECT
- THIS WOULD STOP THE UNIVERSE COLLAPSING
- WHICH FITS WITH THE GENERAL VIEW OF A STATIC AND ETERNAL UNIVERSE

BANG!

⑥ MEANWHILE: FRIEDMANN AND LEMAÎTRE DITCHED THE COSMOLOGICAL CONSTANT AND PROPOSED THAT THE UNIVERSE MIGHT BE DYNAMIC.

THEY PICTURED AN EXPANDING UNIVERSE. LEMAÎTRE DESCRIBED AN ALMIGHTY, COMPACT, PRIMEVAL ATOM, WHICH EXPLODED, EXPANDED AND EVOLVED INTO TODAY'S UNIVERSE.

⇨ WE WOULD NOW CALL THIS A BIG BANG MODEL OF THE UNIVERSE.

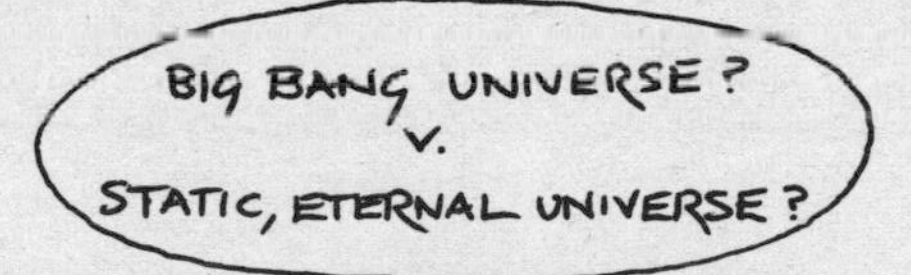

FRIEDMANN AND LEMAÎTRE AND THEIR EXPANDING UNIVERSE ARE IGNORED. WITHOUT ANY OBSERVATIONAL EVIDENCE TO SUPPORT IT, THE BIG BANG MODEL WAS IN THE DOLDRUMS.

THE MAJORITY OF SCIENTISTS CONTINUED TO BELIEVE IN AN ETERNAL, STATIC UNIVERSE.

Chapter 3

THE GREAT DEBATE

The known is finite, the unknown is infinite; intellectually we stand on an islet in the midst of an illimitable ocean of inexplicability. Our business in every generation is to reclaim a little more land.

T.H. HUXLEY

The less one knows about the universe, the easier it is to explain.

LEON BRUNSCHVICG

Errors using inadequate data are much less than those using no data at all.

CHARLES BABBAGE

Theories crumble, but good observations never fade.

HARLOW SHAPLEY

First, get the facts, then you can distort them at your leisure.

MARK TWAIN

Heaven wheels above you displaying to you her eternal glories and still your eyes are on the ground.

DANTE

Science consists of two complementary strands, theory and experiment. While theorists consider how the world works and build models of reality, it is the experimentalists who test these models by comparing them with reality. In cosmology, theorists such as Einstein, Friedmann and Lemaitre had developed competing models of the universe, but testing them was highly problematic: how do you experiment with the entire universe?

When it comes to conducting experiments, astronomy and cosmology stand apart from the rest of science. Biologists can touch, smell, prod, poke and even taste the organisms they study. Chemists can boil, burn and blend chemicals in a test tube to learn more about their properties. And physicists can easily add mass to a pendulum and vary its length to investigate why it swings the way it does. But astronomers can only stand and stare, for the vast majority of celestial objects are so far away that they can be studied only by detecting the rays of light they send towards the Earth. Instead of actively indulging in a wide range of experiments, astronomers can only passively observe the universe. In other words, astronomers can look, but they can't touch.

Despite this severe limitation, astronomers have been able to discover an extraordinary amount about the universe and the objects within it. For instance, in 1967 the British astronomer Jocelyn Bell discovered a new type of star known as a pulsating star or *pulsar*. When she first spotted the regular pulsing light signal on the recording chart, she marked it 'LGM', for 'Little Green Men', because it seemed like a message broadcast by intelligent life. Today, when she lectures on pulsars, Professor Bell Burnell (as she now is) passes a tiny folded slip of paper around the audience. It says: 'In picking up this piece of paper you have used thousands of times more energy than all the world's telescopes have ever received from all the known pulsars.' In other words, these pulsars radiate energy, like any other star, but they are so distant that astronomers have gathered only a tiny amount of energy from them during decades of intense observation. Nevertheless, even though they are so faint, astronomers have been able to deduce several facts about pulsars. For example, they have worked out that pulsars are stars at the end of their life, are made up of subatomic particles called neutrons, are typically 10 km in diameter and are so dense that one teaspoon of pulsar matter weighs a billion tonnes.

Only when as much information as possible has been gleaned by observation can astronomers begin to examine the models put forth by theorists and test whether they are correct. And in order to test the greatest models of all – the competing Big Bang and eternal universe models – astronomers would have to push their observational technology to the limit. They would have to build giant telescopes containing vast mirrors, housed in observatories the size of huge warehouses, sited on remote mountaintops. Before we examine the discoveries made by the major telescopes of the twentieth century, we first need to look at the evolution of the telescope up to 1900 and see how the earlier instruments contributed to the changing view of the universe.

Staring into Space

After Galileo, the next great pioneer in the design and use of the telescope was Friedrich Wilhelm Herschel, born in Hanover in 1738. He started his working life as a musician, following his father into the Hanoverian Guard as a bandsman, but he considered a change in his career at the Battle of Hastenbeck in 1757, at the height of the Seven Years' War. He came under heavy fire and decided to abandon his job and country in favour of a quieter life as a musician abroad. He chose to settle in Britain, because the Hanoverian George Louis had ascended the British throne as George I back in 1714, thus establishing the Hanoverian dynasty, and Herschel thought he would receive a sympathetic welcome. He anglicised his name to William Herschel, bought a house in Bath and earned a comfortable living as an excellent oboist, composer, conductor and music teacher. However, as the years passed, Herschel gradually developed an interest in astronomy which evolved from a minor hobby into a major obsession. He eventually became a full-time professional stargazer and would be recognised by his colleagues as the greatest astronomer of the eighteenth century.

Herschel made his most famous discovery in 1781, observing from his garden and using a telescope that he had built from scratch. He identified a new object in the sky that slowly moved over the course of several nights. He assumed that it was a previously undiscovered comet, until it became clear that the object did not possess a tail, and was in fact a new planet, a momentous addition to the Solar System. For thousands of years astronomers had known only of the five other planets (Mercury, Venus, Mars, Jupiter and Saturn) visible to the naked eye, but now Herschel had identified an entirely new world. He named it Georgium Sidus (George's Star) in honour of his monarch, King George III, a fellow Hanoverian, but French astronomers preferred to call the new planet Herschel after its

Figure 32 William Herschel, the most famous astronomer of the eighteenth century, wrapped up warm for a night of stargazing.

discoverer. In the end the planet was named after Uranus, the father of Saturn and grandfather of Jupiter in Roman mythology.

William Herschel, working in his back garden, had succeeded where the lavish court observatories of Europe had failed. His sister Caroline, who acted as his assistant, played a crucial role in helping him to achieve his success. Although a brilliant astronomer in her own right, discovering eight comets during her career, she devoted herself to supporting William. She worked alongside him during the arduous days that he spent building new telescopes, and she would then assist him during the long, freezing nights of observing. As she wrote: 'Every leisure moment was eagerly snatched at for resuming some work which was in progress, without taking time or changing dress, and many a lace ruffle was torn or bespattered by molten pitch . . . I was even obliged to feed him by putting the vitals by bits into his mouth.'

The pitch mentioned by Caroline Herschel was used by her brother to make tools for polishing mirrors. Indeed, William took great pride in building his own telescopes. As a telescope-maker he was entirely self-taught, yet he constructed what were then the finest telescopes in the world. One of his telescopes could achieve a magnification of ×2,010, whereas the Astronomer Royal's best telescope could manage only ×270.

Magnification is beneficial for any telescope, but even more important is its ability to gather light, and that depends wholly on its *aperture*, the diameter of the main mirror or lens. Only a few thousand stars are bright enough to be seen with the naked eye, but a telescope with a wide aperture opens up entirely new vistas. A very small telescope, such as the one used by Galileo, will show stars slightly below naked-eye visibility, but no fainter than that regardless of the magnification of the eyepiece. A telescope with a wider aperture will capture, focus and intensify a much greater amount of starlight, so that dimmer, more distant and otherwise invisible stars become visible.

In 1789 Herschel constructed a telescope with a 1.2-metre mirror, giving it the widest aperture of any telescope in the world. Unfortunately it was 12 metres in length, making it so unwieldy that valuable observing time was wasted while the telescope was being manoeuvred to point in the right direction. Another problem was that the mirror had to be strengthened with copper to support its own weight, which meant that it tarnished quickly, negating its otherwise excellent light-gathering potential. Herschel abandoned this monster in 1815, and thereafter used a more moderate telescope for most of his observing, with a 0.475-metre aperture and 6 metres long, a compromise between sensitivity and practicality.

Figure 33 Following his discovery of Uranus, Herschel moved to Slough, which had a finer climate than Bath. He was also closer to his patron, King George III, who had granted him an annual pension of £200 and funded his new record-breaking telescope, 1.2 metres in diameter and 12 metres long.

One of Herschel's main research projects was to use his superior telescopes to measure the distances to hundreds of stars, using the rough and ready assumption that all stars emit the same amount of light and the fact that brightness falls away with the square of the distance. For example, if one star is 3 times farther away than another star of the same actual brightness, then it will appear to be $1/3^2$ (or $1/9$) as bright. Conversely, Herschel assumed that a star that was apparently $1/9$ as bright as another star was roughly three times more distant. Using Sirius, the brightest star in the night sky, as his reference star, he defined all his stellar measurements in terms of multiples of the distance to Sirius, a unit he defined as the *siriometer*. Thus, a star that is apparently $1/49$ (or $1/7^2$) as bright as Sirius must be roughly seven times farther away than Sirius, or seven siriometers away. Although Herschel was aware that all stars are probably not equally bright and that his method was therefore inexact, he remained confident that he was building an approximately valid three-dimensional map of the heavens.

While it would be reasonable to expect that the stars would be distributed evenly in all directions and at all distances, Herschel's data strongly implied that the stars are in fact clumped together in a disc, rather like a flat, round pancake. This gigantic pancake was 1,000 siriometers in diameter and 100 siriometers thick. Instead of occupying an infinite extent of space, the stars of Herschel's universe were contained within a close-knit community. One way to imagine the distribution of stars is as a pancake that contains a sprinkling of raisins, each one representing a star.

This view of the universe was completely compatible with one of the most famous features of the night sky. If you imagine that we are embedded somewhere within the pancake of stars, then we would see lots of stars to the left, right, ahead and behind, but we would see fewer stars above and below us because the pancake is thin. Hence, from our vantage point in the cosmos we would expect to see a concentration of starlight around us – and indeed such a band can be seen arching across the night sky (as long as you are far from bright city lights). This feature of the heavens was well known to the ancient astronomers. In Latin this band was called Via Lactea, meaning 'milky way', because it has a hazy, milky quality. Although it was not apparent to the ancients, the first telescopic generation of astronomers could see that the milky band was actually a concentration of individual stars, too remote to be picked out by the naked eye. These stars are positioned around us in the plane of the pancake formation. Once the pancake model of the universe had been accepted, it was not long before the pancake of stars in which we live became known as the Milky Way.

Because the Milky Way supposedly contained all the stars in the universe, the size of the Milky Way was in effect the size of the universe. Although Herschel had estimated the Milky Way's diameter and thickness to be 1,000 siriometers and 100 siriometers respectively, he died in 1822 without knowing how many kilometres were in one siriometer. Therefore he had no idea of the size of the Milky Way in absolute terms. Converting siriometers into kilometres would require someone to measure the distance to Sirius. A major step towards this goal took place in 1838, when the German astronomer Friedrich Wilhelm Bessel became the first person to measure the distance to a star.

The puzzle of stellar distances had plagued generations of astronomers, and their failure to solve it had been a thorn in the side of Copernicus's theory that the Earth orbits the Sun. In Chapter 1 we saw how, if the Earth moves around the Sun, the stars should apparently change their positions when we view them from opposite sides of the Sun, six months apart, an effect known as parallax. Remember, if you hold up your finger and look at it with one eye, then changing your viewpoint by switching to the other eye makes the finger appear to shift against the background. As a rule, as the point of observation shifts, the object being observed seems to shift. However, the stars seemed fixed, a fact that believers in an Earth-centred universe used to support their belief in a fixed Earth. Supporters of the Sun-centred universe countered by pointing out that the stellar parallax effect reduces with distance, so the imperceptible shift in the positions of stars could simply mean that the stars must be incredibly distant.

Friedrich Bessel's efforts to put solid numbers to the vague phrase 'incredibly distant' began in 1810, when the Prussian king, Frederick William III, invited him to construct a new observatory at Königsberg. It would house the finest astronomical instruments in Europe, partly because the British prime minister, William Pitt, had crushed his own country's glass industry with his punitive window tax, thereby allowing Germany to take over as Europe's leading telescope manufacturer. German lenses were finely crafted, and a new triple-lens eyepiece arrangement reduced the problem of *chromatic aberration*, a difficulty in focusing caused by the fact that white light is a combination of colours, each of which is bent differently by glass.

After twenty-eight years at Königsberg, honing and refining his observations, Bessel eventually made his crucial breakthrough. By taking every conceivable error into account and by making painstaking observations six months apart, he was able to state that a star called 61 Cygni shifted its position by an angle of 0.6272 arcseconds, roughly 0.0001742°. This parallax detected by Bessel was minuscule – the equivalent of what you would perceive if you switched between your two eyes when you were observing your forefinger held up at arm's length . . . if your arm were 30 km long!

Figure 34 shows the principle of Bessel's measurement. When he observed 61 Cygni from the Earth at position A, he did so along a particular line of sight. Six months later, when he observed the star from the Earth at position B, he noticed that his line of sight had shifted slightly. The right-angled triangle formed by the Sun, 61 Cygni and the Earth allowed him to use trigonometry to estimate the distance to the star, because he already knew the Earth–Sun distance and now he knew the angle in one corner of the triangle. Bessel's measurements implied that the distance to 61 Cygni was 10^{14} km (100 trillion km). We now know that his measurement was approximately 10% too short, because modern estimates put the distance to 61 Cygni at 1.08×10^{14} km, or 720,000 times as far as the distance to the Sun. As explained in the caption to Figure 34, this is equivalent to 11.4 light years.

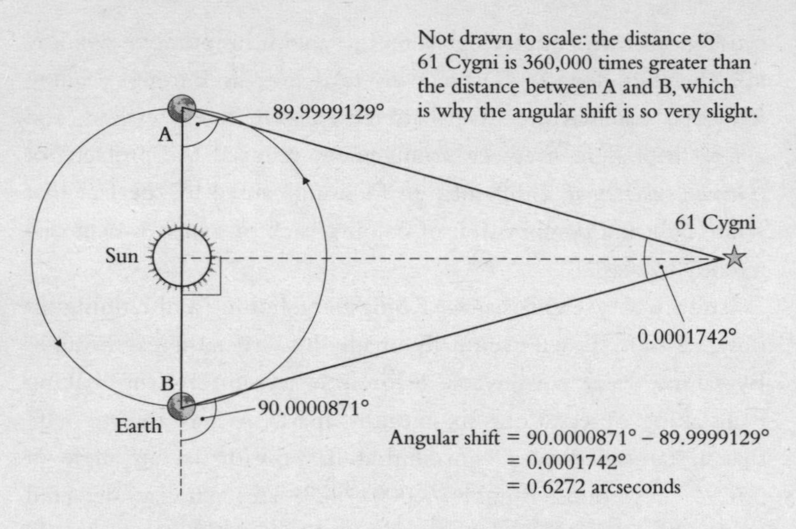

Figure 34 In 1838, Friedrich Bessel made the first measurement of stellar parallax. As the Earth orbits the Sun and moves from point A to point B, so a nearby star (e.g. 61 Cygni) appears in slightly different positions when viewed from A and B. The distance to 61 Cygni can be measured by simple trigonometry. The acute angle in the right-angled triangle = (0.0001742 ° ÷ 2) or 0.0000871 °, and the short side of the triangle is the Earth–Sun distance.

Hence, Bessel estimated the distance to 61 Cygni to be approximately 100,000,000,000,000 km, and now we know it is 108,000,000,000,000 km.

The kilometre is a very small unit of measurement for stellar distances, so astronomers prefer the *light year* as their unit of length, defined as the distance covered by light in one year. One year contains 31,557,600 seconds and light travels at 299,792 km/s, so

$$\begin{aligned}\text{1 light year} &= 31{,}557{,}600\ \text{s} \times 299{,}792\ \text{km/s}\\ &= 9{,}460{,}000{,}000{,}000\ \text{km}\end{aligned}$$

This means that 61 Cygni is 11.4 light years from Earth. The light year reminds us that telescopes act as time machines. Because light takes a finite time to travel any distance, we only ever see celestial objects as they were in the past. It takes 8 minutes for sunlight to reach us, so we only ever see the Sun as it was 8 minutes ago. If the Sun suddenly exploded, it would be 8 minutes before we knew about it. The more distant star 61 Cygni is 11.4 light years away, so we only ever see it as it was 11.4 years ago. The farther that telescopes allow us to look across the universe, the farther back in time we are seeing.

The Copernicans were correct. The stars did move, and the stellar 'jumps' had hitherto been imperceptible because the stars were so incredibly far away. Even though astronomers knew that the stars had to be very remote, they were still shocked by the sheer distance to 61 Cygni, especially bearing in mind that it is one of the closest stars to the Earth. To put this into perspective, if the universe were miniaturised so that our Solar System, everything from the Sun to the outer reaches of Pluto's orbit, could be squeezed inside a house, then our neighbouring stars would still be dozens of kilometres away. It became clear that our Milky Way is exceedingly thinly populated.

Bessel's contemporaries praised his measurement. The German physician and astronomer Wilhelm Olbers said that it 'put our ideas about the universe for the first time on a sound basis'. Similarly, John Herschel, William Herschel's son and himself an acclaimed astronomer, called the result 'the greatest and most glorious triumph which practical astronomy has ever witnessed'.

Not only did astronomers now know the distance to 61 Cygni, but they could also estimate the size of the Milky Way. By comparing the brightness of 61 Cygni to that of Sirius, it was possible to do a ballpark conversion of William Herschel's siriometer unit into light years, whereupon astronomers estimated that the Milky Way was 10,000 light years across and 1,000 light years thick. In fact, they had underestimated the dimensions of the Milky Way by a factor of ten, and we now know that the Milky Way is about 100,000 light years across and 10,000 light years thick.

Eratosthenes had been shocked when he measured the distance to the Sun, and Bessel had been staggered by the distance to the nearest stars, but the size of the Milky Way was truly overwhelming. At the same time, astronomers realised that even the vastness of the Milky Way was insignificant compared with the assumed infinity of the universe. Not surprisingly, some scientists began to wonder what was going on in the space beyond the Milky Way. Was it completely empty, or was it populated by other objects?

Attention turned to the *nebulae*, curious smudges of light in the night sky that looked very different from the sharp pinpricks of light from stars. Some astronomers suggested that these mysterious objects were sprinkled throughout the universe. The majority, however, believed that they were more mundane entities within our own Milky Way. After all, William Herschel had indicated that everything was within our pancake-shaped Milky Way.

The study of nebulae dates back to the ancient astronomers, who had spotted a handful of nebulae using just their naked eyes, but then the invention of the telescope revealed a surprisingly large number of them. The first person to compile a detailed catalogue of nebulae was the French astronomer Charles Messier, who started work on this project in 1764. Previously he had already been successful in tracking down comets, which is why King Louis XV nicknamed him the Comet Ferret, but Messier was continually frustrated because, at first sight, it was easy to confuse a comet with a nebula as both types of object appear as tiny smudges in the sky. Comets move across the sky, so they eventually reveal themselves for what they are, but Messier wanted to compile a list of nebulae so that he did not have to waste time mistakenly staring at a static object waiting in vain for it to move. He published a catalogue of 103 nebulae in 1781, and today these objects are still referred to by their *Messier numbers*; for instance, the Crab Nebula is M1, and the Andromeda Nebula is M31. Messier's sketch of the Andromeda Nebula is shown in Figure 35.

When William Herschel received a copy of the Messier Catalogue, he turned his gaze upon the nebulae, employing his giant telescopes to conduct an exhaustive search of the heavens. Herschel went far beyond Messier and recorded a total of 2,500 nebulae, and during the course of his survey he began to speculate on their nature.

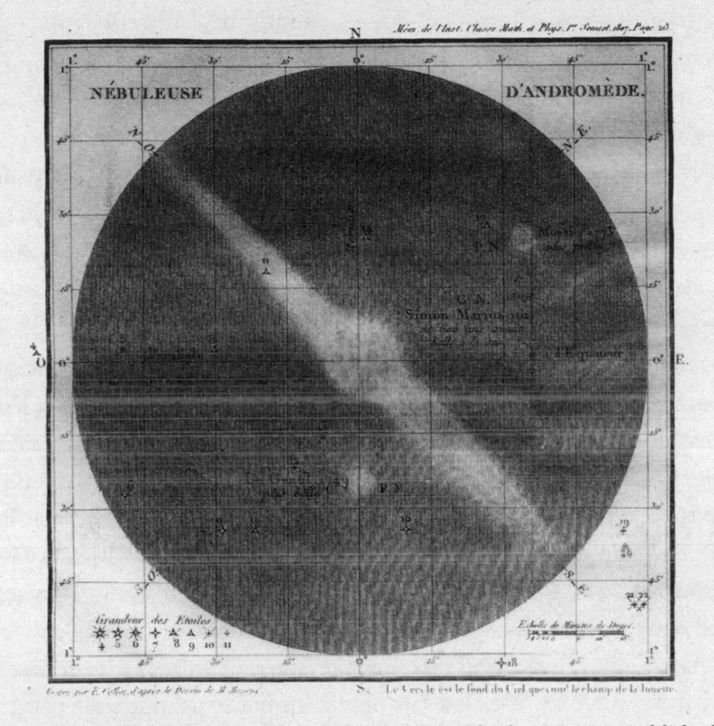

Figure 35 After two decades of observation, Charles Messier published a catalogue of 103 nebulae in 1781. His detailed sketch of the Andromeda Nebula, the 31st entry in his catalogue, illustrates the difference between a nebula, which has a definite extended visible structure, and a star, which appears as a point of light.

Because they looked like clouds (*nebula* means 'cloud' in Latin), he believed that they were indeed large clouds of gas and dust. More specifically, Herschel could discern a single star within some of these clouds, so suggested that the nebulae were young stars surrounded by debris, and this debris was presumably in the process of coalescing to form planets. All in all, it seemed to Herschel that these nebulae were stars in the early phase of their life and that, like all other stars, they existed within the realm of the Milky Way.

While Herschel believed that the Milky Way was the one and only cluster of stars in the whole universe, the eighteenth-century German philosopher Immanuel Kant took the opposite view and argued that at least some of the nebulae were independent groupings of stars, similar to the Milky Way in terms of size, but far beyond its perimeter. According to Kant, the reason why the nebulae looked like clouds was because they contained millions of stars and were so distant that the stars merged into a haze of light. To back his case, he noted that most nebulae had an elliptical appearance, which is exactly what you would expect if they had the same round pancake structure as the Milky Way. Although the Milky Way would look like a circular disc when viewed from above and a thin line when viewed from the side, it would appear elliptical when viewed from an intermediate angle. Kant called the nebulae 'island worlds', because he pictured the universe as an ocean of space populated by separate islands of stars. Our Milky Way was just one such island of stars. Today we refer to any such isolated system of stars as a *galaxy*.

Although Kant's fondness for the idea of nebulae as galaxies beyond the Milky Way had an observational basis, there was also a theological foundation for his belief. He argued that God was omnipotent, so the universe should be both eternal and infinitely rich in content. It seemed absurd to Kant that God's creation should be limited to the finite Milky Way:

> We come no nearer to the infinitude of the creative power of God, if we enclose the space of its revelation within a sphere described with the radius of the Milky Way, than if we were to limit it to a ball an inch in diameter. All that is finite, whatever has limits and a definite relation to unity, is equally far removed from the infinite . . . For this reason the field of the revelation of the Divine attributes must be as infinite as these attributes themselves. Eternity is not sufficient to embrace the manifestations of the Supreme Being, if it is not combined with the infinitude of space.

The battle lines had been drawn up. Herschel's supporters argued that the nebulae were young stars surrounded by clouds of debris and situated within the Milky Way, while the followers of Kant maintained that they were galaxies, independent stellar systems far beyond the Milky Way. The key to settling the debate was better observational evidence, and this began to appear in the middle of the nineteenth century, thanks to the extraordinary William Parsons, the Third Earl of Rosse.

Having married a wealthy heiress and inherited Birr Castle, situated in a large estate in Ireland, Lord Rosse was fortunate in being able to pursue the life of a gentleman scientist. He was determined to build the biggest and best telescope in the world and was not afraid to get his hands dirty. A reporter on the *Bristol Times* wrote:

> I saw the Earl, the telescope maker himself, not in state with his coronet and ermine robe on, but in his shirt sleeves, with his brawny arms bare. He had just quitted the vice at which he had been working and, powdered with steel filings, was washing his hands and face in a coarse ware basin placed on the block of an anvil, while a couple of smiths sledging away on a blazing bar were sending a shower of sparks about his lordship which he little regarded as though he were a Fire King.

Merely casting the mirror for the giant telescope was a major engineering feat in itself. It required 80 cubic metres of peat to melt the ingredients for the 3-tonne mirror, which measured 1.8 metres in diameter. Dr Thomas Romney Robinson, Director of the Armagh Observatory, witnessed the casting:

> The sublime beauty can never be forgotten by those who were so fortunate as to be present. Above, the sky, crowned with stars and illuminated by a most brilliant Moon, seemed to look auspiciously

> on their work. Below, the furnaces poured out huge columns of nearly monochromatic yellow flame, and the ignited crucibles during their passage through the air were fountains of red light.

In 1845, after three years of construction and having spent the equivalent of £1 million of his own money, Lord Rosse completed his gigantic 16.5 metre long telescope, shown in Figure 36, and began making observations. This coincided with the Irish Potato Famine, a tragedy that Rosse had tried to avert when he had earlier advocated new farming practices that would have reduced the risk of potato blight. He quickly halted his survey of the sky and diverted his time and money towards supporting the local community. He also refused to accept rent from his tenants and earned a reputation as an earnest politician who campaigned on behalf of the rural population during this dark period of Irish history.

When Lord Rosse did eventually return to surveying the stars several years later, he would make his observations while precariously perched on the scaffolding that surrounded his magnificent telescope. At the same time, he had to maintain his balance while five labourers worked with mechanical cranks, blocks and pulleys to hoist the telescope to the right elevation. Lord Rosse and his team wrestled with this monster night after night, which is why it was nicknamed the Leviathan of Parsonstown.

Rosse was rewarded for his efforts with spectacular views of the night sky. Johnstone Stoney, Rosse's assistant, assessed the telescope's quality by pointing it at very faint stars: 'Such stars are bright in the great telescope. They are usually seen as balls of light, like small peas, violently boiling in consequence of the atmospheric disturbance . . . the test bordered very closely indeed on theoretical perfection.'

The only problem was that the Leviathan was sited in the middle of Ireland, which does not have a great reputation for clear, cloudless

Figure 36 Lord Rosse's 'Leviathan of Parsonstown', with a mighty aperture of 1.8 metres, was the world's largest telescope when it was built. Parsonstown was the former name of Birr, the town where the telescope was sited.

skies. Apart from the 'fogs from the bogs', there were said to be two types of weather, namely 'just before rain' or 'in rain'. On one occasion the patient lord wrote to his wife, explaining: 'The weather here is still vexatious: but not absolutely repulsive.'

Somehow, in between the clouds, Rosse was able to make extraordinarily detailed observations of the nebulae. Instead of appearing as formless smudges, the nebulae began to show themselves as having a distinct internal structure. The first nebula to succumb to the Leviathan was M51 in Messier's list, which became the subject of an amazingly detailed sketch by Rosse, shown in Figure 37. He could easily discern that M51 had a spiral structure. In particular, he noticed a mini-swirl at the end of one of the spiral arms, which is why M51 was sometimes referred to as Lord Rosse's Question Mark Nebula. Rosse's sketch became well known across Europe, and it has even been suggested that it inspired Vincent Van

Gogh's painting *Starry Night*, which appears to show a spiral nebula with an accompanying swirl.

Its resemblance to a whirlpool gave M51 its other nickname, the Whirlpool Nebula. It also led Rosse to an obvious conclusion: 'That such a system should exist, without internal movement, seems to be in the highest degree improbable.' Also, he believed that the swirling mass was more than merely a gaseous cloud: 'We thus observe, that with each successive increase of optical power, the structure has become more complicated . . . The nebula itself, however, is pretty well studded with stars.'

It was becoming clear that at least some of the nebulae were collections of stars, but this did not necessarily prove Kant's theory that the nebulae were galaxies equivalent to and independent from our own Milky Way. Such nebulae would have to be vast, distinct and remote, but perhaps the Whirlpool Nebula was a relatively small subgrouping of stars within or on the edge of our own Milky Way. The critical issue was distance. If somebody could somehow measure the distances to the nebulae, then it would be easy to decide whether they were in the Milky Way, close to the Milky Way or far beyond the Milky Way. But parallax, the best technique for distance measurement, could not be applied to the nebulae. After all, it was barely possible to measure the angular shifts of the closest stars, so identifying any angular shift associated with a fuzzy nebula on the edge of the Milky Way – or perhaps much more distant – was out of the question. The status of the nebulae remained in limbo.

As each decade passed, astronomers invested more money in building increasingly powerful telescopes, situated in high-altitude locations blessed with cloudless skies (unlike Ireland). Although there were other questions on their agenda, astronomers were particularly anxious to discover the true identity of the nebulae, if not by measuring their distance then by finding some other vital clue that would reveal their nature.

Figure 37 Lord Rosse's drawing of the Whirlpool Nebula (M51), alongside a modern image taken at La Palma Observatory, which shows the power of Rosse's telescope and the accuracy of his observation.

The next great master telescope-builder was the eccentric millionaire George Ellery Hale, who turned out to be even more obsessive than Lord Rosse. Hale was born in 1868, at 236 North LaSalle Street in Chicago, and in 1870 the family moved to the suburb of Hyde Park, just in time to avoid the Great Chicago Fire of 1871, which consumed 18,000 buildings including their old home. The city became a blank slate for architects, and the nine-storey Home Insurance Building became the world's first skyscraper, setting a new trend in building design for Chicago and many other American cities. Hale's father, William, had previously been a struggling salesman, but he was sharp enough to take out a loan and set up a company to supply the elevators necessary for the Chicago skyscrapers. Eventually, he even constructed the elevator for the Eiffel Tower.

The family became wealthy and could afford to indulge young George's interest in microscopes and telescopes. They were unaware that his childhood fascination would evolve into an adult obsession. In fact, Hale grew up to be a serial world-class telescope-builder. His first major project started when he scavenged some redundant lenses from astronomers on the West Coast who had just abandoned their own plans to build a telescope. Hale's ambition was to incorporate these lenses in a 40-inch (1-metre) diameter refracting telescope, and he also wanted to build an entire observatory complex around this telescope.

Hale sought funding for his new telescope and observatory from Charles Tyson Yerkes, a transport tycoon who had made his money building Chicago's elevated rail transit system, which still serves the city today. Yerkes was also a convicted swindler, so Hale tried to persuade him that sponsoring an astronomical observatory would help him to become accepted in Chicago high society. Hale also exploited Yerkes' penchant for one-upmanship by pointing out that the wealthy land investor James Lick had funded the California Lick Observatory. He began to lobby Yerkes with the slogan 'Lick the Lick', because his new telescope would dwarf anything at the Lick Observatory.

Bowled over by Hale's relentless campaigning, it was not long before Yerkes put up half a million dollars, and the Yerkes Observatory was born as part of the University of Chicago. After the dedication ceremony, one newspaper ran a headline highlighting the swindler's new-found status: YERKES BREAKS INTO SOCIETY. Unfortunately for Yerkes, the headline was over-optimistic. He still failed to become accepted by the Chicago elite, so he moved to London, where he played a major role in developing the underground train system, particularly the Piccadilly Line.

The Yerkes Observatory was situated 120 km north of Chicago, near the community of Williams Bay. The town still relied on candles and kerosene lamps for lighting, so the astronomers knew that the faint celestial light would not be polluted by bright electric lamps. Even the resort of Lake Geneva, the nearest community with electric lights, was a safe 10 km away. The telescope, 20 metres in length and weighing 6 tonnes, was finished in 1897. It was guided by 20 tonnes of machinery especially designed to point the telescope in the right direction and then to smoothly synchronise it with the rotation of the Earth. In this way the star or nebula under inspection remained in the instrument's field of view. It was, and still is, the biggest telescope of its type in the world.

Hale, though, was not satisfied. A decade later he raised money from the Carnegie Institute and pushed the limits of engineering even further, building a 60-inch (1.5-metre) telescope at Mount Wilson, near Pasadena in California. This time he would use a mirror rather than a lens, as a 60-inch lens would sag under its own weight. He described his desire for wider, longer and more sensitive telescopes as a symptom of 'Americanitis', namely the insatiable ambition to be the very best. Unfortunately, Hale's compulsive craving for perfection and the responsibility of managing major projects became self-destructive. As a result of the overwhelming stress he suffered periods of psychosis, which ultimately forced him to spend several months in a sanatorium in Maine.

His mental health deteriorated further after he embarked on his third project, a 100-inch (2.5-metre) telescope at Mount Wilson. As the basis for his mirror, Hale ordered a 5-tonne glass disc from France, which the newspapers called the single most valuable piece of merchandise to cross the Atlantic. When it arrived, however, there was concern among Hale's team about the strength and the optical quality of the glass, which turned out to contain tiny air bubbles. Evelina Hale witnessed the suffering caused to her husband by this latest project and came to hate the giant lens that plagued him: 'I wish that glass was in the bottom of the sea.'

Figure 38 Andrew Carnegie and George Ellery Hale at Mount Wilson in 1910, outside the dome housing the 60-inch telescope. The millionaire Carnegie (left) is standing farther up the slope to appear taller – a manoeuvre he often performed when he was being photographed with others.

The project seemed doomed to failure, and during periods of extreme pressure Hale would hallucinate and receive visitations from a green elf, who soon became the only person he would confide in about his plans for the telescope. The elf was usually sympathetic, but occasionally it would taunt him. Hale lamented to a friend: 'How to escape this new form of torment, which is incessant, I do not know.'

Funded by the Los Angeles hardware tycoon John Hooker, the 100-inch Hooker Telescope was eventually completed in 1917. On the night of 1 November, Hale had the honour of being the first person to stare into the eyepiece – and was shocked to see Jupiter overlapped by six ghost planets. Blame for the optical defect was immediately laid on the bubbles in the glass, but calmer minds came up with an alternative theory. Workmen had left the roof of the observatory open that day while they completed installation, so sunlight had been warming the mirror, which had possibly become distorted as a result. The astronomers adjourned until 3 a.m., by which time they hoped that a cooling-off period would have solved the problem. In the chill of the night, Hale's next view of the heavens was clearer than any previous observation in history. The Hooker Telescope was capable of revealing nebulae that previously had been too faint to show up in any other telescope; it was so sensitive that it could have detected a candle at a distance of 15,000 km.

Hale was still not satisfied. Motivated by his guiding principle of 'More light!', he began work on a 200-inch (5-metre) telescope. His obsession became infamous and would later be immortalised on television in an episode of *The X Files*. Mulder explains to Scully that the elf gave Hale advice on fundraising: 'Actually the idea was presented to Hale one night while he was playing billiards. An elf climbed in his window and told him to get money from the Rockefeller Foundation for a telescope.' Scully comments that Mulder must be reassured to know that he is not the only one to see green elves, but Mulder replies: 'In my case, little green men.'

Sadly, Hale would not live to see his 200-inch telescope project completed. He was, however, able to witness the impact of his 40-inch, 60-inch and 100-inch telescopes, each of which revealed further riches in the sheer number and variety of nebulae. Annoyingly, the exact location of these objects remained a mystery. Were they part of our own Milky Way galaxy, or were they far-away galaxies in their own right?

The matter came to a head in April 1920, when the National Academy of Sciences in Washington planned to host what would become known as the Great Debate. The Academy decided it should bring together the two opposing camps on the nature of nebulae to debate the question in front of the most eminent scientists of the age. The view that the Milky Way contains the entire universe, including the nebulae, was strongly championed by the astronomers at the Mount Wilson Observatory, and they sent an ambitious young astronomer, Harlow Shapley, to argue on their behalf. The opposing view, that the nebulae are galaxies in their own right, was popular at the Lick Observatory, who sent Heber Curtis to defend their position.

By chance, the two rival astronomers ended up sharing the train from California to Washington. It was an awkward and uncomfortable journey – two astronomers with directly opposing views trapped in a railway carriage for 4,000 km, each one careful to avoid prematurely engaging in the debate that was intended for later. The situation was made worse by their contrasting personalities.

Curtis had an aura of superiority and a reputation as a distinguished astronomer, well known for speaking with authority and confidence. He relished the battle to come. In contrast, Shapley was nervous and overawed. Having grown up as the son of a poor hay farmer from Missouri, he had stumbled into astronomy more by luck than by judgement. As a teenager he had wanted to study journalism at college, but the course was cancelled, so he had to find a new subject: 'I opened the catalogue of courses and the very first course offered was a-r-c-h-a-e-o-l-o-g-y, and I couldn't pronounce it! . . . I turned over a page and saw a-s-t-r-o-n-o-m-y; I could pronounce that – and here I am!'

By the year of the Great Debate, Shapley had established himself as part of the new generation of promising astronomers, but he still felt very much in the shadow of Curtis, and was grateful for the opportunity to escape his opponent's intimidating personality when their Southern Pacific train broke down in Alabama. Shapley spent the time wandering along the tracks in search of ants, which he had studied and collected for many years.

Figure 39 The two main protagonists in the Great Debate. young Harlow Shapley (left), who believed that the nebulae lay within the Milky Way galaxy; and the more senior Heber Curtis, who put forward the case that the nebulae were independent galaxies far beyond the Milky Way.

When the night of the Great Debate finally arrived, Shapley's nerves grew worse during the long-winded prize-giving ceremony that preceded the main event. The citations honouring the winners and the acceptance speeches seemed to go on for ever. There was not even a drop of wine to help cheer up proceedings, as prohibition had come into force earlier that year. In the audience, Albert Einstein whispered to his neighbour: 'I have just got a new theory of Eternity.'

Eventually, the Great Debate took centre stage and the main event of the evening was under way. It began with Shapley arguing the case that the nebulae were within the Milky Way. In his presentation, he relied upon two pieces of evidence to support his view. First, he discussed the distribution of the nebulae. They were generally found above and below the plane of the pancake-shaped Milky Way, but rarely within the plane itself, a band that became known as the *zone of avoidance*. Shapley explained this situation by claiming that the nebulae were clouds of gas that acted as nurseries for newborn stars and planets. He believed that such clouds existed only in the upper and lower reaches of the Milky Way, drifting towards the central plane as the stars and planets matured. Hence, he could explain the zone of avoidance in terms of the Milky Way being the only galaxy. He then turned to his opponents and claimed that the zone of avoidance was incompatible with their model of the universe: if the nebulae represented galaxies that were peppered throughout the entire universe, they should appear all around the Milky Way.

Shapley's second piece of evidence was a nova that had appeared in the Andromeda Nebula in 1885. A nova is not, as the name suggests, a new star, but a very dim star that has suddenly increased in brightness, fuelled by material stolen from a companion star. The 1885 nova was one-tenth as bright as the entire Andromeda Nebula, which was perfectly sensible if Andromeda was just a smattering of stars situated within the boundaries of our home galaxy. However, if Andromeda was a galaxy in its own right, as his opponents argued, then it would consist of billions of stars, and the nova (one-tenth as bright as Andromeda) would have been as bright as hundreds of millions of stars! Shapley argued that this was preposterous, and that the only sensible conclusion was that the Andromeda Nebula was not a separate galaxy, but merely part of our Milky Way galaxy.

For some, this level of evidence was more than sufficient. Agnes Clerke, a historian of astronomy, was already aware of Shapley's arguments and had previously written: 'No competent thinker, with the whole of the available evidence before him, can now, it is safe to say, maintain any single nebula to be a star system of coordinate rank with the Milky Way.'

However, for Curtis, the matter was far from settled. As far as he was concerned, Shapley's case was weak, and he attacked his two main arguments. Both men had 35 minutes to present their case, but their styles differed. While Shapley had given a largely non-technical talk, aimed at scientists who came from a variety of disciplines, Curtis presented his riposte with ruthless attention to detail.

With respect to the zone of avoidance, Curtis believed that this was an illusion. He argued that the nebulae, being galaxies, were sprinkled symmetrically all around and way beyond the Milky Way. According to Curtis, the only reason that astronomers could not see many nebulae in the plane of the Milky Way was because their light was blocked by all the stars and interstellar dust that occupy the galactic plane.

When it came to the other pillar of Shapley's case, the nova of 1885, Curtis dismissed it as abnormal. There were many other novae that had been observed within the spiral arms of nebulae, and they had all been inordinately fainter than the notorious Andromeda nova. In fact, most of the novae observed in nebulae were so extremely faint that, Curtis claimed, this proved that the nebulae must be incredibly distant and beyond the Milky Way. In short, Curtis was not prepared to abandon his cherished model just because of a single bright nova observed thirty-five years earlier. Curtis once said of his unproven multiple galaxy model:

> Few greater concepts have ever been formed in the mind of thinking man than this one. Namely that we, the microbic inhabitants of a minor satellite of one of millions of suns which form our galaxy, may look out beyond its confines and behold other similar galaxies, tens of thousands of light-years in diameter, each composed, like ours, of a thousand million or more suns, and that, in so doing, we are penetrating the greater cosmos to distances of from half a million to a hundred million light years.

Curtis put forward various other arguments during his presentation, some supporting his own theory, some attacking Shapley's. He was confident that he had presented a convincing case and wrote to his family shortly afterwards: 'Debate went off fine in Washington, and I have been assured that I came out considerably in front.' The truth is that there was no clear-cut winner, and if there was any slight swing towards Curtis's point of view, then Shapley attributed it to style rather than substance: 'As I remember it, I read my paper and Curtis presented his paper, probably not reading much because he was an articulate person and was not scared.'

The Great Debate did little more than focus attention on a question that was far from being resolved. It keenly illustrated the nature of conducting research at the frontiers of science, where competing theories square up to each other, armed only with the feeblest of hard data. The observations used by each side to prop up its own view lacked rigour, detail and volume, and it was far too easy for the opposition to label any data as flawed, inaccurate or open to interpretation. Unless somebody could establish some concrete observations, in particular something that would firmly establish the distance to the nebulae, then the rival theories were nothing more than speculations. The popularity of the theories seemed to depend on the personality of their supporters rather than on any real evidence.

The Great Debate was all about humankind's place within the cosmos, and settling the matter would require a major breakthrough in astronomy. Some scientists, such as the popular astronomy writer Robert Ball, believed that such a breakthrough was impossible. In *The Story of the Heavens*, he was of the opinion that astronomers were at the limits of knowledge: 'We have already reached a point where man's intellect begins to fail to yield him any more light, and where his imagination has succumbed in the endeavour to realise even the knowledge he has gained.'

Similar statements had probably been made by some ancient Greeks dismissing the possibility of measuring the size of the Earth or the distance to the Sun. However, the first generation of scientists, including Eratosthenes and Anaxagoras, invented techniques that allowed them to span the globe and the Solar System. Then Herschel and Bessel used brightness and parallax to measure the size of the Milky Way and the distance to the stars. Now it was time for someone to invent a yardstick that could cross the cosmos, one that would resolve the true nature of the nebulae.

Now You See It, Now You Don't

Nathaniel Pigott came from a wealthy and well-connected Yorkshire family, and was a gentleman astronomer of the first order. A close friend of William Herschel, Pigott made careful observations of two solar eclipses and the 1769 transit of Venus. He also constructed one of only three private observatories that existed in England in the late 1700s. Consequently, his son Edward was brought up surrounded by telescopes and other astronomical instruments. Edward developed a fascination with the night sky and in due course he would surpass his father in both his enthusiasm for and expertise in astronomy.

Edward Pigott's main interest was variable stars. Novae are considered to be a class of variable star, because they flare up suddenly after a long period of being relatively faint, and then they gradually fade back to their former dimness. Other stars brighten and fade more regularly, such as Algol in the constellation Perseus, nicknamed the Winking Demon. These variable stars were significant in astronomy because they directly contradicted the ancient view that the stars were immutable, and as a result there was a concerted effort to understand what was driving their fluctuations.

In his twenties, Edward Pigott befriended the teenager John Goodricke. He was a deaf-mute who had developed a keen interest in science, having grown up during a period when educationalists were for the first time addressing the issue of schooling deaf children. He attended Britain's first school for the deaf, opened in Edinburgh in 1760 by Thomas Braidwood. The school had such an excellent reputation that the author and lexicographer Samuel Johnson paid a visit in 1773, when he may well have encountered Goodricke, who would have been a nine-year-old student at the time. Johnson was particularly interested in educating deaf children, because he had contracted tuberculosis from his wet nurse and had suffered from scarlet fever as a baby, the combined effect of which left him permanently deaf in one ear and partially sighted. Johnson was so impressed with Braidwood Academy that he mentioned it in his *Journey to the Western Islands of Scotland*:

> This school I visited, and found some of the scholars waiting for their master, whom they are said to receive at his entrance with smiling countenances and sparkling eyes, delighted with the hope of new ideas. One of the young Ladies had her slate before her, on which I wrote a question consisting of three figures, to be multiplied by two figures. She looked upon it, and quivering her fingers in a manner which I thought very pretty, but of which I know not whether it was art or play, multiplied the sum regularly in two lines, observing the decimal place.

Then, at the age of fourteen, Goodricke moved from Braidwood to Warrington Academy, where he was able to learn alongside hearing students. His teachers described him as 'a very tolerable classic and an excellent mathematician'. When he returned home to York he continued his studies under the guidance of Edward Pigott, who taught him about astronomy, and in particular the significance of variable stars.

Goodricke proved to be an extraordinary astronomer. He had developed an unparalleled visual acuity and sensitivity, and was able to evaluate with great precision how the brightness of a variable star changed from night to night. This was an amazing achievement, because he had to take into consideration the effects of atmospheric conditions and the varying level of moonlight to obtain a sufficient degree of accuracy. To help him gauge the brightness of a variable star, Goodricke compared it with the fixed brightnesses of surrounding non-variable stars. One of his first research projects was to observe the subtle winks of Algol from November 1782 to May 1783, carefully plotting a graph of brightness versus time, showing that it reached minimum brilliance every 68 hours and 50 minutes. The variation of Algol is shown in Figure 40.

Goodricke's brain was as sharp as his sight. By studying the pattern of variation in Algol's brightness, he deduced that it was not a lone star, but a binary star – a pair of stars orbiting each other, which we now know to be a relatively common situation for stars. In the case of Algol, Goodricke proposed that one star was much dimmer than the other and that the variability in overall brightness was a result of the dim star passing in front of the bright star and blocking its light during their mutual orbiting. In other words, the variability was an eclipsing effect.

Goodricke was just eighteen years old, and absolutely correct in his analysis of Algol – the pattern was symmetric and an eclipse is a symmetrical process, and the star system was generally bright and with a relatively short dim phase, which again was typical of an eclipsing system. In fact, a large proportion of variable stars can be explained in this way. His work was recognised by the Royal Society, which awarded him the prestigious Copley Medal for the year's most significant discovery in science. Three years earlier it had been won by William Herschel, and in later years it would be awarded to Dmitri Mendeleev for developing the periodic table, to

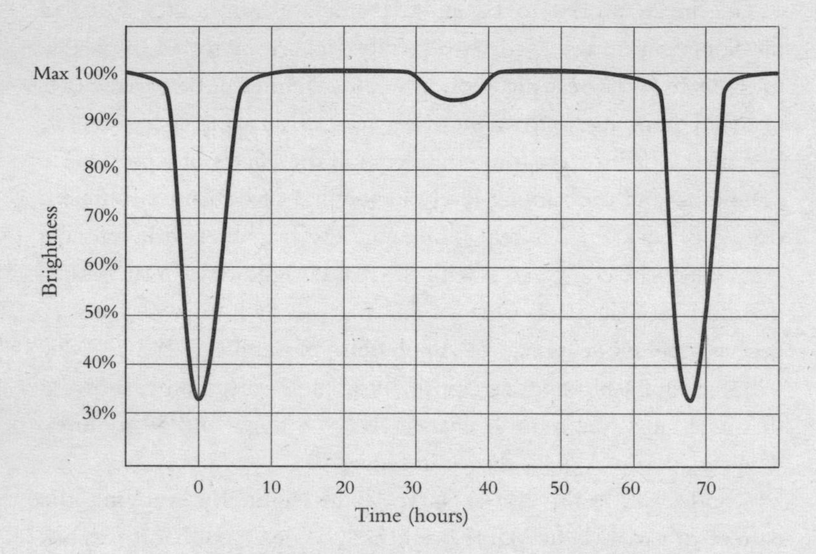

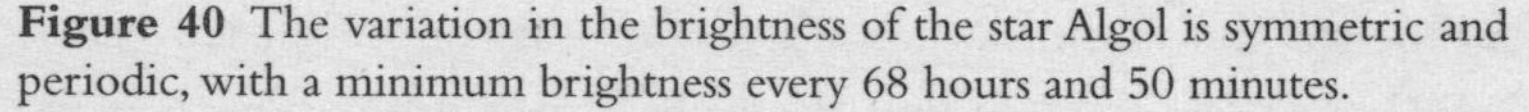
Figure 40 The variation in the brightness of the star Algol is symmetric and periodic, with a minimum brightness every 68 hours and 50 minutes.

Einstein for his work on relativity, and to Francis Crick and James Watson for unravelling the secret of DNA.

The phenomenon of eclipsing binary stars was a major discovery in the history of astronomy, but it would play no role in the drama of the nebulae. Instead, it was a set of observations made by Goodricke and Pigott in 1784 which would ultimately resolve the Great Debate that was to come. On the night of 10 September, Pigott observed that the star Eta Aquilae varied in brightness. A month later, on 10 October, Goodricke spotted that Delta Cephei was also varying. Nobody had previously noticed the variability of these stars, but Pigott and Goodricke had a knack for detecting subtle changes in brightness. Goodricke plotted the variation of both stars with time and showed that Eta Aquilae repeated its pattern every seven days, whereas Delta Cephei took just five days, so both

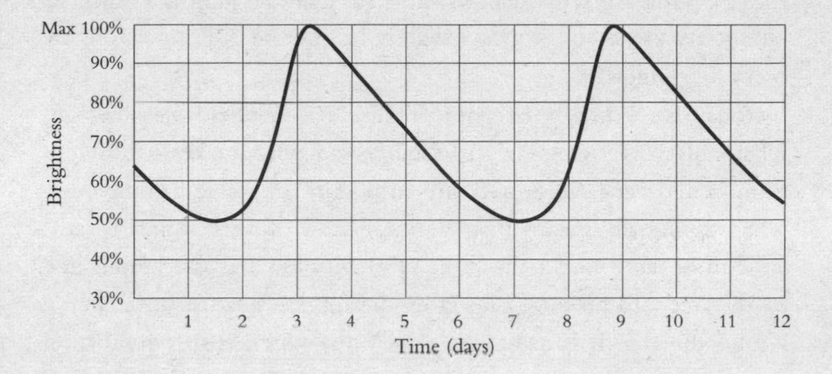

Figure 41 The variable brightness of the star Delta Cephei The variation is asymmetric, increasing in brightness quickly and decreasing slowly.

had a distinctly longer period of variation compared to Algol. What made Eta Aquilae and Delta Cephei even more remarkable was the overall shape of their variations in brightness.

Figure 41 shows a plot of Delta Cephei's variation. The most striking feature is the lack of symmetry. Whereas the Algol plot (Figure 40) displays a series of thin, symmetric valleys, Delta Cephei ramps up to peak brightness in just a day and then gradually fades to a minimum over the course of four days. Eta Aquilae showed a similar sawtooth or shark's-fin pattern. This pattern cannot be explained by any sort of eclipse effect, so the two young men assumed that there must be something intrinsic to the two stars that was causing the variation. They decided that Eta Aquilae and Delta Cephei belonged to a new class of variable star, which we now call *Cepheid variables*, or simply Cepheids. Some Cepheids are very subtle, such as Polaris, the North Star, which is our closest Cepheid. William Shakespeare was completely unaware of the star's variable nature, and in *Julius Caesar* he has Caesar proclaim: 'But I am constant as the Northern Star.'

Although this star is constant inasmuch as it always indicates north, its luminosity varies and it grows slightly brighter and dimmer roughly every four nights.

Today we know what goes on inside a Cepheid variable star, what causes its asymmetric variability and what makes it different from other stars. Most stars are in a state of stable equilibrium, which essentially means that the huge mass of a star wants to collapse in on itself under the force of gravity, but this is counteracted by the outward pressure caused by the intense heat of the material within the star. It is a bit like a balloon, which is in equilibrium because the rubber skin on the outside wants to contract inwards, while the air pressure on the inside wants to push outwards. Put the balloon in a fridge overnight, and the air in the balloon cools, the air pressure inside the balloon decreases and the balloon contracts to find a new equilibrium state.

However, Cepheid variable stars are not in a stable equilibrium, but fluctuate. When a Cepheid is relatively cool, it is unable to counteract the gravitational force, which will then cause the star to contract. This compresses the fuel in the stellar core and encourages more energy to be generated, which heats the star, forcing it to expand. Energy is released during and after the expansion, whereupon the star cools and contracts, and the process repeats itself all over again. Crucially, the contraction phase compresses the outer layer of the star, which causes it to become more opaque, resulting in the dimming phase of the Cepheid.

Although Goodricke was unaware of the explanation behind the variability of Cepheids, the discovery of this new type of star was in itself a great achievement. At the age of just twenty-one, a new honour was bestowed on him: he was made a Fellow of the Royal Society. Then, just fourteen days later, the life of this brilliant young astronomer was cut short. Goodricke died of pneumonia, contracted during long freezing nights spent staring at the stars. His friend and collaborator Pigott lamented: 'This worthy young man exists no more; he is not only regretted to many friends, but will prove a loss to astronomy, as the discoveries he so rapidly made evince.' In a career lasting just a few years, Goodricke had made an outstanding contribution to astronomy. Although he did not realise it, his discovery of Cepheid variables would prove pivotal to the Great Debate and to the development of cosmology.

Over the next century, Cepheid spotters would discover thirty-three stars with the distinctive shark's-fin variation. Each one increased and decreased its brightness, sometimes over the course of less than a week, sometimes taking more than a month. However, one problem plagued the study of Cepheids, namely subjectivity. Indeed, this major problem was common throughout astronomy. If observers saw something in the sky, they would inevitably interpret it with some level of bias, especially if the phenomenon was fleeting and the interpretation relied on memory. Also, the observation could only be recorded in words or a sketch, neither of which could be relied upon for perfect accuracy.

Then, in 1839, Louis Daguerre released details of the *daguerreotype*, a process for chemically imprinting an image on a metal plate. Suddenly, daguerreomania swept the world, with people queuing up to be photographed. As with every new technology, there were some critics, as demonstrated by this extract from the *Leipzig City Advertiser*: 'The wish to capture evanescent reflections is not only impossible . . . but the mere desire alone, the will to do so, is blasphemy. God created man in His own image, and no man-made machine may fix the image of God. Is it possible that God should have abandoned His eternal principles, and allowed a Frenchman to give to the world an invention of the Devil?'

John Herschel, son of William and now president of the Royal Astronomical Society, was one of the first people to adopt this new technology. Within a few weeks of Daguerre's announcement, he was able to replicate the process and took the first photograph on glass (Figure 42), which showed his father's biggest telescope shortly before it was dismantled. He went on to make enormous contributions to improving the photographic process, and coined the words 'photograph' and 'snapshot', along with other photographic terms such as 'positive' and 'negative'. In fact, Herschel was just one of many astronomers who pushed photography to the limit and developed new photographic technology in an effort to capture the very faintest celestial objects.

Photography provided astronomers with the objectivity that they had been searching for. When Herschel tried to describe the brightness of a star, he had previously had to write: 'Alpha Hydrae much inferior to Gamma Leonis, rather inferior to Beta Aurigae.' Such vague jottings could now be replaced with a more objective and accurate photograph.

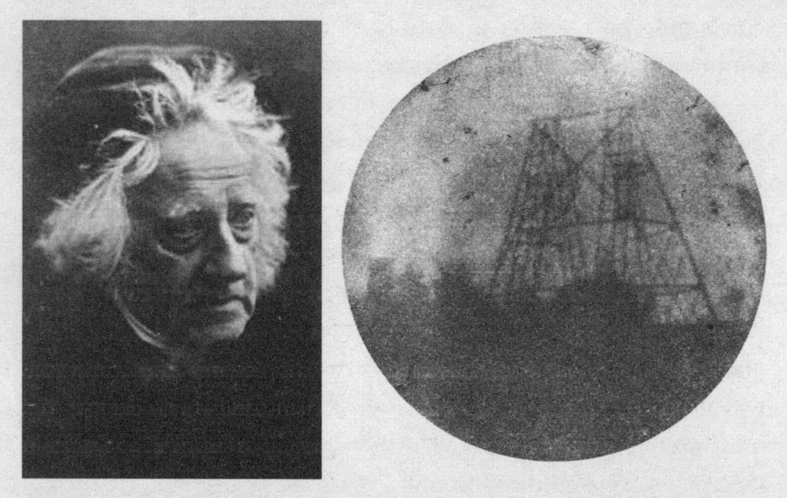

Figure 42 Sir John Herschel, son of William Herschel, by the celebrated portrait photographer Julia Margaret Cameron. Alongside is the very first photograph on glass, taken by John Herschel himself in 1839. It shows an image of his father's telescope, also shown as an etching in Figure 33 (p. 172).

Despite the advantages of photography, there was a certain level of suspicion from traditionalists who worried about the implications of this new technology. Sketching astronomers were wary that the technology would introduce new features into the night sky that were merely artefacts of the chemical process. For example, might some chemical residue be misinterpreted as a nebula? Henceforth, any reported observation had to be labelled either 'visual' or 'photographic' so that its provenance was unambiguous.

Once the technique had matured and natural conservatism had subsided, it was generally accepted that photographs were the best method for recording observations. In 1900, an astronomer at the Princeton Observatory argued that photographs provided 'a record that is permanent, authentic, and free from the personal bias of an imagination and hypothesis, which so seriously impairs the authority of many ocular observations'.

Photography proved to be an invaluable technology for recording observations accurately and objectively, but equally important was its power to detect previously invisible objects. If a telescope is pointed at a very distant object, then the light that reaches the human eye might be too feeble to be perceived, even if the telescope has a wide aperture. If, however, the eye is replaced with a photographic plate, then it can be exposed for several minutes or even hours, capturing more and more light as time goes by. The human eye absorbs light, processes it and disposes of it in an instant, and then it starts from scratch all over again, whereas the photographic plate keeps on accumulating light, building up an image that gets stronger over time.

In summary, the eye has a limited sensitivity, a telescope with a wide aperture boosts that sensitivity, and that same telescope coupled with a photographic plate is even more sensitive. For example, the Pleiades (or Seven Sisters) star cluster contains seven stars visible to the naked eye, but Galileo with his telescope could see forty-seven stars in this region. In the late 1880s, the French brothers Paul and Prosper Henry took a long photographic exposure of that part of the sky and counted 2,326 stars.

At the centre of the photographic revolution in astronomy was the Harvard College Observatory, partly thanks to its first director, William Cranch Bond, who had taken the first daguerreotype of a star at night, Vega, back in 1850. Also, the amateur astronomer Henry Draper, whose father John Draper had taken the first photograph of the Moon, bequeathed his personal fortune to Harvard in order to photograph and catalogue all the observable stars.

This allowed Edward Pickering, who became director of the observatory in 1877, to initiate a relentless programme of celestial photography. The observatory would take half a million photographic plates in the decades to come, so one of Pickering's biggest challenges was to establish an industrial-scale system for analysing the photographs. Each plate contained hundreds of stars, and each speck would need to have its brightness evaluated and its location measured. Pickering recruited a team of young men to work as *computers*, a term that was originally used to describe people who manipulated data and performed calculations.

Unfortunately, he soon became frustrated because of his team's lack of concentration and failure to pay attention to detail. One day, when his patience had been exhausted, he blurted out that his Scotch maid could do a better job. To prove his point, he sacked his all-male team, hired women computers to replace them and put his maid in charge. Williamina Fleming had been a teacher in Scotland before emigrating to America, where she had been abandoned by her husband when pregnant, forcing her to take a job as a housekeeper. Now she was leading a team nicknamed 'Pickering's harem' and scrutinising the world's largest set of astronomical images.

Pickering is generally respected for his liberal recruitment policy, but to some extent he was motivated by practical issues. The women

Figure 43 The Harvard 'computers' at work, busy examining photographic plates while Edward Pickering and Williamina Fleming watch over them. On the back wall are two plots that show the oscillating brightness of stars.

were generally more accurate and meticulous than the men they replaced, and they also tolerated being paid between 25 and 30 cents per hour, whereas the men had demanded 50 cents. Also, the women were restricted to the role of computers and were denied the opportunity to make observations themselves. This was partly because the telescopes were housed in cold, dark observatories, which were considered unsuitable for the fairer sex, and partly because Victorian sensitivities would have been offended by the thought of a man and a woman working together late into the night, staring up at the romantic array of stars. But at least the women could now examine the photographic results of night-time observations and contribute to astronomy, a discipline that had largely excluded them in the past.

Although Williamina Fleming's team of women computers were supposed to focus on the drudgery of harvesting data from the photographs so that the male astronomers could conduct the research, it was not long before they were reaching their own scientific conclusions. Endless days spent staring at the photographic plates had given them an intimate familiarity with the stellar objects that they were surveying.

For example, Annie Jump Cannon catalogued roughly 5,000 stars per month between 1911 and 1915, calculating the location, brightness and colour of each one. She drew upon her hands-on experience to make a major contribution to the system of stellar classification, dividing stars into seven classes (O, B, A, F, G, K, M). Today's astronomy undergraduates still learn this system of stellar classification, usually according to the mnemonic 'Oh, Be A Fine Guy – Kiss Me!' In 1925 Cannon became the first woman to receive an honorary doctorate from Oxford University, in recognition of this insightful and painstaking work. She was voted one of the twelve greatest American women in 1931, and in the same year became the first woman to receive the prestigious Draper Gold Medal from the American National Academy of Sciences.

Cannon had been struck down by scarlet fever as a child, which left her almost completely deaf, just like the Cepheid pioneer John Goodricke. It seems likely that they had both compensated for their loss of hearing by sharpening their sense of sight, thus allowing them to pick up fine details that had been missed by others. The most famous member of Pickering's team, Henrietta Leavitt, was also profoundly deaf. It was Leavitt who spotted features in the photographic plates that would settle the Great Debate once and for all. She would enable astronomers to measure the distance to the nebulae, and her discovery would influence cosmology for decades to come.

Leavitt was born in 1868 in Lancaster, Massachusetts, the daughter of a Congregational minister. Professor Solon Bailey, who knew her at the Harvard College Observatory, recalled how her character was shaped by her religious upbringing:

> She was a devoted member of her intimate family circle, unselfishly considerate in her friendships, steadfastly loyal to her principles, and deeply conscientious and sincere in her attachment to her religion and church. She had the happy faculty of appreciating all that was worthy and lovable in others, and was possessed of a nature so full of sunshine that, to her, all of life became beautiful and full of meaning.

In 1892, Leavitt graduated from Harvard University's Radcliffe College, which at the time was known as the Society for the Collegiate Instruction of Women. For the next two years she remained housebound, recovering from a serious illness, possibly meningitis, that caused her loss of hearing. Once she had regained her strength she became a volunteer at the Harvard College Observatory, sifting through the plates and searching for variable stars, which she had been designated to catalogue. Photography had transformed the study of variable stars, because two photographic glass plates taken on different nights could be overlaid and directly compared, making it much easier to spot any variations in brightness. Leavitt made the most of this burgeoning technology and would discover more than 2,400 variable stars, about half of the total known in her day. Professor Charles Young of Princeton University was so impressed that he called her 'a variable-star fiend'.

Of the various types of variable star, Leavitt developed a particular passion for Cepheids. After months spent measuring and cataloguing Cepheid variables, she yearned to gain some understanding of what determined the rhythm of their fluctuations. In an effort to solve the mystery she turned her attention to the only two firm

Figure 44 Henrietta Leavitt, who rose from being an unpaid volunteer at Harvard College Observatory to make one of the most important breakthroughs in twentieth-century astronomy.

pieces of information available for any Cepheid variable: its period of variation and its brightness. Ideally, she wanted to see whether there was any relationship between period and brightness – perhaps brighter stars might prove to have a longer period of variation than dimmer stars, or vice versa. Unfortunately, it seemed virtually impossible to make any sense of the brightness data. For example, an apparently bright Cepheid might actually be a dim star that was close by, while an apparently dim Cepheid might actually be a bright star that was far away.

Astronomers had long ago realised that they could perceive only the apparent brightness of a star, as opposed to its actual brightness.

The situation seemed hopeless, and most astronomers would have given up, but Leavitt's patience, dedication and concentration led her to a rather cunning and brilliant idea. She made her breakthrough by focusing her attention on the stellar formation known as the Small Magellanic Cloud, named after the sixteenth-century explorer Ferdinand Magellan, who recorded it when he sailed the southern oceans while circumnavigating the globe. Because the Small Magellanic Cloud is visible only from the southern hemisphere, Leavitt had to rely on photographs taken at Harvard's southern station at Arequipa in Peru. Leavitt managed to identify twenty-five Cepheid variables within the Small Magellanic Cloud. She did not know the distance from the Earth to the Small Magellanic Cloud, but she suspected that it was relatively far away and that the Cepheids within it were relatively close together. In other words, all twenty-five Cepheids were more or less at the same distance from the Earth. Suddenly, Leavitt had exactly what she needed. If the Cepheids in the Small Magellanic Cloud were all roughly the same distance away, then if one Cepheid was brighter than another it was because it was intrinsically more luminous, not just apparently brighter.

The assumption that the stars in the Small Magellanic Cloud were roughly equidistant from the Earth was something of a leap of faith, but a very reasonable one. Leavitt's line of thinking was akin to an observer seeing a flock of twenty-five birds in the sky and assuming that the distance between each one is relatively small compared with the distance between the observer and the entire flock. Hence, if one bird *seems* smaller than the others, then it probably *is* genuinely smaller. However, if you saw twenty-five birds spread around the sky and one seemed smaller than the others, then you could not be sure whether that bird was genuinely smaller or just farther away.

Leavitt was now ready to explore the brightness versus period relationship for Cepheids. Building on the assumption that the apparent brightness of each Cepheid in the Small Magellanic Cloud was a true indication of its actual brightness in relation to the other Cepheids in the Cloud, Leavitt plotted a graph of the apparent brightness against the period of variation for the twenty-five Cepheid stars. The result was astonishing. Figure 45(a) shows how Cepheids that fluctuate over a longer period are typically brighter, and even more importantly, the data points generally seem to follow a smooth curve. Figure 45(b) shows the same data but with a change of scale for the period of variation, which reveals more clearly the relationship between brightness and period. In 1912 Leavitt announced her conclusion: 'A straight line can be readily drawn among each of the two series of points corresponding to maxima and minima, thus showing that there is a simple relation between the brightnesses of the variables and their periods.'

Leavitt had discovered a strict mathematical relationship between the true luminosity of a Cepheid and the period of its variations in apparent brightness: the higher the luminosity of the Cepheid, the longer the period between the peaks in brightness. Leavitt was confident that this rule could be applied to any Cepheid variable star in the universe, and that her graph could be extended to include Cepheids with very long periods. This was a staggering result, pregnant with cosmic repercussions, but it was published with the understated title 'Periods of 25 Variable Stars in the Small Magellanic Cloud'.

The power of Leavitt's discovery was that it was now possible to compare any two Cepheids in the sky and work out their relative distances from the Earth. For example, if she could find two Cepheids in different parts of the sky that both varied with very similar periods, then she knew that they would be shining approximately as brightly as each other – the plot in Figure 45 predicted that a certain period implied a certain inherent brightness. So, if one of those stars appeared to be 9 times fainter than the other,

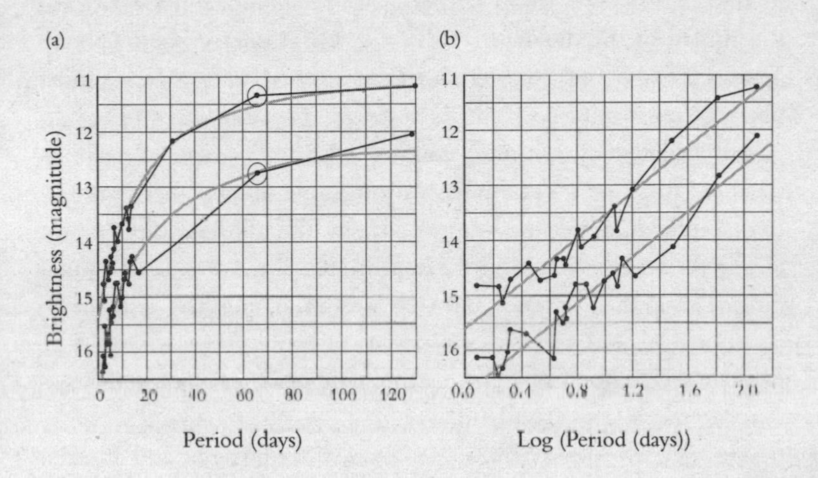

Figure 45 These two graphs show Henrietta Leavitt's observations of Cepheid variable stars in the Small Magellanic Cloud. Graph (a) is a plot of brightness (on the vertical axis) against period, measured in days (on the horizontal axis), and each point represents a Cepheid. There are two lines in the plot: one represents the maximum brightness and the other the minimum brightness of each variable star.

To help interpret the graph, the points that are circled represent a Cepheid with a period of roughly 65 days and its brightness varies between 11.4 and 12.8. A pair of smooth curves can be drawn through the data points. Not every point sits exactly on its curve, but if allowance is made for errors, the curves do seem to be a valid fit to the data.

Stellar brightness is measured in terms of magnitude, which is an unusual unit of measurement because the brighter the star, the lower the magnitude, which is why the vertical scale runs from 16 up to 11. Also, magnitude is measured on a *logarithmic scale*. It is not necessary for our purposes to define a logarithmic scale; all we need to know is that the relationship between brightness and period of variation becomes clearer if the period is also plotted on a logarithmic scale, as in graph (b). The points now all lie reasonably close to a pair of straight lines, which indicates that there is a simple mathematical relationship between a Cepheid's period of variability and its brightness.

then it must be farther away. Indeed, if it was 9 times fainter, then it must be exactly 3 times farther away, because brightness fades as the square of the distance and $3^2 = 9$. Or if one of the Cepheids appeared to be 144 times fainter than another with a very similar period, then it must be 12 times as distant, because $12^2 = 144$.

But although astronomers could use Leavitt's graph to calibrate Cepheid brightness and establish the relative distance between any two Cepheids, as yet they did not know the absolute distance for any of them. They could prove that one Cepheid was, say, 12 times farther away than another, but that was all. If only the distance to just one Cepheid variable star could be found, then it would be possible to anchor Leavitt's measurement scale and gauge the distance to every single Cepheid.

The decisive observations that made this possible and thereby calibrated the Cepheid distance scale were achieved thanks to a team effort by astronomers who included Harlow Shapley and Denmark's Ejnar Hertzsprung. Together they used a combination of techniques, including parallax, to measure the distance to one Cepheid variable, which then transformed Leavitt's research into the ultimate distance guide for the cosmos. Cepheid variables could act as a yardstick for the universe.

In summary, an astronomer could now measure the distance to any Cepheid by a simple three-step process. First, see how quickly it varies, which reveals how bright it really is. Second, see how bright it appears to be. And third, work out what distance would turn the actual brightness into the apparent brightness.

As a crude analogy, picture the pulsing Cepheid stars as flashing lighthouses. Imagine that the speed at which a lighthouse flashes depends on its brightness (just like a Cepheid star), so a 3 kW lighthouse flashes three times per minute and a 5 kW lighthouse flashes five times per minute. If a sailor at sea on a dark night sees a lighthouse flashing in the distance, he can gauge the distance to it by the same three-step process. First, he counts the frequency of the flashing, which immediately gives him the true brightness of the lighthouse. Second, he sees how bright it appears to be. And third, he works out what distance would turn the actual brightness into the apparent brightness.

Also, the sailor can estimate the distance from his ship to a seaside village which is along the same line of sight as the lighthouse, because he can assume that the village is roughly as far away as the distance he has already worked out for the lighthouse. It could be that the village is set back a long way from the coast and far from the lighthouse, or that the lighthouse is located some way out to sea on a rocky outcrop and some distance from the village, but in general the lighthouse will be close to the village and the estimate will be fairly accurate. Similarly, an astronomer who works out the distance to a Cepheid variable also knows the rough distance to any other stars in its vicinity. The method is not foolproof, but it is effective in most cases.

Professor Gösta Mittag-Leffler of the Swedish Academy of Sciences was so impressed by Leavitt and the power of her Cepheid yardstick that in 1924 he started on the paperwork that would be needed to nominate her for a Nobel prize. However, when he began to research Leavitt's current scientific interests he was shocked to find that she had died of cancer three years earlier, on 12 December 1921, at the age of just fifty-three. Leavitt was not an astronomer with a high profile who travelled the world giving seminars, but rather a humble researcher who quietly and diligently studied her photographic plates, so her passing went virtually unnoticed in Europe. Not only did she not live long enough to receive the recognition she deserved, she never witnessed the decisive impact of her work on the Great Debate over the nature of the nebulae.

The Titan Astronomer

The astronomer who would fully exploit the potential of Leavitt's discovery was Edwin Powell Hubble, arguably the most famous astronomer of his generation. He was born in Missouri in 1889, the son of John and Jennie Hubble, who had met when John was seriously injured in a farming accident and Jennie, the local doctor's daughter, had the job of nursing him back to health. He was so bloody and battered that she said that she 'never wanted to see John Hubble again'. But as he recovered she fell in love with him, and they married in 1884.

Edwin had a largely happy childhood, except for one traumatic incident when he was seven years old. He and his brother Bill had come to resent their fourteen-month-old attention-grabbing sister Virginia, and they decided to get their own back by deliberately stepping on her fingers to make her cry. A few days later she came down with a severe undiagnosed illness, which proved to be fatal. Confused and distraught, Edwin blamed himself, even though Virginia's illness was unrelated to his earlier actions. As one of his siblings recalled: 'Edwin became psychologically ill and had it not been for his very understanding and intelligent parents, this paranoia might have caused another tragedy in the family.' Edwin was particularly close to his mother, and it was she who helped him through this disturbing episode in his childhood.

Edwin also developed a close relationship with his grandfather, Martin Hubble, who introduced him to astronomy by building him a telescope for his eighth birthday. Martin would persuade the boy's parents to let Edwin stay up late into the night to stare at the myriad stellar specks in the black Missouri sky. He became so fascinated by the stars and planets that he was inspired to write an article about Mars, which was published in his local newspaper while he was still a high-school student. His teacher, Miss Harriet Grote, recognised

Edwin's escalating enthusiasm for astronomy: 'Edwin Hubble will be one of the most brilliant men of his generation.' Probably every teacher says much the same about their favourite pupil, but in Edwin's case he would truly fulfil Miss Grote's prediction.

Hubble went on to study at Wheaton College, hoping to earn a scholarship to a major university. At the graduation ceremony, where such scholarships were announced, the superintendent shocked Hubble by proclaiming: 'Edwin Hubble, I have watched you for four years and I have never seen you study for ten minutes.' After a dramatic pause worthy of the greatest of thespians, he continued: 'Here is a scholarship to the University of Chicago.'

Hubble had planned to study astronomy at Chicago, but his forceful father compelled him to pursue a degree in law because of the steady income it would guarantee. As a young man, John Hubble had struggled to earn a decent wage, and he gained financial security only later in life when he became an insurance salesman. He took great pride in the profession that had made the Hubbles a respectable middle class family: 'The best definition we have found for civilisation is that a civilised man does what is best for all, while the savage does what is best for himself. Civilisation is but a huge mutual insurance company against human selfishness.'

Edwin resolved the conflict between his own ambition and his father's pragmatism by formally studying law to pacify his father, while also completing enough courses in physics to keep alive his dream of becoming an astronomer. The Chicago physics department was headed by Albert Michelson, who had dispensed with the ether and won America's first Nobel Prize for Physics in 1907. The university was also home to Robert Millikan, who would go on to become America's second Nobel Laureate in physics, and who took on Hubble as his part-time laboratory assistant while Edwin was still an undergraduate. This was a brief but pivotal relationship,

Figure 46 Edwin Powell Hubble, the greatest observational astronomer of his generation, puffing at his trademark briar pipe.

because Millikan helped to propel Hubble towards his next goal, a Rhodes scholarship to study at Oxford University.

The Rhodes scholarships were established in 1903 and funded from the fortune of the Victorian empire-builder Cecil Rhodes, who had died the previous year. They were awarded to young Americans who displayed both strength of character and intellect. George Parker, who helped to administer the scheme, said that the thirty-two scholarships were for those 'likely to become President of the United States, Chief Justice of the Supreme Court, or American Ambassador to Great Britain'. Millikan duly gave Hubble a first-class recommendation: 'I find Hubble a man of magnificent physique, admirable scholarship, and worthy and loveable character . . . Seldom have I known a man who seemed to be better qualified to meet the conditions imposed by the founder of the Rhodes scholarships than is Mr. Hubble.' Thanks to this endorsement from one of America's best-known scientists, Hubble achieved his goal of a Rhodes scholarship and left for England in September 1910. The only disappointment for Hubble was that, through paternal pressure, his main subject at Oxford was still supposed to be law.

During his two years at Oxford, Hubble became an extreme Anglophile, adopting everything from an English dress sense to an aristocratic accent. Fellow Rhodes scholar Warren Ault was unpleasantly surprised when he encountered Hubble towards the end of his time in Britain: 'He was dressed in plus-fours, a Norfolk jacket with leather buttons, and a huge cap. He also sported a cane and spoke in a British accent I could scarcely understand . . . Those two years had transformed him, seemingly, into a phoney Englishman, as phoney as his accent.' Jakob Larsen of Iowa, who was with Hubble at Queen's College, was similarly unimpressed: 'We laughed at his effort to acquire an extreme English pronunciation while the rest of us tried to keep the pronunciation we brought from home. We always claimed that he could not be consistent, so that he might take a bāth in a băth tub.'

Hubble's time in England came to an abrupt end when his father became seriously ill and died on 19 January 1913. He was forced to return home, still sporting his Oxford cape and fake English accent, and took on the responsibility of supporting his mother and four siblings, whose suffering had been compounded by a collapse in the family's financial investments. Hubble worked as a high-school teacher and managed to get some part-time legal work for the next eighteen months, which was enough to put the family's finances back on a firm footing. Thereafter, having done his duty to his family, and now liberated from his misguided, domineering father,

Hubble was suddenly free to follow his childhood dream of becoming an astronomer. 'Astronomy is something like the ministry,' he once said. 'No one should go into it without a call. I got that unmistakable call, and I knew that even if I were second rate or third rate, it was astronomy that mattered.' He reiterated the point in a remark that seemed aimed at his late father: 'I would much rather be a second-rate astronomer than a first-rate lawyer.'

Hubble began to make up for the time he had wasted in legal lectures and set off on the long road to becoming a professional astronomer. Thanks to his scientific connections at the University of Chicago, he obtained a graduate position at the nearby Yerkes Observatory, the site of Hale's first great telescope. He went on to complete his Ph.D., a survey of nebulae, which he sometimes called by their German name, *nebelflecken*. Hubble knew that his thesis was a solid piece of work but not an inspired one: 'It does not add appreciably to the sum total of human knowledge. Some day I hope to study the nature of these *nebelflecken* to some purpose.'

To achieve this particular goal, Hubble realised that he had to obtain a research post at whichever observatory had the best telescopes. He once said: 'Equipped with his five senses, man explores the universe around him and calls the adventure science.' The key sense for astronomers is vision and whoever had access to the best telescope would see farthest and clearest. Mount Wilson was therefore the place to be: it already boasted the great 60-inch telescope, and the even greater 100-inch telescope would soon be completed. As it happened, the California observatory was already aware of Hubble's potential and was keen to headhunt him, so he was delighted when he received a job offer from Mount Wilson in November 1916. The appointment was delayed, because by this time America had entered the First World War, and Hubble felt duty bound to help defend Britain, the country he loved so much. He arrived in Europe too late to be involved in combat, but stayed on for four months after the war as part of the occupation forces in Germany. He postponed his return to America to undertake a long tour of his beloved England, and eventually arrived at the Mount Wilson Observatory in the autumn of 1919.

Although he was still a junior astronomer with relatively little experience, Hubble was soon a conspicuous figure at the observatory. One of his assistants gave a vivid description of Hubble as he stood taking photographs with the 60-inch telescope:

> His tall, vigorous figure, pipe in mouth, was clearly outlined against the sky. A brisk wind whipped his military trench coat around his body and occasionally blew sparks from his pipe into the darkness of the dome. 'Seeing' that night was rated as extremely poor on our Mount Wilson scale, but when Hubble came back from developing his plate in the dark room he was jubilant. 'If this is a sample of poor seeing conditions,' he said, 'I shall always be able to get usable photographs with the Mount Wilson instruments.' The confidence and enthusiasm which he showed on that night were typical of the way he approached all his problems. He was sure of himself – of what he wanted to do, and of how to do it.

When it came to the Great Debate, Hubble sympathised with the view that the nebulae were independent galaxies. This was slightly embarrassing, because Mount Wilson was dominated by astronomers who believed that the Milky Way was the only galaxy and that the nebulae lay within it. In particular, Harlow Shapley, who had defended the single galaxy theory in Washington, took great exception to the new boy, to his views and his demeanour. Shapley's own humble manner was completely at odds with a man who was fixated by the English aristocracy, who sported an Oxford tweed jacket and who called out 'By Jove!' and 'What ho!' several times a

Figure 47 Edwin Hubble (left) next to the 100-inch Hooker Telescope at Mount Wilson Observatory. Figure 48 shows the whole telescope.

day. Hubble liked to be the centre of attention. He took great delight in being able to light a match, flip it in the air through 360°, catch it and light his briar pipe. He was the consummate showman, whereas Shapley was quite the opposite and disdained such exhibitionism. Worst of all for Shapley, who had argued vehemently against America entering the war, Hubble persisted in wearing his army trench coat around the observatory.

The constant clash of personalities ended in 1921, when Shapley left Mount Wilson to become director of the Harvard Observatory. This was definitely a promotion for Shapley, partly in recognition of his leading role in the as yet unresolved Great Debate, but moving to the East Coast turned out to be a disaster. Although he had escaped Hubble and taken up a prestigious directorship, Shapley had also left behind the observatory that would dominate astronomy for four decades. Mount Wilson possessed the world's most powerful telescopes, and was destined to be the observatory that would make the next great breakthrough in astronomy.

Hubble moved up the pecking order, gradually obtained more telescope time and committed himself to taking the best possible pictures of the nebulae. Whenever his name was on the observing schedule, he would make the journey up the steep, winding road that led to the 1,740-metre peak of Mount Wilson, where he would spend a few days living in the aptly named Monastery, the male-only residence for those who had abandoned contact with the outside world to devote themselves to staring into space.

This might give the impression of astronomers as a meditative breed who spend their nights in contemplation and wonder, but in reality observing was hard work. It required hours of intense concentration, as the gnawing pain of sleep deprivation increased over the course of the night. To make matters worse, temperatures at Mount Wilson were often freezing, which meant that delicate adjustments to the telescope's orientation had to be performed with fingers numb with pain, while eyelashes could become glued to the eyepiece with frozen tears. The observatory logbook offered a few words of caution: 'When tired, cold and sleepy never make any movement of telescope or dome without pausing and thinking.' Only the most diligent and determined observers would succeed. In a demonstration of supreme mental and physical discipline, the hardiest astronomers were capable of suppressing their own shivers so as not to vibrate the photographic equipment as it captured priceless images of the cosmos.

On the night of 4 October 1923, four years after his arrival at Mount Wilson, Hubble was observing with the 100-inch telescope. The viewing conditions were rated as 1, which was as poor as it was allowed to get before the dome was closed, but he managed to take a 40-minute exposure of M31, the Andromeda Nebula. After developing and studying the photograph in the clear light of day, he spotted a new speck, which he assumed was either a photographic glitch or a nova. On the next night, the last of his observing run, the weather was much clearer and he repeated the exposure, adding an extra five minutes in the hope that it would confirm the nova. The speck was there again, and this time two other potential novae joined it. He marked the plate with an 'N' next to each candidate nova and, once his time at the telescope was over, he returned to his office and the photographic plate library in Santa Barbara Street, Pasadena.

Figure 48 The 100-inch Hooker Telescope in its dome at the Mount Wilson Observatory. It was the most powerful telescope in the world when Hubble made his historic observation in 1923.

Hubble was anxious to compare his new plate with previous plates of the same nebula to see whether his novae were genuine. All the observatory's photographic plates were stored in an earthquake-proof vault, with each image carefully catalogued and filed, so it was a simple matter to find the appropriate plates and check the candidate novae. The good news was that two of the specks were indeed new novae. The even better news was that the third one was not a nova, but a Cepheid variable star. This third star had been recorded on some of the earlier plates but not on others, indicating its variability. Hubble had made the greatest discovery of his career. He quickly crossed out the 'N' and scribbled, triumphantly, 'VAR!', as shown in Figure 49.

This was the first Cepheid to be discovered in a nebula. What made the discovery so important was that Cepheids could be used to measure distance, so Hubble could now measure the distance to the Andromeda Nebula and thereby conclusively settle the Great Debate. Were the nebulae entities within our own Milky Way, or were they galaxies in their own right and much farther away? The new Cepheid brightened and dimmed over a 31.415-day period, so Hubble could use Leavitt's research to calculate the absolute brightness of the star. It turned out that the Cepheid was 7,000 times more luminous than the Sun. By comparing its absolute brightness and apparent brightness, Hubble deduced its distance.

The result was staggering. The Cepheid variable star, and therefore the Andromeda Nebula which it inhabited, appeared to be roughly 900,000 light years from the Earth.

The Milky Way was roughly 100,000 light years in diameter, so Andromeda was clearly not part of our galaxy. And if Andromeda was so far away, it must be incredibly bright because it was still visible to the naked eye. Such brightness implied a system containing hundreds of millions of stars. The Andromeda Nebula just had to be a galaxy in its own right. The Great Debate was over.

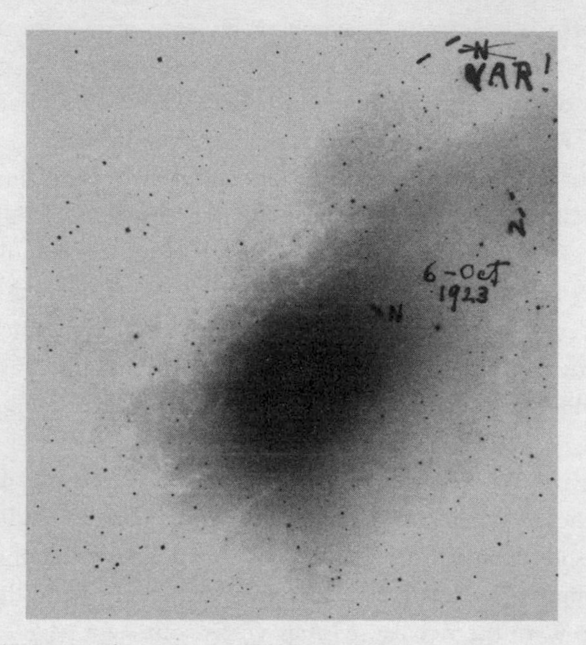

Figure 49 In October 1923 Hubble located three candidate novae in the Andromeda Nebula, each marked with an 'N'. One of these novae turned out to be a Cepheid variable, a star that changes predictably in brightness, so the 'N' was crossed out and the star relabelled 'VAR!'. Cepheids can be used to measure distance, so Hubble could now measure the distance to the Andromeda Nebula and settle the Great Debate.

The Andromeda Nebula was now the Andromeda Galaxy, because it and the majority of other nebulae were indeed separate galaxies, as mighty and magnificent as our own Milky Way and positioned far beyond it. Hubble had proved that Curtis was right and Shapley was wrong.

The huge distance to Andromeda was such a shock that Hubble decided not to go public until he had more proof. At Mount Wilson he was surrounded by believers in the single galaxy theory, so he was wary of making a fool of himself. He exercised enormous

Figure 50 Galaxies are no longer classed as nebulae, so the Andromeda Nebula is today known as the Andromeda Galaxy. This photograph was taken at La Palma Observatory in 2000. It shows that Andromeda is composed of millions of stars and is a galaxy in its own right.

self-discipline and patience, taking several more photographs of Andromeda and discovering a second, dimmer Cepheid, which corroborated his initial result.

At last, in February 1924, he broke his silence by revealing his results in a letter to Shapley, the spokesman for the single galaxy theory. Shapley had helped to calibrate Leavitt's Cepheid distance scale, and now it had undermined his position in the Great Debate. When Shapley read Hubble's note, he remarked: 'Here is the letter that has destroyed my universe.'

Shapley tried to attack Hubble's data by suggesting that Cepheid stars with periods longer than 20 days were unreliable indicators because very few long-period Cepheids had been studied. He also argued that the supposed variability of Hubble's Andromeda stars might be nothing more than a quirk of the photographic development process or the exposure time. Hubble knew that his observations were not perfect, but there was no error that was significant enough to bring Andromeda back into the Milky Way. So Hubble was confident that Andromeda was roughly 900,000 light years from the Earth, and in the years ahead it would become clear that the vast majority of other galaxies are even farther away. The only exceptions to this are a small number of dwarf galaxies, such as the Small Magellanic Cloud studied by Henrietta Leavitt. This is now known to be a small, satellite galaxy gravitationally attached to and on the periphery of our Milky Way galaxy.

The term 'nebula' had originally been used for any celestial object with a cloud-like appearance, but now the bulk of these nebulae were relabelled as galaxies. However, it would turn out that a few nebulae were nothing more than mere clouds of gas and dust within the Milky Way, and in due course the term 'nebulae' came to refer specifically to such clouds. Despite the existence of these relatively small, local nebulae of gas and dust, this did not alter the fact that many of the original nebulae, such as Andromeda, were actually galaxies in their own right and lay far beyond the Milky Way. The central question in the Great Debate was whether the universe was full of such galaxies, and Hubble had shown that this was indeed the case.

But what about the nova of 1885 in the Andromeda Galaxy? Shapley had argued that its brightness proved that Andromeda could not be a distant, independent galaxy, because the nova would have had to be impossibly bright. In fact, we now know that the 1885 event was not a nova but a *supernova*, which is indeed an 'impossibly' bright event. A supernova is a cataclysmic phenomenon on an altogether different scale than an ordinary nova, and it occurs when a single star blasts itself to oblivion, outshining for a brief time the combined output of billions of stars. Supernovae are rare events, and their brilliance had not been appreciated when Curtis and Shapley argued their cases in 1920.

And what of the other pillar in Shapley's counter-argument? If the universe was populated with galaxies, then they should be visible in all directions. However, there were plenty to be seen above and below the plane of the Milky Way, but very few in the plane itself, which was dubbed the zone of avoidance. It turned out that Curtis had been right in claiming that the zone of avoidance was the consequence of interstellar dust in the plane of the pancake-shaped Milky Way, obscuring our view of galaxies beyond. Modern telescope technology has since been able to penetrate the dust, and we now know that there just as many galaxies in this 'empty' zone as there are visible in other directions.

As news of Hubble's discovery emerged, his peers began to applaud his success in resolving one of the longest-running disputes in astronomy. Henry Norris Russell, director of the Princeton Observatory, wrote to Hubble: 'It is a beautiful piece of work, and you deserve all the credit that it will bring you, which will undoubtedly be great. When are you going to announce the thing in detail?'

Hubble's result was formally announced at the 1924 meeting of the American Association for the Advancement of Science, held in Washington, where he shared the $1,000 prize for the most exceptional paper – the co-winner was Lemuel Cleveland, for his groundbreaking work on intestinal protozoa found in termites. A letter drafted by the Council of the American Astronomical Society highlighted the implications of Hubble's work: 'It opens up depths of space previously inaccessible to investigation and gives promise of still greater advances in the near future. Meanwhile, it has already expanded one hundred fold the known volume of the material universe and has apparently settled the long-mooted question of the nature of the [spiral nebulae], showing them to be gigantic agglomerations of stars almost comparable in extent with our own galaxy.'

With a single observation, captured on a single photographic plate, Hubble had changed our view of the universe and forced us to re-evaluate our position within it. Our tiny Earth now seemed more insignificant than ever – just one of many planets, orbiting one of many stars, within one of many galaxies. Indeed, it would later become clear that our galaxy is just one of billions of galaxies, each containing billions of stars. Also, the scale of the universe was much greater than previously imagined. Shapley had argued that all the matter in the universe was contained within the disc of the Milky Way, of the order of 100,000 light years across, but Hubble had proved that there were other galaxies more than a million light years from the Milky Way and beyond. Today we know of galaxies that are billions of light years away.

Astronomers were already aware of the huge gap between the planets and our Sun, and they were also familiar with the even greater gaps between the stars, but now they had to consider the gigantic emptiness between galaxies. Hubble used his observations to work out that if all the matter in the stars and planets was smeared out evenly across space, then the average cosmic density would be a single gram of matter in a volume the size of one thousand Earths. This density, which is not far from modern estimates, shows that we inhabit a very rich patch of space within a generally empty universe. 'No planet or star or galaxy can be typical, because the Cosmos is mostly empty,' wrote the astronomer Carl Sagan. 'The only typical place is within the vast, cold, universal vacuum, the everlasting night of intergalactic space, a place so strange and desolate that, by comparison, planets and stars and galaxies seem achingly rare and lovely.'

The implications of Hubble's measurement were truly sensational, and Hubble himself soon became the subject of popular debate and newspaper coverage. One paper called him 'the titan astronomer'. He also received numerous prizes and awards, both in his own country and abroad, and his colleagues were quick to praise him. Herbert Turner, Savilian Professor of Astronomy at Oxford University, was of the opinion: 'It will be years before Edwin realizes the magnitude of what he has done. Such a thing can come only once to most men and they are fortunate.'

But Hubble was destined to shake astronomy again in the years to come, this time with an even more revolutionary observation, one that would force cosmologists to reassess their assumption of an eternal static universe. In order to achieve this next breakthrough, he would need to exploit a relatively new piece of technology, one that would make full use of the power of the telescope and the sensitivity of photography. This piece of equipment, known as a *spectroscope*, would allow astronomers to drain every last piece of information from the meagre light that reached their giant telescopes. It was an instrument that had its origins in the hopes and ambitions of nineteenth-century science.

World in Motion

In 1842, the French philosopher Auguste Comte tried to identify the areas of knowledge which would remain forever beyond the wit of scientific endeavour. For example, he thought that some qualities of the stars could never be ascertained: 'We see how we may determine their forms, their distances, their bulk, and their motions, but we can never know anything of their chemical or mineralogical structure.'

In fact, Comte would be proved wrong within two years of his death, as scientists began to discover which types of atom exist in our closest star, the Sun. To understand how astronomers would unravel the chemistry of the stars, it is first necessary to understand the nature of light at a basic level. In particular, there are three key points to appreciate.

First, physicists think of light as a vibration of electric and magnetic fields, which is why light and related forms of radiation are known as *electromagnetic radiation*. Second, and more simply, we can think of electromagnetic radiation or light as a wave. The third key point is that the distance between two neighbouring peaks in a light wave (or two successive troughs), the *wavelength*, tells us almost everything we need to know about a light wave. Examples of wavelengths are illustrated in Figure 51.

For example, light is a form of energy, and the amount of energy carried by a particular light wave is inversely proportional to the wavelength. In other words, the longer the wavelength, the lower the energy of the light wave. At a human level we are much less concerned with the energy of a light wave, and instead use colour as the basic feature to distinguish one light wave from another. The colours blue, indigo and violet correspond to light waves of shorter wavelengths and higher energies, whereas orange and red correspond to light waves of longer wavelengths and lower energies. Green and yellow correspond to intermediate wavelengths and energies.

In particular, violet light has a wavelength of roughly 0.0004 mm and red has a wavelength of roughly 0.0007 mm. There are waves with shorter and longer wavelengths, but our eyes are not sensitive to them. Most people use the word 'light' to describe only those waves that we can see, but physicists use the term loosely to describe any form of electromagnetic radiation, visible or invisible to the human eye. Light with even shorter wavelengths and higher energies than violet light includes ultraviolet radiation and X-rays, while light with even longer wavelengths and lower energies than red light includes infrared radiation and microwaves.

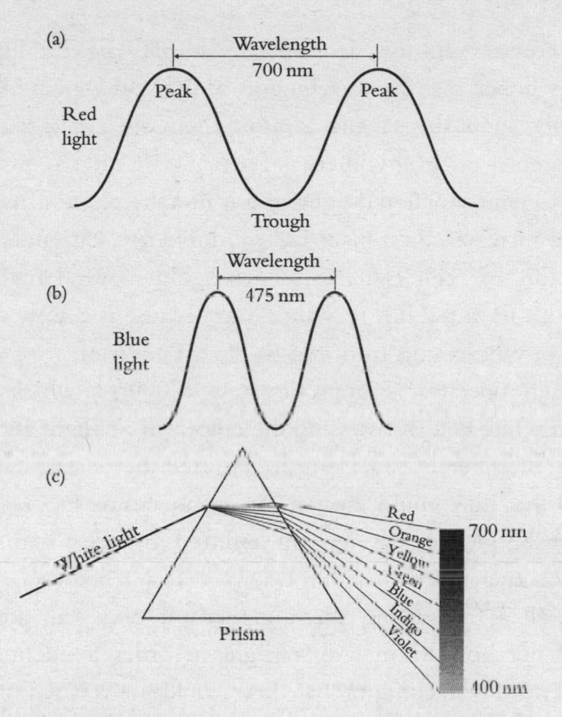

Figure 51 Light can be pictured as a wave. The wavelength of a light wave is the distance between two successive peaks (or troughs), and it tells us almost everything we need to know about the light wave. In particular, the wavelength is related to the colour and energy of the light wave. Diagram (a) shows a longer-wavelength, lower-energy wave of red light. Diagram (b) shows a shorter-wavelength, higher-energy wave of blue light. The wavelengths for visible light are less than one-thousandth of a millimetre, ranging from roughly 0.0004 mm for violet to 0.0007 mm for red. Usually wavelengths are measured in nanometres (nm); 1 nm is one-billionth of a metre. So red light has a wavelength of roughly 700 nm.

There are light waves with wavelengths that are shorter than blue light (e.g. ultraviolet radiation, X-rays) and longer than red light (e.g. infrared radiation, microwaves), but these are invisible to the human eye.

A beam of white light is a mixture of colours and wavelengths, which becomes apparent when it is passed through a glass prism, because the beam is split into a rainbow, as shown in diagram (c). This happens because each wavelength behaves differently. In particular, each wavelength bends at a different angle as it enters and leaves the glass.

The crucial point for astronomers was that stars emit light waves, and they hoped that the wavelengths of the starlight could tell them something about the star that emitted them, such as its temperature. For example, once an object reaches 500°C it has just enough energy to emit visible red light, and is literally red hot. As the temperature increases, the object has more energy and emits a greater proportion of higher-energy, shorter, bluer wavelengths and it transforms from red hot to white hot, because it is now emitting a variety of wavelengths from red to blue. The filament of a standard light bulb operates at approximately 3,000°C, which certainly makes it white hot. By assessing the colour of starlight and the proportion of different wavelengths emitted by a star, astronomers realised that they could estimate its temperature. Figure 52 shows the distribution of wavelengths emitted by stars with differing surface temperatures.

As well as measuring the temperature of a star, astronomers worked out how to analyse starlight in order to identify a star's ingredients. The technique that they would use is based on research dating back to 1752, when the Scottish physicist Thomas Melvill made a curious observation. He subjected various substances to a flame and noticed that each one produced a characteristic colour. For example, table salt gave off a bright orange flash of colour. You can easily observe the orange signature of salt by sprinkling a tiny amount over a gas cooker flame.

The distinctive colour associated with salt can be traced to its structure at the atomic level. Salt is otherwise known as sodium chloride, and the orange light is generated by the sodium atoms within the sodium chloride crystals. This also explains why sodium streetlamps are orange. By passing the light from sodium through a prism, it is possible to analyse exactly which wavelengths are emitted, and the two dominant emissions are both in the orange region of the spectrum, as shown in Figure 53.

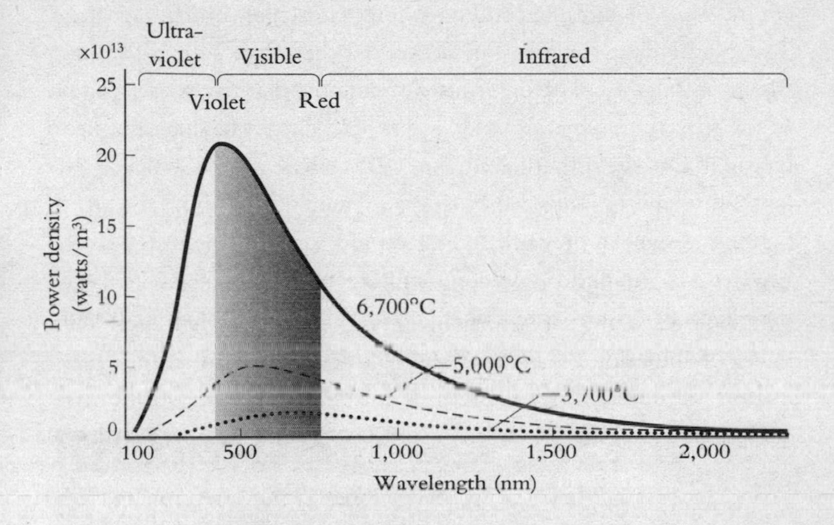

Figure 52 This graph shows the range of light wavelengths emitted by three stars with different surface temperatures. The main curve shows the distribution of wavelengths emitted by a star with a surface temperature of 6,700°C. The distribution peaks at blue and violet wavelengths, but it also emits other colours in the visible spectrum. This star also radiates an ample fraction of infrared and a large amount of ultraviolet radiation, wavelengths that are, respectively, longer and shorter than the visible wavelengths. The middle curve represents the wavelength distribution emitted by a star with a cooler surface temperature of 5,000°C. It peaks at a longer wavelength in the middle of the visible region, so the star emits a good mix of colours. The lowest curve represents the wavelength distribution emitted by an even cooler star (3,700°C). Its distribution peaks at even longer wavelengths, giving off a significant amount of red light and a large amount of invisible infrared radiation. This star has an orange-red appearance.

By looking at the range of wavelengths emitted by a star, an astronomer on Earth can deduce the star's temperature. The wavelength distribution acts as a signature for temperature. In summary, the cooler the star, the greater its tendency to emit long wavelengths and the redder it appears. Conversely, the hotter the star, the greater its tendency to emit short wavelengths and the bluer it appears.

Each type of atom has the ability to emit particular wavelengths (or colours) of light, depending on its particular atomic structure. The emitted wavelengths for elements other than sodium are also shown in Figure 53. Neon emits wavelengths that are at the red end of the spectrum, which is what you would expect having seen neon lighting. On the other hand, mercury emits several bluer wavelengths, which explains the blueness of mercury lighting. As well as lighting designers, firework manufacturers are also interested in the wavelengths emitted by different substances and use them to create the effects that they desire. For example, fireworks containing barium emit green light, while those containing strontium emit red.

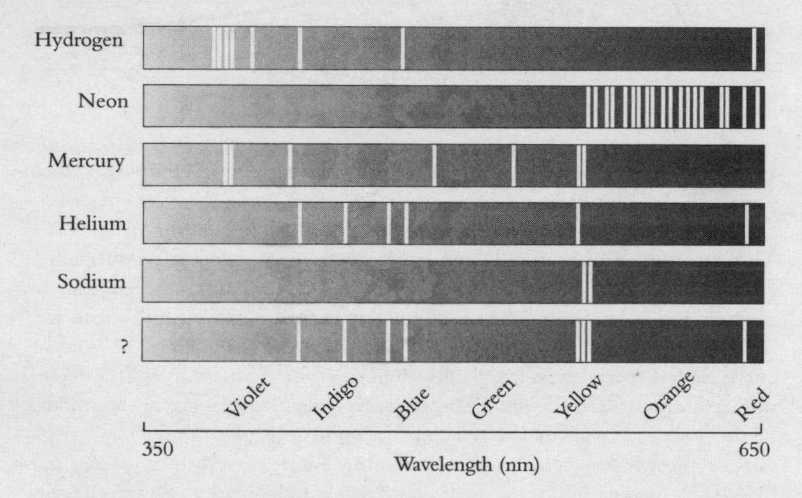

Figure 53 The main visible light emitted by sodium is shown in the fifth spectrum chart. There are two dominant wavelengths at roughly 0.000589 mm (589 nm), which corresponds to an orange colour. This chart represents a fingerprint for sodium. Indeed, each atom has its own fingerprint, which is apparent from the different wavelength charts. An atom may exhibit a slightly different fingerprint according to its environment, such as when the atom is subjected to high pressure. The lowest chart is for an unknown gas. By comparing the emitted wavelengths against the other charts it becomes obvious that the gas contains helium and sodium.

The exact wavelengths emitted by each atom act as a fingerprint. So by studying the wavelengths emitted by a heated substance, it is possible to identify the atoms in that substance. The lowest spectrum in Figure 53 is from an unknown hot gas, but by matching its emitted wavelengths against the other spectra then it is possible to see that the gas contains helium and sodium.

This science of atoms, light, wavelengths and colour is known as *spectroscopy*. The process by which a substance emits light is called spectroscopic *emission*. The opposite process, spectroscopic *absorption*, also exists, and this is when specific wavelengths of light are absorbed by an atom. So, if a whole range of wavelengths of light were directed at vaporised salt, then most of the light would pass through unaffected, but a few key wavelengths would be absorbed by the sodium atoms in the salt, as shown in Figure 54. The absorbed wavelengths for sodium are exactly the same as the emitted wavelengths, and this symmetry between absorption and emission is true for all atoms.

In fact, it was absorption rather than emission that attracted the attention of astronomers, which then took spectroscopy out of the chemistry laboratory and into the observatory. They realised that absorption could give clues to the make-up of stars, starting with the Sun. Figure 55 shows how sunlight can be passed through a

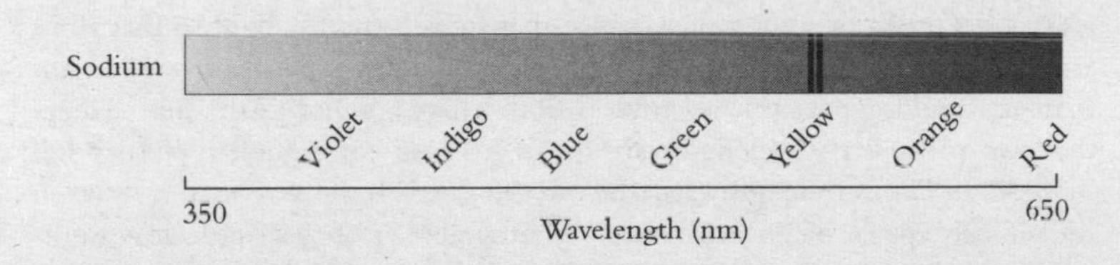

Figure 54 Spectroscopic absorption is the opposite process to spectroscopic emission. This absorption chart for sodium is identical to the one shown in Figure 53, except that it is black on grey, not white on grey, because we are seeing all the wavelengths, except the two wavelengths absorbed by sodium.

prism so that the complete range of wavelengths can be studied. The Sun is hot enough to emit wavelengths over the entire range of visible light, but physicists at the start of the nineteenth century noticed that specific wavelengths were missing. These wavelengths revealed themselves as fine black lines in the solar spectrum. It was not long before somebody realised that the missing wavelengths had been absorbed by atoms in the Sun's atmosphere. Indeed, the missing wavelengths could be used to identify the atoms that make up the Sun's atmosphere.

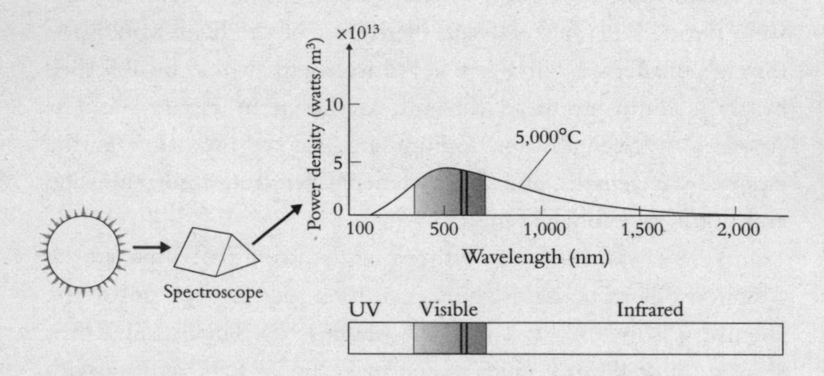

Figure 55 The Sun is hot enough to emit the complete range of visible wavelengths from red through to violet, as well as ultraviolet and infrared. Sunlight can be studied by passing it through a spectroscope, which incorporates a glass prism or some other device that spreads out the light so that all its wavelengths are discernible. The graph shows the distribution of wavelengths that we would expect to see emitted from a body as hot as the Sun, except that two particular wavelengths are missing. These correspond to absorption by sodium. The wavelength chart below the graph is the way that absorption lines usually appear on an astronomer's photographic plate, except real measurements may be much less distinct. In reality, detailed studies of sunlight showed that there were hundreds of missing wavelengths in the solar spectrum. These wavelengths had been absorbed by various atoms in the Sun's atmosphere, and by measuring the wavelengths of these dark absorption lines it was possible to identify the atoms that make up the Sun.

Although much of the groundwork was done by Joseph von Fraunhofer, a German pioneer in optics, it was Robert Bunsen and Gustav Kirchhoff who made the crucial breakthrough in around 1859. Together they built a *spectroscope*, a specially designed instrument for accurately measuring the wavelengths of light emitted by an object. They used it to analyse sunlight and were able to identify two of the missing wavelengths as ones associated with sodium, thereby concluding that sodium must exist in the Sun's atmosphere.

'At present Kirchhoff and I are engaged in a common work which doesn't let us sleep,' wrote Bunsen. 'Kirchhoff has made a wonderful, entirely unexpected discovery in finding the cause of the dark lines in the solar spectrum . . . thus a means has been found to determine the composition of the Sun and fixed stars with the same accuracy as we determine sulphuric acid, chlorine, etc., with our chemical reagents.' Comte's assertion that humans would never identify the constituents of the stars was thus shown to be wrong.

Kirchhoff went on to search for evidence of other materials, such as the heavy metals, in the Sun's atmosphere. His bank manager was not very impressed, and asked him, 'Of what use is gold in the Sun if I cannot bring it down to earth?' Many years later, when he was awarded a gold medal for his research, Kirchhoff paid a triumphant visit to the narrow-minded banker and said, 'Here is gold from the Sun.'

This technique of stellar spectroscopy was so powerful that in 1868 the Englishman Norman Lockyer and the Frenchman Jules Janssen independently discovered an element in the Sun before it was discovered on Earth. They identified an absorption line in sunlight that could not be matched with any known atom, so Lockyer and Janssen took this as evidence for a completely new type of atom. It was named helium, after Helios, the Greek sun-god.

Although helium accounts for a quarter of the Sun's mass, it is very rare on Earth and it would be over twenty-five years before it was detected here, whereupon Lockyer was knighted.

William Huggins was another scientist who appreciated the power of spectroscopy. As a young man he had been forced to take over the running of his father's draper's shop, but he later decided to sell the family business and pursue his scientific dream, using the money to set up an observatory on Upper Tulse Hill, now a suburb of London. When he heard about Bunsen and Kirchhoff's spectroscopic discoveries, Huggins was overjoyed: 'This news was to me like the coming upon a spring of water in a dry and thirsty land.'

During the 1860s, he applied spectroscopy to the stars beyond the Sun and confirmed that they too contained the same elements that existed on Earth. For example, he saw that the spectrum of the star Betelgeuse contained dark lines that appeared at the wavelengths absorbed by atoms such as sodium, magnesium, calcium, iron and bismuth. The ancient philosophers had argued that the

Figure 56 Mr and Mrs Huggins, who pioneered the use of spectroscopy in astronomy to measure the velocity of stars.

stars were made of *quintessence*, a fifth element beyond the mundane terrestrial elements of air, earth, fire and water, but Huggins had succeeded in showing that Betelgeuse, and presumably the entire universe, was made of the same materials as those found on Earth. Huggins concluded: 'One important object of this original spectroscopic investigation of the light of the stars and other celestial bodies, namely to discover whether the same chemical elements as those of our Earth are present throughout the universe, was most satisfactorily settled in the affirmative; a common chemistry, it was shown, exists throughout the universe.'

Huggins continued to study the stars for the rest of his life, accompanied by his wife Margaret and his dog Kepler. Margaret Huggins was an accomplished astronomer in her own right and twenty-four years his junior. So when William was aged eighty-four and getting towards the end of his career as an astronomer, he relied on his sprightly sixty-year-old wife to clamber around the telescope and make the necessary adjustments. 'Astronomers need universal joints and vertebrae of India rubber,' she complained. Together, Mr and Mrs Huggins developed an entirely new application for spectroscopy, one that would transform our view of the universe. In addition to assessing the ingredients of a star, they demonstrated how spectroscopy could be used to measure a star's velocity.

Following Galileo, astronomers had assumed that the stars were stationary. Although the stars all moved across the sky every night, astronomers realised that this apparent motion was caused by the Earth's rotation. In particular, they assumed that the stars' positions relative to one another remained the same. In fact, this was false, as pointed out in 1718 by the English astronomer Edmund Halley. Even after taking into account the motion of the Earth, he became aware of subtle discrepancies in the recorded positions of the stars Sirius, Arcturus and Procyon compared with measurements made by

Ptolemy many centuries earlier. Halley realised that these differences were not down to inaccurate measurements, but were the result of genuine shifts in the positions of these stars over time.

With infinitely precise measuring tools and infinitely powerful telescopes, astronomers would have been able to detect the so-called *proper motion* of every star, but in reality the stars change position so gradually that even modern astronomers can barely detect shifts in stellar positions. In general, detecting proper motion has required careful observations of the closest stars taken across several years, as shown in Figure 57. In other words, it has been a struggle to measure proper motion even in our closest stellar neighbours. Another limitation of studying proper motion is that it is a measure of motion across the sky only, and says nothing about motion towards or away from the Earth, known as *radial velocity*. All in all, the detection of proper motion has given only a limited insight into stellar velocities.

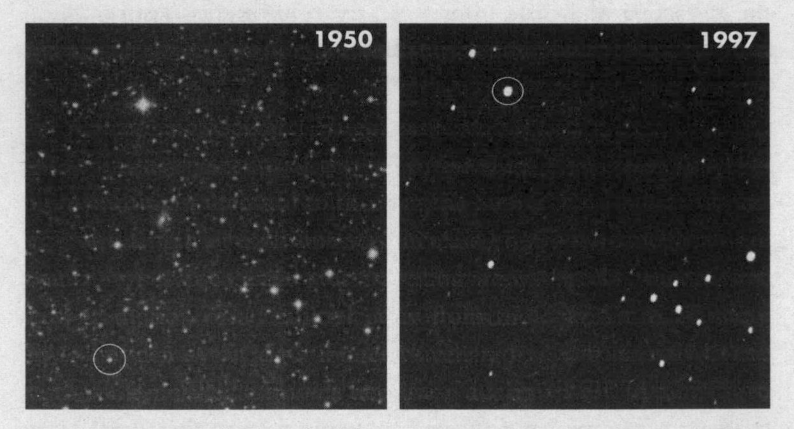

Figure 57 Barnard's Star (circled) is the second nearest star to our Solar System and the one with the greatest proper motion. It moves across the sky at 10 arcseconds each year. These pictures were taken almost half a century apart and show that the star has significantly shifted relative to all the other stars. To help appreciate the shift, the stars forming a < shape in the bottom right quarter provide a useful landmark.

William Huggins, however, realised that he was able to exploit spectroscopy to make up for the twin inadequacies of proper motion measurements. His new spectroscopic technique could be used to accurately measure the radial velocity of any star, and it could be applied to even the most distant stars. His idea relied on coupling the spectroscope with a piece of physics that had been discovered by the Austrian scientist Christian Doppler.

In 1842 Doppler announced that the movement of an object would affect any waves it was emitting, whether they were water waves, sound waves or light waves. For a simple illustration of this *Doppler effect*, picture a frog relaxing on a lily pad and tapping his webbed foot in the water each and every second, generating a series of waves that are 1 metre apart and which travel at 1 m/s, as shown in Figure 58. If we were looking from above and if the lily pad was not moving, then we would see the peaks of the waves forming a series of concentric symmetric rings, as shown in column (a) of Figure 58. Observers on either bank would see the waves arriving spaced 1 metre apart.

But things change if the frog is moving, as shown in column (b). Imagine that the lily pad and frog drift towards the right bank at a speed of 0.5 m/s and that the frog continues to generate one wave per second, and the waves still travel across the water at 1 m/s. This time the result is a clumping of the waves in the direction in which the frog is moving, and an increased spacing of waves in the opposite direction. An observer on the right bank sees the waves arriving only 0.5 metres apart, whereas the other observer sees a spacing of 1.5 metres. One observer sees a decreased wavelength, the other sees an increased wavelength. This is the Doppler effect.

In summary, when an object emitting waves moves towards an observer, then the observer perceives a decrease in the wavelength, whereas when the emitter moves away from the observer, then the observer perceives an increase in the wavelength. Alternatively, the

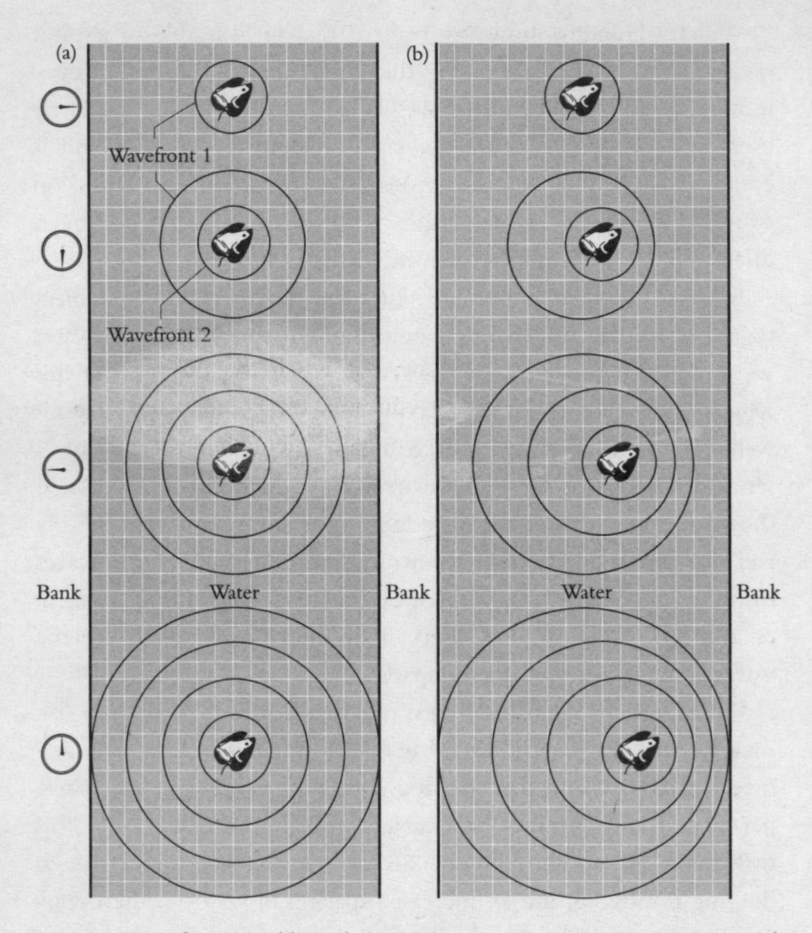

Figure 58 A frog on a lily pad is emitting water waves once every second which are 1 metre apart. When the frog is stationary, as shown in the series of diagrams in column (a), the observers on both banks see the water waves arriving 1 metre apart. However, when the frog drifts towards the right bank at a steady rate of 0.5 m/s, column (b), then the observers see two different effects. In the direction in which the frog is moving, the waves appear to bunch up, whereas the waves become more spaced out in the opposite direction. This is a consequence of the frog moving towards or away from different parts of the wavefront in the process of emitting the next wave, and is an example of the Doppler effect in water waves.

emitter might be stationary and the observer might be moving, in which case the same effects are apparent.

The Doppler effect was tested for sound waves in 1845 by the Dutch meteorologist Christoph Buys-Ballot, who was actually trying to disprove its existence. Trumpeters were split into two groups and asked to play the note E-flat. One group of trumpeters played from an open-top railway carriage on a piece of newly opened track between Utrecht and Maarsen, while the other trumpeters remained on the platform. When both groups were stationary then both notes were the same, but when the rail carriage was approaching, then a musically educated ear could detect that the note became higher, and it became even higher as the speed of the carriage increased. When the carriage moved away, the note became deeper. This change in pitch is associated with a change in the wavelength of the sound waves.

Today we can hear the same effect with an ambulance siren, which seems to have a higher pitch (shorter wavelength) as the ambulance approaches, and then a lower pitch (longer wavelength) as it moves away. The transition between the higher and lower pitch as the ambulance passes us at speed is quite noticeable. Formula 1 cars, because of their higher speed, demonstrate an even clearer Doppler effect as they pass by – the engine makes a distinct 'eeeeeeeey-oooooow' noise, going from higher to lower pitch.

The shift in wavelength and pitch is highly predictable thanks to an equation developed by Doppler. The received wavelength (λ_r) depends on the initial emitted wavelength (λ), and the ratio between the speed of the emitter (v_e) and the speed of the wave (v_w). If the emitter is travelling towards the observer, then v_e is reckoned as positive, and it if is travelling away from the observer then it is negative:

$$\lambda_r = \lambda \times \left(1 - \frac{v_e}{v_w}\right)$$

We can now perform a rough calculation to work out the perceived change in wavelength of a siren as an ambulance races past. The speed of the sound waves (v_w) in air is roughly 1,000 km/h, and the speed of an ambulance (v_e) might be 100 km/h, so the wavelength increases or decreases by 10% depending on the direction of the ambulance.

A similar calculation gives us the change in wavelength of the ambulance's blue flashing light. This time, the waves travel at the speed of light, so v_w is roughly 300,000 km/s, which is 1,000,000,000 km/h, and the speed of the ambulance (v_e) is still 100 km/h. Therefore the wavelength changes by 0.00001%. This difference in wavelength and colour would be imperceptible to the human eye. In fact, at an everyday level, we never perceive any form of Doppler shift in connection with light because even our fastest vehicles are extremely slow compared with the speed of light. However, Doppler predicted that the optical Doppler shift was a genuine effect and could be detected, as long as the emitter was moving fast enough and the detection equipment was sufficiently sensitive.

Sure enough, in 1868 William and Margaret Huggins succeeded in detecting a Doppler shift in the spectrum of the star Sirius. The absorption lines of Sirius were almost identical to those in the Sun's spectrum, except that the wavelength of each line was increased by 0.015%. This was presumably because Sirius was travelling away from the Earth. Remember, motion of an emitter away from the observer causes its light to be seen as having a longer wavelength. An increase in wavelength is often called a *redshift*, because red is at the longer-wavelength end of the visible spectrum. Similarly, a decrease in the wavelength caused by an approaching emitter is called a *blueshift*. Both types of shift are shown in Figure 59.

Although Doppler's equation would later need to be modified to conform with Einstein's theory of relativity, the nineteenth-century

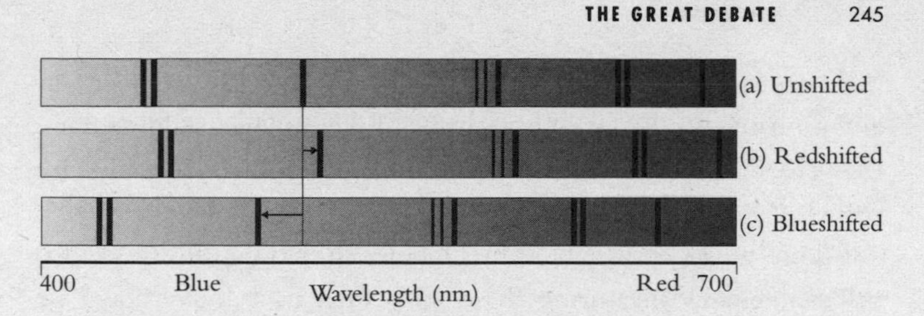

Figure 59 The three spectra show how the light emitted by a star depends on its radial motion. Spectrum (a) shows the wavelengths of some absorption lines from a star (e.g. the Sun) which is neither moving closer to nor farther from the Earth. Spectrum (b) shows redshifted absorption lines from a star which is moving away from the Earth – the lines are identical, except they have all been shifted to the right. Spectrum (c) shows blueshifted absorption lines from a star which is moving towards the Earth – again, the lines are identical, except this time they have all been shifted to the left. The blueshifted star is moving towards us faster than the redshifted star is moving away, because the blueshift is larger than the redshift.

version was satisfactory for Huggins' purposes and he could calculate the speed at which Sirius was receding from the Earth. He had measured the wavelengths from Sirius as being increased by 0.015%, so the relationship between the received and standard wavelengths was such that $\lambda_r = \lambda \times 1.00015$. And he knew that the speed of the waves was the speed of light, so v_w was 300,000 km/s. By rearranging the equation and plugging in the numbers, he could show that Sirius was receding at a speed of 45 km/s:

$$\text{We know that } \lambda_r = \lambda \times \left(1 - \frac{v_e}{v_w}\right) \text{ and } \lambda_r = \lambda \times 1.00015$$

$$\text{So, } 1.00015 = \left(1 - \frac{v_e}{v_w}\right)$$

$$\begin{aligned} v_e &= -0.00015 \times v_w \\ &= -0.00015 \times 300{,}000 \text{ km/s} \\ &= -45 \text{ km/s} \end{aligned}$$

William Huggins, the ex-cloth merchant who had pursued his ambition to practise astronomy, had proved that he could measure the velocities of the stars; each star contained ordinary Earthly elements (e.g. sodium), which emitted specific standard wavelengths, but these wavelengths would be Doppler shifted by the star's radial velocity, and by measuring these shifts then its velocity could be calculated. His method had huge potential, because any visible star, or nebula, could be analysed with a spectroscope and thus have its Doppler shift measured and its velocity determined. In addition to a star's proper motion across the sky, it was now possible to measure its radial velocity, towards or away from the Earth.

Using Doppler shifts to measure velocities is an unfamiliar technique for most people, but it really does work. Indeed, it is so reliable that some police forces use Doppler shifts to identify speeding motorists. The police officer fires a pulse of radio waves, an invisible part of the light spectrum, at an approaching car and then detects it after it has been reflected back from the car. The returning pulse has effectively been emitted by a moving object, the car, so its wavelength is shifted by an amount that depends on the car's speed. The faster the car, the greater the shift and the higher the speeding fine.

One tall tale explains how an astronomer driving to his observatory tried to use the Doppler effect to outwit the police. Having been caught jumping a red traffic light, the astronomer argued that the light had appeared green to him because he was moving towards it and consequently it was blueshifted. The police officer excused him the ticket for running a red light, and instead doubled the fine and gave him a speeding ticket. To achieve such a dramatic wavelength shift, the astronomer would have had to be driving at roughly 200,000,000 km/h.

By the beginning of the twentieth century, spectroscopes had become a mature technology, and could be coupled with the newly built giant telescopes and the latest, highly sensitive photographic plates. This technological trinity offered astronomers an unparalleled opportunity to explore the make-up of stars and their velocities. By identifying the many missing wavelengths from a particular star, astronomers could identify its ingredients, which turned out to be mainly hydrogen and helium. Then, by measuring how these lines were shifted, astronomers could see that some stars were moving towards the Earth and some were moving away, with the slowest dawdling at a few kilometres per second and the fastest zipping along at 50 km/s. To put this speed into context, if a plane could fly as quickly as the fastest star, it would be able to cross the Atlantic in a couple of minutes.

In 1912, an ex-diplomat turned astronomer took velocity measurement into uncharted territory. Vesto Slipher became the first astronomer to successfully measure the Doppler shift of a nebula. He used the Clarke Telescope, a 24-inch refractor at the Lowell Observatory in Flagstaff, Arizona. The telescope had been funded by a donation from Percival Lowell, a wealthy Boston aristocrat, who was obsessed with the belief that Mars was home to intelligent life and who was desperate to find proof of a Martian civilisation. Slipher's interests were more mainstream than Lowell's, and whenever possible he would point the telescope at the nebulae.

Slipher took a 40-hour exposure across several nights of the faint light from the Andromeda Nebula (which would later be confirmed as a galaxy), and measured a Doppler blueshift equivalent to 300 km/s, six times faster than any star. In 1912, the majority opinion was that Andromeda was within our own Milky Way, so astronomers could not believe that such a local object could have such a high velocity. Even Slipher doubted his own measurement, and to check that he had not made a mistake, he trained his telescope on the nebula now known as the Sombrero Galaxy. This time he discovered a redshift, not a blueshift, and the Doppler effect was even more extreme. The Sombrero was redshifted to such an extent that it had to be moving away from the Earth at 1,000 km/s. This is approaching 1% of the speed of light. If a plane could travel this fast, it would fly from London to New York in six seconds.

Over the next few years, Slipher measured the velocities of an increasing number of galaxies, and it became clear that they did indeed travel at phenomenally high velocities. However, a new puzzle began to emerge. The first two measurements had shown that one galaxy was approaching (blueshifted) and one was receding (red-shifted), but the first dozen measurements showed that many more galaxies were receding than were approaching. By 1917 Slipher had measured twenty-five galaxies, twenty-one of which were receding and only four of which were approaching. Over the next decade twenty more galaxies were added to the list, and every single one was receding. Virtually every galaxy appeared to be racing away from the Milky Way, as if our galaxy had a bad case of cosmic body odour.

Some astronomers had expected that the galaxies would be roughly static, effectively floating in the void; this was clearly not the case. Others thought that the distribution of their velocities would be balanced, with some approaching and some receding; this did not seem to be the case either. The fact that the galaxies had a distinct tendency to recede rather than approach confounded all expectations. Slipher and others attempted to make sense of the picture that was emerging. Various weird and wonderful explanations were put forward, but there was no consensus.

The case of the receding galaxies remained a mystery until Edwin Hubble applied his mind and his telescope to the problem. When he entered the debate he saw no point in wild theorising, particularly when the power of the mighty 100-inch Mount Wilson telescope held the promise of new data. His mantra was simple: 'Not until the empirical results are exhausted need we pass on to the dreamy realms of speculation.'

It would not be long before Hubble made the vital observations that would allow astronomers to slot Slipher's measurements into a new coherent model of the universe. Hubble was about to unwittingly provide the first major evidence to back up Lemaître and Friedmann's model of cosmological creation.

Hubble's Law

In the years after he measured the distance to the nebulae and proved that many of them were independent galaxies, Edwin Hubble had stamped his authority on the world of astronomy. At the same time, there was a major development in his personal life, because he had met and fallen in love with Grace Burke, the daughter of a local millionaire banker. According to Grace, she became infatuated with Hubble when she visited Mount Wilson and saw him staring intently at a photographic plate showing a field of stars. Later she would recall that he looked like 'an Olympian, tall, strong, and beautiful, with the shoulders of the Hermes of Praxiteles . . . There was a sense of power, channelled and directed in an adventure that had nothing to do with personal ambition and its anxieties and lack of peace. There was hard concentrated effort and yet detachment. The power was controlled.'

Grace was already married when she first met Hubble, but she was widowed in 1921 when her husband Earl Leib, a geologist, fell to his death while collecting mineral samples in a vertical mineshaft. After resuming her acquaintance and a period of courting, Edwin married Grace on 26 February 1924.

Thanks to Hubble's resolution of the Great Debate and the publicity that followed, Edwin and Grace found themselves catapulted onto the celebrity A-list. Mount Wilson was just 25 km from Los Angeles, and they became regulars on the Hollywood social circuit. The Hubbles dined with actors such as Douglas Fairbanks and mixed with the likes of Igor Stravinsky, while famous names such as

Leslie Howard and Cole Porter visited Mount Wilson and brought a touch of glamour to the observatory.

Hubble revelled in his cult status as the world's most famous astronomer, and he enjoyed regaling guests, students and journalists with stories of his colourful past. Having been dominated by his father throughout his youth, Hubble now enjoyed showing off to an adoring public. For example, he would often tell the tale of how he had duelled with swords while in Europe. His friends loved to hear this story, but when his father had heard of his duelling exploits, he had merely rebuked him and reminded Edwin that 'the duellist scar is not a badge of honour'.

Despite his fame and celebrity lifestyle, Hubble never forgot that he was first and foremost a pioneering astronomer. He considered himself a giant standing on the shoulders of giants, a natural successor to the throne previously occupied by Copernicus, Galileo and Herschel. While on honeymoon in Italy, he even took Grace to the tomb of Galileo to pay homage to the man whose work had provided the foundation for his own great discovery.

Naturally, when Hubble heard about Slipher's preponderance of redshifted galaxies, he felt compelled to enter the fray and resolve the mystery. He saw it as his duty as the greatest astronomer of the day to make sense of the fleeing galaxies. He set to work at Mount Wilson, where the 100-inch telescope gathered seventeen times more light than Slipher's telescope at the Lowell Observatory. He spent night after night working in almost continual darkness so that his eyes would become sensitised to the darkness of the night sky. The only illumination that was allowed to break the monotonous blackness inside the great observatory dome was the occasional gentle glow from his briar pipe.

Hubble's assistant was Milton Humason, who had risen from humble beginnings to become the world's finest astronomical photographer. Humason had dropped out of school at the age of fourteen and then worked as a bellboy at the Mount Wilson Hotel, which provided accommodation for visiting astronomers. He was then appointed as the observatory's mule driver, helping to take provisions and equipment to the top of the mountain. He next obtained a job as a janitor at the observatory, and as each night passed he learned more and more about what the astronomers were up to and about the photographic techniques they employed. He even persuaded one of the students to give him tutorials in mathematics. Word spread that Mount Wilson had a curious janitor with a rapidly growing knowledge of astronomy, and within three years of joining the observatory he was appointed to the photographic division. Two years later he became a fully fledged assistant astronomer.

Hubble took a liking to Humason, and the two men struck up an unlikely partnership. Hubble maintained the persona of a distinguished English gentleman, while Humason would spend cloudy nights playing cards and drinking an illicit alcoholic brew known as panther juice. Their relationship relied on Hubble's belief that 'the history of astronomy is a history of receding horizons', and Humason was capable of delivering the images that allowed Hubble to penetrate farther into the universe than anybody else in the world. While Humason was photographing a galaxy he would keep his fingers permanently on the buttons that steered the telescope, keeping the galaxy fixed in the field of view and compensating for any errors in the tracking mechanism. Hubble admired Humason's patience and meticulous attention to detail.

To explore Slipher's redshift mystery, the duo divided the work between them. Humason would measure the Doppler shifts of numerous galaxies, and Hubble set about measuring their distances. The telescope was fitted with a new camera and spectroscope so that photographs that previously would have taken several nights of exposure could be snapped in just a few hours. They began by confirming the galactic redshifts first measured by Slipher, and by 1929 Hubble and Humason had gauged the redshifts and distances for forty-six galaxies. Unfortunately, the margin of error in half of these measurements was too large. Being cautious, Hubble took only those galactic measurements of which he was confident and plotted velocity versus distance for each galaxy, as shown in Figure 60.

In almost every case the galaxies were redshifted, implying that they were receding. Also, the points on the graph seemed to indicate that the velocity of a galaxy strongly depended on its distance. Hubble drew a straight line through the data, suggesting that the velocity of a given galaxy was proportional to its distance from the Earth. In other words, if one galaxy was twice as far away as another galaxy, then it seemed to be moving away at roughly twice the velocity. Or if a galaxy was three times farther away, then it seemed to be receding three times as fast.

If Hubble was right, the repercussions were immense. The galaxies were not randomly dashing through the cosmos, but instead their speeds were mathematically related to their distances, and when scientists see such a relationship they search for a deeper significance. In this case, the significance was nothing less than the realisation that at some point in history all the galaxies in the universe had been compacted into the same small region. This was the first observational evidence to hint at what we now call the Big Bang. It was the first clue that there might have been a moment of creation.

The link between Hubble's data and a moment of creation was simple. Take a galaxy which is travelling away from the Milky Way at some velocity today, and let us see what happens if we wind the clock backwards. Yesterday the galaxy must have been closer to the Milky Way than it currently is, and last week it would have been closer still, and so on. In fact, by dividing the current distance to the galaxy by its speed, we can deduce when the galaxy would have been sat on top of our Milky Way (assuming that its velocity has remained

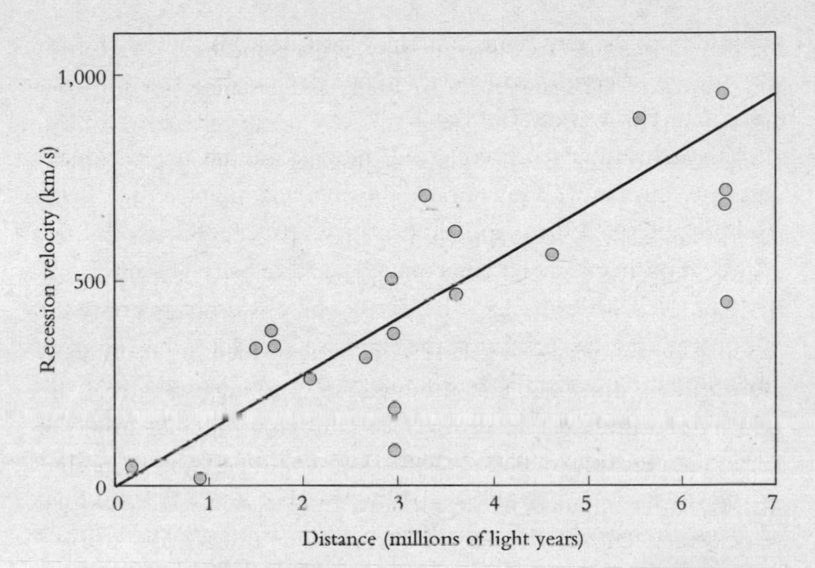

Figure 60 This plot shows Hubble's first set of data (1929) showing galactic Doppler shifts. The horizontal axis represents distance and the vertical axis represents recession velocity, and each dot represents the measurements for a single galaxy. While the points do not all fall on the line, there is a general trend. This suggests that the speed of a galaxy is proportional to its distance.

constant). Next, we pick a galaxy that is twice as far away as the first one and go through the same process, working out when it would also have been on top of our Milky Way. The graph suggests that a galaxy that is twice as far away as the first galaxy is travelling twice as fast. So, if we run the clock backwards the second galaxy will take exactly the same time as the first galaxy to return to the Milky Way. In fact, if every galaxy has a speed proportional to its distance from our Milky Way, then at some point in the past they would all have been simultaneously positioned on top of our own Milky Way, as shown in Figure 61.

So everything in the universe apparently emerged from a single dense region during a moment of creation. And if the clock is run forward from the zero hour, then the consequence is an evolving and expanding universe. This is exactly what Lemaître and Friedmann had theorised. This was the Big Bang.

Although Hubble had collected the data, he did not personally instigate, promote or encourage the implication of a Big Bang. Hubble published his graph in a six-page paper modestly entitled 'A Relation between Distance and Radial Velocity among Extra-Galactic Nebulae'. The hard-headed Hubble was not interested in speculating on the origin of the universe or addressing the great philosophical questions of cosmology. He just wanted to make good observations and get accurate data. It was the same when he made his previous breakthrough. He had proved that certain nebulae existed far beyond the Milky Way, but it was left to others to draw the conclusion that these nebulae were galaxies in their own right. Hubble seemed pathologically unable to engage with the deeper meaning of his data, so his colleagues were the ones who interpreted his graph of velocity versus distance.

But before anybody would speculate seriously about Hubble's observations, they first had to believe that his measurements were accurate. This was a major hurdle, because many of his fellow astronomers were not convinced by Hubble's graph. After all, many of the points were quite far from his superimposed line. Perhaps the points did not really lie on a straight line, but rather along a curve? Or perhaps there was no line or curve at all, and the points were actually random? The evidence had to be concrete, because the implications were potentially momentous. Hubble needed better measurements and more of them.

For two years Hubble and Humason continued to put in gruelling nights at the telescope, pushing the technology to the limit. Their efforts paid off, and they managed to measure galaxies that were twenty times as distant as any they had reported in their 1929 paper. In 1931 Hubble published another paper containing a new plot,

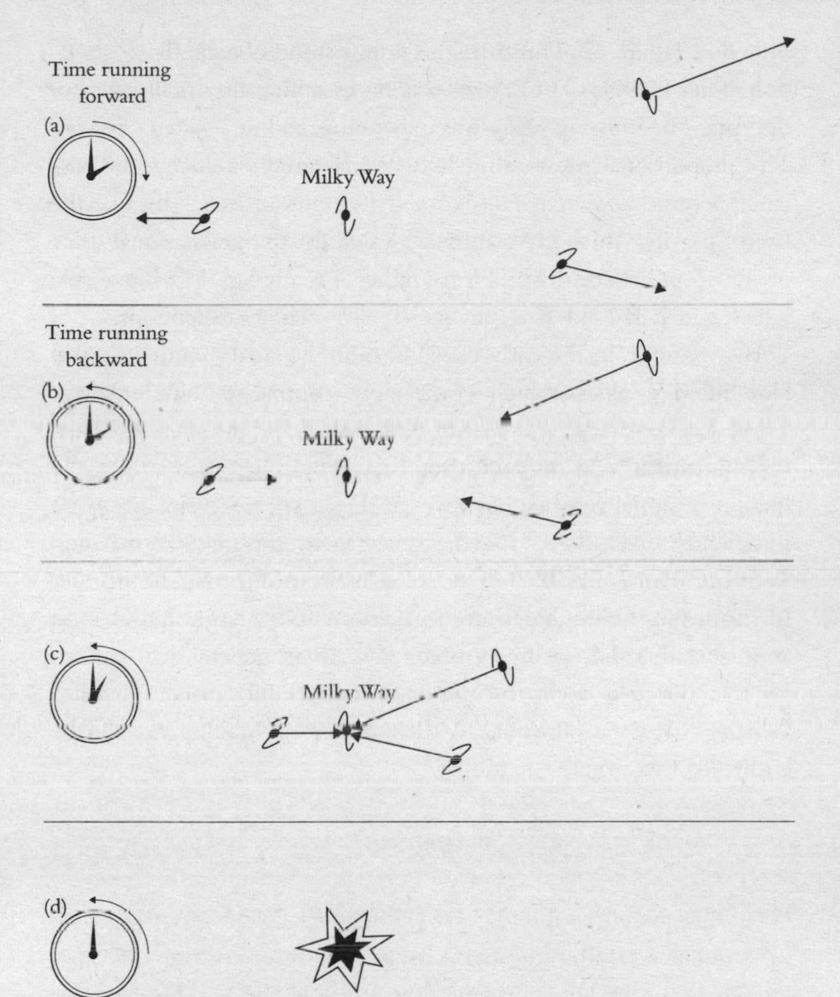

Figure 61 Hubble's observations implied a moment of creation. Diagram (a) represents the universe today, labelled 2 o'clock, with just three other galaxies for simplicity. The farther the galaxy, the faster it is receding, as indicated by the length of the arrows. However, if we run the clock backwards, diagram (b), then the galaxies seem to be approaching. At 1 o'clock, diagram (c), the galaxies will be closer to us. At midnight, diagram (d), they will all be on top of us. This would have been when the Big Bang started.

shown in Figure 62. This time the points stood obediently to attention along Hubble's line. There was no escaping the implications of the data. The universe really was expanding, and in a systematic way. The proportional relationship between a galaxy's velocity and distance became known as *Hubble's law*. It is not an exact law, like the law of gravity, which gives an exact value for the gravitational force of attraction between two objects; rather it is a broad descriptive rule which generally holds true, but which also tolerates exceptions.

For example, in the early days Vesto Slipher had identified a few blueshifted galaxies, which completely contradicts Hubble's law. These galaxies were close to our Milky Way, and if a galaxy's speed is proportional to its distance, then they should have had a relatively small recessional velocity. However, if their expected velocities were sufficiently small, they could be reversed by the gravitational pull from our own Milky Way or other galaxies in our neighbourhood. In short, the slightly blueshifted galaxies could be ignored as local anomalies that did not fit Hubble's law. So, in general, it is true to say that the galaxies in the universe are receding from us with a velocity that is proportional to their distance. Hubble's law can be enshrined in a simple equation:

$$v = H_0 \times d$$

What this says is that the velocity (v) of any galaxy is generally equal to its distance (d) from the Earth multiplied by a fixed number (H_0), known as the Hubble constant. The value of the Hubble constant depends on the units that are used for distance and velocity. Velocity is usually measured in the familiar unit of kilometres per second, but for technical reasons astronomers often prefer to measure distance in megaparsecs (Mpc), such that 1 Mpc equals 3,260,000 light years, or 30,900,000,000,000,000,000 km. Using the megaparsec unit, Hubble calculated that his constant had a value of 558 km/s/Mpc.

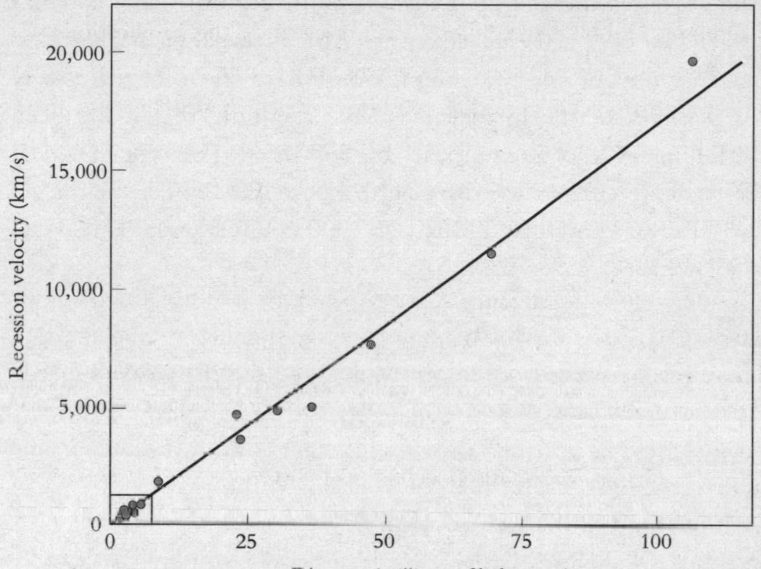

Figure 62 As in his 1929 graph (Figure 60), each point on Hubble's 1931 graph represented the measurements for one galaxy. The measurements were much improved compared to the 1929 paper. In particular, Hubble was able to measure galaxies at far greater distances, so much so that all the data points from the 1929 paper are contained in the small box in the bottom left corner. This time it was much more obvious that the points lay on a straight line.

The value of the Hubble constant has two implications. First, if a galaxy is 1 Mpc from the Earth then it should be travelling at roughly 558 km/s, or if a galaxy is 10 Mpc from Earth then it should be travelling at roughly 5,580 km/s, and so on. In fact, if Hubble's law is correct then we can deduce the speed of any galaxy just by measuring its distance, or conversely we can work out its distance from its speed.

The second implication of Hubble's constant is that it tells us the age of the universe. How long ago was it that all the matter in the universe emerged from a single dense region? If the constant is 558 km/s/Mpc, then a galaxy at 1 Mpc is travelling at 558 km/s, so we can work out how long it would have taken for that galaxy to have reached a distance of 1 Mpc assuming that it has been travelling at a constant speed of 558 km/s. The calculation is easier if we convert the distance to kilometres, which we can do because we know that 1 Mpc = 30,900,000,000,000,000,000 km.

$$\text{time} = \frac{\text{distance}}{\text{velocity}}$$

$$\text{time} = \frac{30{,}900{,}000{,}000{,}000{,}000{,}000\,\text{km}}{558\,\text{km/s}}$$

$$\text{time} = 55{,}400{,}000{,}000{,}000{,}000\,\text{s}$$

$$\text{time} = 1{,}800{,}000{,}000 \text{ years}$$

So, according to Hubble and Humason's observations, all the matter in the universe was concentrated into a relatively small region roughly 1.8 billion years ago and has been expanding outwards ever since. This picture completely contradicted the established view of an eternal unchanging universe. It reinforced the notion put forward by Lemaître and Friedmann that the universe began with a Big Bang.

Astronomers had already been obliged to tolerate a minimal level of evolution in the universe, because they had witnessed changes with their own eyes, such as the appearance of novae and supernovae. But astronomers had assumed that a dying star was compensated for by the emergence of a newborn star elsewhere, maintaining the overall stability and balance of the universe. In other words, the occasional nova would not change the overall character of the universe. However, this latest data implied continual evolution on a grand cosmic scale. Hubble's observations and

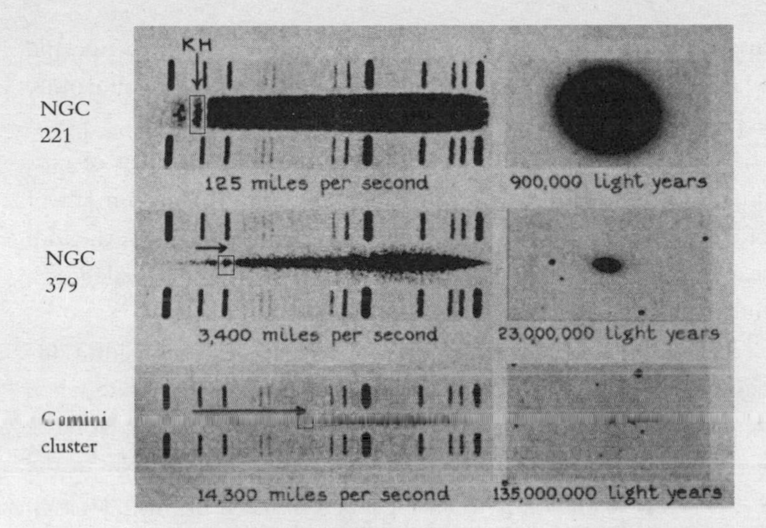

Figure 63 Unlike the idealised absorption spectrum in Figure 54, these spectra show some real measurements taken by Hubble and Humason. Although hard to interpret, each row shows the absorbed wavelengths for one galaxy, accompanied by an image of the galaxy on the right.

The first galaxy, NGC 221, is 0.9 million light years away. Humason's spectroscopic measurements provide the speed of the galaxy. The central horizontal strip shows the light from the galaxy, and the vertical line enclosed within a box represents a wavelength of light that has been absorbed by calcium in the galaxy. This vertical bar is actually farther to the right than it ought to be, representing a redshift (see Figure 59), implying a recessional velocity of 125 miles/s (200 km/s). The extent of the shift is measured relative to the calibration scale that runs above and below the data for NGC 221.

The second set of measurements relate to the galaxy NGC 379, which is 23 million light years away, which is why it appears smaller in the photograph than NGC 221. The key point is that the calcium absorption line (boxed) has been shifted farther to the right, which means a higher redshift – indeed, its recessional velocity is 1,400 miles/s (2,250 km/s). NGC 379 is 26 times as distant as NGC 221 and is travelling 27 times faster. Hence, the increase in velocity is roughly proportional to the increase in distance.

The third set of measurements relate to a galaxy in the Gemini cluster at a distance of 135 million light years. The calcium line (boxed) is shifted even farther to the right, which is an even higher redshift, implying a speed of 14,300 miles/s (23,000 km/s). It is roughly a hundred times farther away than NGC 221 and is travelling roughly a hundred times faster.

his expansion law meant that the whole universe was dynamic and evolving, with distances increasing and the universe's overall density decreasing with time.

Naturally, innate conservatism meant that the majority of cosmologists rejected the idea of an expanding universe and a moment of creation, just as there were those who had fought against the idea that the nebulae were distant galaxies, or that light travelled at a finite velocity, or that the Earth travelled around the Sun.

As far as the ex-mule driver was concerned, such highfalutin discussions did not trouble him. Humason's work was complete when he had measured the redshifts, and their interpretation was none of his concern: 'I have always been rather happy that my part in the work was, you might say, fundamental, it can never be changed – no matter what the decision is as to what it means. Those lines are always where I measured them and the velocities, if you want to call them that or redshifts or whatever they are going to be called eventually, will always remain the same.'

It is worth stressing again that Hubble also steered clear of any speculation. He may have provided the measurements, but he took no part in the cosmological debate. Hubble and Humason's scientific paper contained the following statement: 'The writers are constrained to describe the "apparent velocity-displacements" without venturing on the interpretation and its cosmological significance.'

So instead of getting involved in the next Great Debate, Hubble luxuriated in his ever-increasing fame. In 1937 he was Frank Capra's guest of honour at the Motion Picture Academy Awards. Capra, president of the academy, opened the Oscars evening by introducing the world's greatest astronomer. The Hollywood glitterati were playing supporting roles to Hubble, who stood up to accept his applause, illuminated by three brilliant spotlights. He had spent his life staring at the stars in wonder, and now the stars were staring at him in equal awe.

Everyone in the auditorium appreciated the magnitude of Hubble's achievements. Here was the man whose distance measurements had enlarged our view of the universe from a single finite Milky Way to an infinite space peppered with other galaxies. Here was the man who had shown that the cosmos was expanding and, whether Hubble himself acknowledged it or not, this seemed to imply that the universe had a limited history and that it had once been an embryo of compact matter ready to explode and evolve. Edwin Hubble had unwittingly discovered the first real evidence in favour of creation. At last the Big Bang model was more than just a theory.

CHAPTER 3 · THE GREAT DEBATE
SUMMARY NOTES

① ASTRONOMERS BUILT BIGGER AND BETTER TELESCOPES – THEY EXPLORED THE SKY AND MEASURED THE DISTANCES TO THE STARS.

② 1700s HERSCHEL SHOWS THAT THE SUN IS EMBEDDED WITHIN A GROUPING OF STARS – THE MILKY WAY. THIS WAS OUR GALAXY – PERHAPS THE ONLY GALAXY?

③ 1781 MESSIER CATALOGUED THE NEBULAE (FAINT SMUDGES) THAT APPEAR NOT TO BE STARS (SHARP POINTS OF LIGHT). THE GREAT DEBATE IS ABOUT THE NATURE OF THESE NEBULAE:
- ARE THEY OBJECTS WITHIN OUR MILKY WAY OR
- ARE THEY SEPARATE GALAXIES?

IS OUR MILKY WAY THE ONLY GALAXY?
OR
IS THE UNIVERSE PEPPERED WITH GALAXIES THROUGHOUT?

④ 1912 HENRIETTA LEAVITT STUDIED CEPHEID VARIABLE STARS AND SHOWED HOW THEIR PERIOD OF VARIABILITY CAN BE USED TO INDICATE THEIR ACTUAL BRIGHTNESS AND ESTIMATE THEIR DISTANCE.

ASTRONOMERS NOW HAD A RULER FOR MEASURING THE UNIVERSE.

⑤ 1923 EDWIN HUBBLE IDENTIFIED A CEPHEID VARIABLE STAR IN A NEBULA AND PROVED THAT IT WAS FAR BEYOND THE MILKY WAY! THEREFORE (MOST) NEBULAE WERE SEPARATE GALAXIES, EACH COMPOSED OF BILLIONS OF STARS, JUST LIKE OUR MILKY WAY.

THE UNIVERSE WAS FULL OF GALAXIES.

⑥ SPECTROSCOPY - DIFFERENT ATOMS EMIT/ABSORB SPECIFIC WAVELENGTHS OF LIGHT

WAVELENGTH

SO ASTRONOMERS STUDIED STARLIGHT TO SEE WHAT STARS ARE MADE OF:

ASTRONOMERS NOTICED THAT THE WAVELENGTHS IN STARLIGHT WERE SLIGHTLY SHIFTED. THIS COULD BE EXPLAINED BY THE DOPPLER EFFECT: - AN APPROACHING STAR HAS ITS LIGHT SHIFTED TO SHORTER WAVELENGTHS (BLUESHIFT) AND
- A RECEDING STAR HAS ITS LIGHT SHIFTED TO LONGER WAVELENGTHS (REDSHIFT).

THE MAJORITY OF GALAXIES SEEMED TO BE RACING AWAY (REDSHIFTED) FROM THE MILKY WAY!

⑦ 1929 HUBBLE SHOWED THAT THERE IS A DIRECT RELATION BETWEEN A GALAXY'S DISTANCE AND VELOCITY.
THIS IS KNOWN AS HUBBLE'S LAW:

IF THE GALAXIES ARE RECEDING THEN:
① TOMORROW THEY WILL BE FARTHER AWAY FROM US
② BUT YESTERDAY THEY WERE CLOSER TO US
③ AND LAST YEAR THEY WERE CLOSER STILL
④ AT SOME POINT IN THE PAST ALL GALAXIES MUST HAVE BEEN RIGHT ON TOP OF US.

HUBBLE'S MEASUREMENTS SEEMED TO IMPLY THAT THE UNIVERSE STARTED IN A SMALL CONDENSED STATE AND THEN EXPANDED OUTWARDS. IT IS STILL EXPANDING TODAY.

IS THIS EVIDENCE FOR A BIG BANG?

Chapter 4

MAVERICKS OF THE COSMOS

The super-system of the galaxies is dispersing as a puff of smoke disperses. Sometimes I wonder whether there may not be a greater scale of existence of things, in which it is no more than a puff of smoke. **ARTHUR EDDINGTON**

Nature shows us only the tail of the lion. But I have no doubt that the lion belongs to it even though he cannot totally reveal himself all at once because of his large size. We can see him only the way a louse that is sitting on him would.

ALBERT EINSTEIN

Cosmologists are often in error, but never in doubt.

LEV LANDAU

Albert Michelson, having banished the ether a few years earlier, delivered a speech at the University of Chicago in 1894. He proclaimed: 'The most important fundamental laws and facts of physical science have all been discovered, and these are now so firmly established that the possibility of their ever being supplemented in consequence of new discoveries is exceedingly remote . . . Our future discoveries must be looked for in the sixth place of decimals.'

The second half of the nineteenth century had indeed been a glorious time for physics, with many great mysteries solved, but to suggest that the only remaining task was to increase the accuracy of measurements was to prove patently absurd. Michelson would live to see his bold statement crumble. Within a few decades, the development of quantum and nuclear physics would shake the very foundations of science. Moreover, cosmologists would have to completely reassess their understanding of the universe.

The late-nineteenth-century view of the universe had been of an eternal and largely unchanging cosmos. But, while flappers flapped and stock markets crashed, the scientists of the 1920s were forced to consider a rival cosmic model which described an expanding universe that had been born a billion or two years ago.

This kind of upheaval in scientific thinking can be initiated in two ways. One involves theorists, who might reach a surprising conclusion by applying the laws of physics in a new direction. The other way involves experimenters or observers, who might measure something or see something that causes them to question previous assumptions. The upheaval in cosmology that took place in the 1920s was unusual because the established model of an eternal universe came under simultaneous attack on both fronts. Georges Lemaître and Alexander Friedmann had used theory to develop the idea of an expanding universe, as described in Chapter 2. In parallel, Edwin Hubble was independently observing galactic redshifts, which also implied an expanding universe, as described in Chapter 3.

Friedmann was no longer alive to hear about Hubble's observations, having died without receiving any recognition for his ideas. Lemaître, however, was more fortunate. In his 1927 paper, in which he proposed the Big Bang model of the universe, he predicted that the galaxies should be racing away at speeds that were proportional to their distances. Initially, his work was ignored because there was no evidence to support it, but two years later Hubble published his observations which showed that the galaxies were indeed receding, and Lemaître was vindicated at last.

Lemaître had previously written to Arthur Eddington about his Big Bang model, but had received no reply. When Hubble's discovery hit the headlines, Lemaître wrote to Eddington again, hoping that this time the distinguished astrophysicist would realise that his theory tied in perfectly with the emerging data. George McVittie was Eddington's student at the time and recalled his supervisor's reaction to the persistent priest: 'Eddington, rather shamefacedly, showed me a letter from Lemaître which reminded Eddington of the solution to the problem which Lemaître had already given. Eddington confessed that although he had seen Lemaître's paper in 1927 he had forgotten completely about it until that moment. The oversight was quickly remedied by Eddington's letter to the prestigious journal *Nature* in June 1930, in which he drew attention to Lemaître's brilliant work of three years before.'

He had overlooked Lemaître's research in the past, but now it seemed that Eddington was prepared to give it his blessing by promoting it. In addition to his letter to *Nature*, Eddington also translated Lemaître's paper and published it in the *Monthly Notices of the Royal Astronomical Society*. He called it a 'brilliant solution' and 'a complete answer to the problem', meaning that Lemaître's model perfectly explained Hubble's measurements.

Gradually word spread through the scientific community, and there was a slowly growing admiration for the perfect match between Lemaître's theoretical predictions and Hubble's observa tions. Until this point, all cosmologists had focused their attention on Albert Einstein's eternal static model of the universe, but now a significant minority considered Lemaître's model to be far more powerful.

To recap: Lemaître had argued that general relativity (in its purest form) implied that the universe is expanding. If the universe is expanding today, then in the past it must have been more compact. Logically, the universe must have started from a highly compact state, the so-called primeval atom of small but finite size. Lemaître thought that the primeval atom might have existed for eternity before there was some 'rupture of the equilibrium', whereupon the atom decayed and ejected all its fragments. He defined the beginning of this decay process as the start of our universe's history. This was effectively the moment of creation – in Lemaître's words, 'a day without a yesterday'.

Friedmann's view of the moment of creation had been slightly different from Lemaître's. Instead of picturing the universe as emerging from a primeval atom, Friedmann's Big Bang model had argued that everything emerged from a point. In other words, the entire universe had been squeezed into nothing. Either way, primeval atom or single point, theories about the actual moment of creation were clearly highly speculative and would remain so for some time. With other aspects of the Big Bang model, however, there was a greater degree of confidence and broad agreement among its advocates.

For example, Hubble had observed that the galaxies were receding from the Earth, just as predicted by the Big Bang model, but Big Bang theorists unanimously believed that the galaxies were not actually moving through space, but were moving along with space. Eddington explained this subtle point by comparing space to the surface of a balloon, simplifying the three spatial dimensions of the universe onto a two-dimensional closed rubber sheet, as shown in Figure 64. The balloon's surface is covered with dots, which represent the galaxies. If the balloon is inflated to twice its original diameter, then the distance between the dots will double in size, so the dots are effectively moving away from one another. The crucial point is that the dots are not moving across the surface of the balloon – instead, it is the surface itself that is expanding, thereby increasing the distance between the dots. Similarly, the galaxies are not moving through space, rather it is the space between the galaxies that is expanding.

Although the redshifting of galactic light was explained in Chapter 3 simply in terms of the recession of galaxies, it now becomes clear that the actual cause of the redshift is the stretching of space. As the light waves leave a galaxy and travel towards the Earth, they are stretched because the space in which they are travelling is itself being stretched, which is why the wavelengths grow longer and the light appears redder. Although this cosmological redshifting of light has a different cause than the usual Doppler shifting of waves, the description of the Doppler effect in Chapter 3 remains a useful way to think about the redshifts of galaxies.

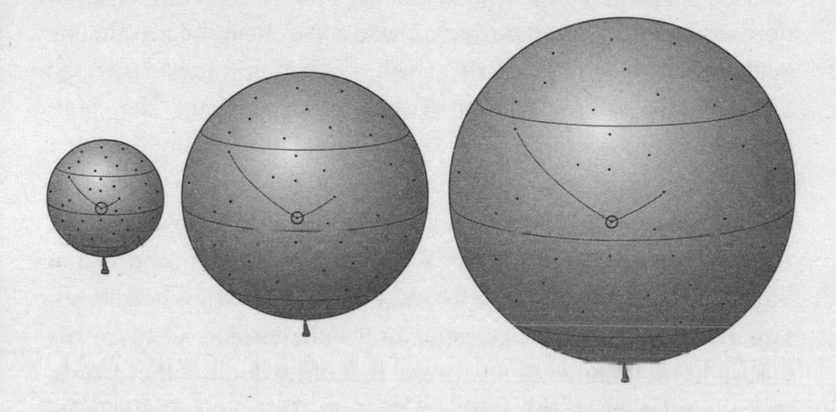

Figure 64 The universe is represented here as the surface of a balloon. Each dot represents a galaxy, and the circled dot represents our own Milky Way galaxy. As the balloon inflates (i.e. as the universe expands), the other dots appear to recede from us, just as Hubble observed that all the galaxies are receding from us. The more distant the galaxy, the farther it moves in a given time interval, so the faster it moves – which is Hubble's law. This effect is highlighted by the distances marked to two galaxies, one near and one far.

If all of space is expanding and the galaxies sit in space, then you might think that the galaxies would also be expanding. In theory this could happen, but in practice the huge gravitational forces that exist within galaxies mean that this effect is insignificant. Therefore expansion is relevant on a cosmic intergalactic level, but not on a local intragalactic level. In a flashback at the start of the Woody Allen film *Annie Hall*, Mrs Singer takes her son Alvy to see a therapist because he is depressed. The boy explains to the doctor that he has read that the universe is expanding, so he thinks that everything around him will eventually be torn apart. His mother interrupts: 'What has the universe got to do with it? You're here in Brooklyn! Brooklyn is not expanding!' Mrs Singer was absolutely correct.

Now that the balloon analogy has been introduced, this is a good time to clear up a common misunderstanding. If all the galaxies are getting farther away from the Earth, doesn't this imply that the Earth is at the centre of the universe? It seems as though the entire universe emerged from where we now live. Do we really occupy a special place in the cosmos? In fact, no matter where an observer is situated, there is the illusion of centrality. Returning to Figure 64, we can imagine that our Milky Way is one of the dots, and that as the balloon inflates, all the other dots seem to move away from us. However, from the vantage point of a different dot, all the other dots appear to be moving away from that other dot. In other words, that other dot thinks that it is at the centre of the universe. There is no centre to the universe – or perhaps every galaxy can claim to be at the centre of the universe.

Albert Einstein had lost interest in cosmology in the mid-1920s, but he re-engaged with the subject after Hubble's observations reinforced the idea of a Big Bang. In 1931, while on a sabbatical at the California Institute of Technology (Caltech), he and his second wife, Elsa, paid a visit to the Mount Wilson Observatory as Hubble's guests of honour. They were given a guided tour of the giant 100-inch Hooker Telescope, and the astronomers explained how this gigantic machine was essential for exploring the universe. To their surprise, Elsa was not particularly impressed: 'Well, well, my husband does that on the back of an old envelope.'

However, Einstein's efforts were restricted to theorising, and theories can be wrong. That is why investing in expensive experiments and vast telescopes is so worthwhile, because they alone make it possible for us to differentiate between a good theory and a bad theory. Einstein's earlier envelope jottings had argued for a static universe, which Hubble's observations now seemed to contradict, thus illustrating the power of observation to judge theory.

While at Mount Wilson, Einstein spent time with Milton Humason, Hubble's assistant, who showed him various photographic plates and pointed out the galaxies they had probed. He also showed Einstein the galaxies' spectra, which had revealed a systematic redshift. Einstein had already read Hubble and Humason's published papers, but now he could see the data for himself. The conclusion seemed to be unavoidable. The observations indicated that the galaxies were receding and that the universe was expanding.

On 3 February 1931, Einstein made an announcement to journalists gathered in the library of the Mount Wilson Observatory. He publicly renounced his own static cosmology and endorsed the Big Bang expanding universe model. In short, he found Hubble's observations to be convincing, and admitted that Lemaître and Friedmann had been right all along. With the world's most famous genius changing his mind and now backing the Big Bang, the expanding universe was official as far as the newspapers were concerned. Hubble's hometown paper, the *Springfield Daily News*, ran the headline YOUTH WHO LEFT OZARK MOUNTAINS TO STUDY STARS CAUSES EINSTEIN TO CHANGE HIS MIND.

Not only did Einstein abandon his static universe model, but he also reconsidered his equation for general relativity. Remember, Einstein's original equation had accurately explained the familiar force of gravitational attraction, but this attractive force would eventually cause the entire universe to collapse. Because the universe was supposed to be eternal and static, he added the cosmological constant – in effect, a fudge – to his equation in order to simulate a repulsive force that acted over large distances, thereby preventing collapse. Now that the universe no longer appeared to be static, Einstein ditched the cosmological constant and returned to his original equation for general relativity.

Einstein had always felt uncomfortable about the cosmological constant, having inserted it into his equation only to comply with the establishment view of a static and eternal universe. Convention and compliance, it turned out, had led him astray. Throughout his early life as a physicist, when he was at his intellectual peak, he had always followed his instinct and ignored authority. On the single occasion on which he had bowed to peer pressure, he was proved to be wrong. Later he would call the cosmological constant the greatest blunder of his entire life. As he wrote in a letter to Lemaître: 'Since I have introduced this term I had always a bad conscience . . . I am unable to believe that such an ugly thing should be realised in nature.'

Although Einstein was keen to abandon his cosmic fudge factor, cosmologists who still believed in an eternal, static universe were convinced that the cosmological constant was an essential and valid part of general relativity. Even some Big Bang cosmologists had become quite fond of it and were reluctant to lose it. By retaining the cosmological constant and varying its value, they could tweak their theoretical models of the Big Bang and modify the universe's expansion. The cosmological constant represented an anti-gravity effect, so it made the universe expand faster.

The value and validity of the cosmological constant generated some conflict among the supporters of the Big Bang theory, but Lemaître and Einstein showed a united front when they met at a seminar at Mount Wilson's base camp in Pasadena in January 1933, nearly two years after Einstein's first visit to the observatory. Lemaître presented his vision of the Big Bang model to the seminar's distinguished audience of astronomers and cosmologists, including Edwin Hubble. Although this was an academic gathering, Lemaître wove some poetic imagery in among the physics. In particular, he returned to his favourite firework analogy: 'In the beginning of everything we had fireworks of unimaginable beauty. Then there was an explosion followed by the filling of the heavens with smoke. We come too late to do more than visualise the splendour of creation's birthday!'

Figure 65 Albert Einstein and Georges Lemaître at Pasadena in 1933 for the seminar on Hubble's observations and the Big Bang model of the universe.

Even though Einstein had probably hoped for more mathematical detail and less embroidery, he still paid tribute to Lemaître's pioneering efforts: 'This is the most beautiful and satisfactory explanation of creation to which I have ever listened.' Praise indeed, especially from a man who just six years earlier had called Lemaître's physics 'abominable'.

Einstein's endorsement marked the start of Lemaître's life as a celebrity, within science and beyond. After all, here was the man who had proved Einstein wrong and who had such great foresight that he had predicted the expansion of the universe before telescopes were powerful enough to detect the fleeing galaxies. Lemaître was invited to speak all over the world and he received numerous international awards – indeed, he could lay claim to that rare honour of being a famous Belgian. Part of his popularity, charm and iconic status came from his dual role as a priest and a physicist. Duncan Aikman of the *New York Times*, who covered the 1933 Pasadena meeting, wrote: 'His view is interesting and important not because he is a Catholic priest, not because he is one of the leading mathematical physicists of our time, but because he is both.'

Like Galileo, Lemaître believed that God had blessed humans with an enquiring mind and that He would look fondly upon scientific cosmology. At the same time, Lemaître kept his physics and his religion separate, declaring that his religious beliefs certainly did not motivate his cosmology. 'Hundreds of professional and amateur scientists actually believe the Bible pretends to teach science,' he said. 'This is a good deal like assuming that there must be authentic religious dogma in the binomial theorem.'

Nevertheless, some scientists continued to believe that theology had negatively influenced the priest's cosmology. This anti-religious faction complained that his primeval atom theory of creation was nothing more than a pseudo-scientific justification of a master creator, a modern version of the Book of Genesis. In order to undermine Lemaître's position, these critics highlighted a serious flaw in the Big Bang hypothesis, namely its estimate for the age of the universe. According to Hubble's observations, the distance and velocity measurements implied a universe less than 2 billion years old. Given that contemporary geological research had estimated the age of some Earth rocks to be 3.4 billion years, there was an embarrassing age gap of at least 1.4 billion years. The Big Bang model seemed to imply that the Earth was older than the universe.

As far as the Big Bang critics were concerned, the fundamental problem with Lemaître's model was that the universe did not have a finite age. They argued that the universe was eternal and unchanging, and that the Big Bang model was nonsense. This was still the establishment view.

However, the establishment could not merely sit back and attack the Big Bang – they also had to explain the latest observations in the context of their preferred eternal universe model. Hubble's observations clearly indicated that the galaxies were redshifted and receding, so the Big Bang critics had to demonstrate that this did not necessarily imply a moment of creation in the past.

The Oxford astrophysicist Arthur Milne was one of the first to come up with an alternative way of explaining Hubble's law that was consistent with an eternal universe. In his theory, dubbed *kinematic relativity*, galaxies had a wide range of speeds, some moving slowly through space, some moving very quickly. Milne argued that it was only natural for the more distant galaxies to be the faster ones, as observed by Hubble, because it was only thanks to their speed that they had got so far away. According to Milne, the fact that the galaxies receded with a speed in proportion to their distance was not a consequence of some exploding primeval atom, but emerged naturally when randomly moving entities were allowed to move unhindered. This argument was far from watertight, but it did

Figure 66 Fritz Zwicky, inventor of the flawed theory of tired light, which attempted to explain Hubble's galactic redshift observations.

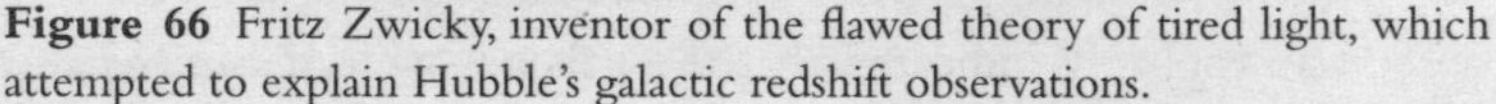

encourage other astronomers to think creatively about Hubble's redshifts in the framework of an eternal universe.

One of the fiercest critics of the Big Bang model was the Bulgarian-born Fritz Zwicky, infamous among cosmologists for his eccentricity and recalcitrance. He had been invited to Caltech and Mount Wilson in 1925 by the Nobel Laureate Robert Millikan, and Zwicky repaid the favour by announcing on one occasion that Millikan had never had a good idea in his life. All of his colleagues were targets of his abuse, and many of them were subjected to his favourite insult – 'spherical bastard'. Just as a sphere looks the same from every direction, a spherical bastard was someone who was a bastard whatever way you looked at them.

Zwicky examined Hubble's data and questioned whether the galaxies were even moving at all. His alternative explanation for the galactic redshifts was based on the accepted notion that anything emitted from a planet or star loses energy. For example, if you throw a stone high into the air, it leaves the Earth's surface with energy and speed, but the gravitational force of the dense Earth reduces the stone's kinetic energy, slowing it down until it stops and falls back to Earth. Similarly, light escaping from a galaxy will have its energy sapped by the galaxy's gravitational force. The light cannot slow down because the speed of light is constant, so instead the loss of energy manifests itself as an increase in the light's wavelength, making it appear redder. In other words, here was another possible explanation for Hubble's redshift observations, one that did not involve universal expansion.

Zwicky's argument that the redshifts were caused by galactic gravity draining light of its energy was called the *tired light theory*. The main problem with the tired light theory was that it was not supported by the known laws of physics. Calculations showed that gravity would have some effect on light and cause a redshift, but only at a very minor level and certainly not enough to account for Hubble's observations. Zwicky countered by criticising the observations and claiming that they might be exaggerated. True to form, he even questioned Hubble and Humason's integrity, implying that their team might have abused their privilege of controlling the world's best telescope. Zwicky claimed: 'Sycophants among their young assistants were thus in a position to doctor their observational data, to hide their shortcomings.'

Although this sort of outspoken behaviour certainly turned many scientists against Zwicky, there were still a few who joined his tired light brigade. They were not even dissuaded by his apparently faulty physics, because Zwicky had an impeccable track record in research. Indeed, during the course of his career he would go on to do groundbreaking work on supernovae and neutron stars. He even predicted the existence of *dark matter*, a mysterious invisible entity which was initially derided, but which is now widely accepted as real. The tired light theory seemed equally laughable, but perhaps it too would turn out to be right.

The Big Bang supporters, however, rejected the notion of tired light completely. At best, they argued, it could account for only a tiny fraction of the observed redshift. On behalf of the Big Bang camp, Arthur Eddington summarised what he thought was wrong with Zwicky's theory: 'Light is a queer thing – queerer than we imagined twenty years ago – but I should be surprised if it is as queer as all that.' In other words, Einstein's theory of relativity had transformed our understanding of light, but there was still no room for tired light in terms of explaining Hubble's redshifts.

Although Eddington had attacked Zwicky's tired light theory and promoted Lemaître's original paper, he still kept a relatively open mind on the question of the origin of the universe. Eddington thought that Lemaître's ideas were important and worthy of a wider audience, which is why he wrote about them in major journals and helped to translate the Belgian's work, but he was not wholly convinced by the thought of the entire universe being suddenly born out of the decay of a primeval atom: 'Philosophically the notion of a beginning of the present order of Nature is repugnant to me. I should like to find a genuine loophole . . . As a scientist I simply do not believe the Universe began with a bang . . . it leaves me cold.' Eddington felt that Lemaître's model of creation was 'too unaesthetically abrupt'.

In the end, Eddington developed his own variation of Lemaître's model. He was content to start off with a small compact universe, not unlike Lemaître's primeval atom. Then, instead of a sudden expansion, he favoured a very gradual expansion, which eventually accelerated to arrive at the expansion that we see today. Lemaître's expansion was like a bomb exploding suddenly and violently; Eddington's expansion was more like the gradual build-up of an avalanche. A mountain covered in snow might be stable for many months. Then a faint puff of wind causes a snowflake to dislodge an ice crystal, which topples onto another crystal, which rolls and forms a crumb of snow and then a mini-snowball, which gathers more weight, knocking more ice and snow down the slope until sheets of snow start collapsing and a full-blown avalanche is under way.

Eddington explained why he preferred his more gradual build-up to the Big Bang. 'There is at least a philosophical satisfaction in regarding the world as beginning to evolve infinitely slowly from a primitive uniform distribution in unstable equilibrium.'

Eddington also claimed that his version of events could explain something emerging from nothing, thanks to some rather dubious logic. His train of thought began with the premise that the universe had always existed, and if we went back in time far enough then we would discover a perfectly smooth, compact universe, which had itself lasted for an eternity. Next, Eddington argued that such a universe was equivalent to nothing: 'To my mind *undifferentiated sameness* and *nothingness* cannot be distinguished philosophically.' The tiniest imaginable fluctuation in the universe – the equivalent of a snowflake starting an avalanche – would then have fractured the symmetry of the cosmos and triggered a chain of events that led to the full-blown expansion that we see today.

In 1933 Eddington wrote a popular primer, *The Expanding Universe*, which was intended to explain the latest ideas in cosmology in a mere 126 pages. He covered general relativity, Hubble's observations, Lemaître's primeval atom and his own ideas, maintaining a whimsical touch throughout. For example, because all the galaxies are fleeing, Eddington urged astronomers to quickly build better telescopes before the galaxies got too far away to see. In another tongue-in-cheek aside, Eddington turned Hubble's observations inside out: 'All change is relative. The universe is expanding relatively to our common material standard; our material standards are shrinking relatively to the size of the universe. The theory of the "expanding universe" might also be called the theory of the "shrinking atom" . . . Is not the expanding universe another example of distortion due to our egocentric outlook? Surely the universe should be the standard and we should measure our own vicissitudes by it.'

In a more serious vein, Eddington gave an honest summary of the state of the Big Bang model. He pointed out that there were important theoretical reasons and persuasive observational evidence in favour of a moment of creation, but also that there was still a huge amount of work to be done before the Big Bang model could be widely accepted. He called Hubble's redshifts 'too slender a thread on which to hang far-reaching conclusions'. The burden of proof was clearly upon the proponents of the Big Bang model, and he encouraged them to seek out more evidence with which to defend their position.

While the scientific establishment still held to its traditional view of an eternal and largely static universe, the Big Bang supporters prepared themselves for the battle ahead, buoyed to some extent by the knowledge that they were now in a position to hold a mature debate with the conservatives. Cosmology was no longer dominated by myth, religion and dogma, and it was less susceptible to fashion or the force of personality, because the power of twentieth-century telescopes held the promise of observations that might help shore up one theory and destroy another.

Eddington himself was optimistic that some version of the Big Bang model would eventually triumph. Towards the end of his book, he crafted a simple yet compelling image to illustrate the state of the Big Bang model in the early 1930s:

How much of the story are we to believe? Science has its showrooms and its workshops. The public to-day, I think rightly, is not content to wander round the showrooms where the tested products are exhibited; they demand to see what is going on in the workshops. You are welcome to enter; but do not judge what you see by the standards of the showroom. We have been going round a workshop in the basement of the building of science. The light is dim, and we stumble sometimes. About us is confusion and mess which there has not been time to sweep away. The workers and their machines are enveloped in murkiness. But I think that something is being shaped here – perhaps something rather big. I do not quite know what it will be when it is completed and polished for the showroom.

From the Cosmic to the Atomic

In order for the Big Bang model to be accepted, there was one seemingly innocuous question that could not be ignored: why are some substances more common than others? If we look at our own planet, we find that the Earth's core is made of iron, its crust is dominated by oxygen, silicon, aluminium and iron, the oceans are largely made of hydrogen and oxygen (i.e. H_2O, water), and the atmosphere is mainly nitrogen and oxygen. If we venture slightly farther afield, then we find that this distribution is not typical on a cosmic scale. By using spectroscopy to study starlight, astronomers realised that hydrogen was by far the most abundant element in the universe. This conclusion was celebrated by updating a famous nursery rhyme:

> Twinkle, Twinkle little star,
> I don't wonder what you are;
> For by spectroscopic ken,
> I know that you are hydrogen;
> Twinkle, Twinkle little star,
> I don't wonder what you are.

The next most abundant element in the universe is helium, and together hydrogen and helium overwhelmingly dominate the universe. These are also the two smallest and lightest elements, so astronomers were confronted by the fact that the universe consists predominantly of small atoms rather than large atoms. The extent of this bias is highlighted by the following list of cosmic abundances according to the number of atoms. These values are based on current measurements, which are not far from the values estimated in the 1930s:

Element	**Relative abundance**
Hydrogen	10,000
Helium	1,000
Oxygen	6
Carbon	1
All others	less than 1

In other words, hydrogen and helium together accounted for roughly 99.9% of all the atoms in the universe. The two lightest elements were extremely abundant, then the next batch of light or medium-weight atoms were much less common, and finally the heaviest atoms such as gold and platinum were rare indeed.

Scientists began to wonder why there should be these extremes of cosmic abundance between the light and heavy elements. The supporters of the eternal universe model were unable to give a clear answer; their fallback position was that the universe had always contained the elements in their present proportions, and always would. The range of abundances was simply an inherent property of the universe. It was not a very satisfactory answer, but it had a certain self-consistency.

However, the mystery of the abundances was more problematic for supporters of the Big Bang. If the universe had evolved from a moment of creation, why had it evolved in such a way as to generate hydrogen and helium rather than gold and platinum? What was it about the process of creation that preferentially created light elements rather than heavy elements? Whatever the explanation, the Big Bang supporters had to find it and show that it was compatible with the Big Bang model. Any reasonable cosmological theory had to accurately explain how the universe came to be the way it is today, otherwise it would be considered a failure.

Addressing this problem would require a very different approach to any previous cosmic investigation. In the past, cosmologists had concentrated on the very large. For example, they had studied the universe using general relativity, the theory that described the long-range force of gravity between giant celestial bodies. And they used giant telescopes to look at very big galaxies that were very far away. But to tackle the problem of cosmic abundances, scientists would need new theories and new equipment to describe and probe the very, very small.

Before embarking on this part of the Big Bang story, it is first necessary to take a short step back in time and examine the modern history of the atom. The rest of this section tells the story of the physicists who laid the foundations for atomic physics, whose work enabled the Big Bang supporters to investigate why the universe was full of hydrogen and helium.

Attempts to understand the atom took off when chemists and physicists became intrigued by the phenomenon of *radioactivity*, which was discovered in 1896. It became apparent that some of the heaviest atoms, such as uranium, are radioactive, which means that they are capable of spontaneously emitting very high amounts of energy in the form of radiation. For a while, nobody could understand what this radiation was or what caused it.

Marie and Pierre Curie were at the forefront of research into radioactivity. They discovered new radioactive elements, including radium, which is a million times more radioactive than uranium. Radium's radioactive emissions are eventually absorbed by whatever surrounds it, and the energy is converted into heat. In fact, 1 kilogram of radium generates enough energy to boil a litre of water in half an hour and, more impressively, the radioactivity continues almost unabated – so a single kilogram could continue to boil a fresh litre of water every thirty minutes for thousands of years. Although radium releases its energy very slowly compared with an explosive, it eventually releases a million times more energy than the equivalent weight of dynamite.

For many years nobody fully appreciated the dangers associated with radioactivity, and substances such as radium were looked upon with naive optimism. Sabin von Sochocky of the US Radium Corporation even predicted that radium would be used as a domestic power source: 'The time will doubtless come when you will have in your own house a room lighted entirely by radium. The light, thrown off by radium paint on walls and ceiling, would in color and tone be like soft moonlight.'

The Curies both suffered from lesions, but carried on with their research regardless. Their notebooks became so radioactive after years of exposure to radium that today they have to be stored in a lead-lined box. So often were Marie's hands covered with radium dust that her fingers have left invisible radioactive traces on the pages of her notebooks, and a photographic film slipped between the pages can actually record her fingerprints. Marie eventually died of leukaemia.

In many ways, the great sacrifices made by the Curies in their cramped Parisian laboratory served only to highlight the huge lack of understanding as to what was going on inside the atom. Scientists seemed to have gone backwards in their knowledge – just a few decades earlier they had claimed to fully comprehend the building blocks of matter thanks to the periodic table. In 1869,

1 H																	2 He
3 Li	4 Be											5 B	6 C	7 N	8 O	9 F	10 Ne
11 Na	12 Mg											13 Al	14 Si	15 P	16 S	17 Cl	18 Ar
19 K	20 Ca	21 Sc	22 Ti	23 V	24 Cr	25 Mn	26 Fe	27 Co	28 Ni	29 Cu	30 Zn	31 Ga	32 Ge	33 As	34 Se	35 Br	36 Kr
37 Rb	38 Sr	39 Y	40 Zr	41 Nb	42 Mo	43 Tc	44 Ru	45 Rh	46 Pd	47 Ag	48 Cd	49 In	50 Sn	51 Sb	52 Te	53 I	54 Xe
55 Cs	56 Ba	57 La	72 Hf	73 Ta	74 W	75 Re	76 Os	77 Ir	78 Pt	79 Au	80 Hg	81 Tl	82 Pb	83 Bi	84 Po	85 At	86 Rn
87 Fr	88 Ra	89 Ac	104 Rf	105 Db	106 Sg	107 Bh	108 Hs	109 Mt	110 Uun								
		58 Ce	59 Pr	60 Nd	61 Pm	62 Sm	63 Eu	64 Gd	65 Tb	66 Dy	67 Ho	68 Er	69 Tm	70 Yb	71 Lu		
		90 Th	91 Pa	92 U	93 Np	94 Pu	95 Am	96 Cm	97 Bk	98 Cf	99 Es	100 Fm	101 Md	102 No	103 Lr		

Figure 67 The periodic table displays all the chemical elements, the building blocks of matter. They could have been put in a single line, from lightest to heaviest (1 hydrogen, 2 helium, 3 lithium, 4 beryllium, etc.), but this tabular arrangement is far more illuminating. The periodic table groups elements to reflect common properties. For example, the column on the far right contains the so-called noble gases (helium, neon, etc.), whose atoms very seldom react with other atoms to form molecules. Despite its role in helping to understand how the elements reacted with one another, the periodic table did not offer any insight into the cause of radioactivity.

the Russian chemist Dmitri Mendeleev had drawn up a chart that listed all the elements then known, from hydrogen to uranium. By combining the atoms of different elements in the periodic table in various ratios, it was possible to build molecules and explain every material under the Sun, inside the Sun and beyond the Sun. For example, two atoms of hydrogen plus one atom of oxygen made one molecule of water, H_2O. This much still remained true, but the Curies demonstrated that there was a mighty energy source within some atoms, and the periodic table could not explain this phenomenon. Nobody really had a clue about what was actually going on deep inside the atom. Nineteenth-century scientists had pictured atoms as simple spheres, but there had to be something more complicated about the atomic structure to account for radioactivity.

One of the physicists drawn to this problem was a New Zealander, Ernest Rutherford. He was much loved by his colleagues and students, but he was also known as a gruff authoritarian who was prone to temper tantrums and displays of arrogance. For example, according to Rutherford, physics was the only important science. He believed that it provided a deep and meaningful understanding of the universe, whereas all the other sciences were preoccupied with mere measuring and cataloguing. He once stated: 'All science is either physics or stamp collecting.' This blinkered comment backfired when the Nobel Committee awarded him the 1908 chemistry prize.

Figure 68 The portrait of Ernest Rutherford was taken when he was in his mid-thirties. He had a disdain for chemists, which was not uncommon among physicists. For example, Nobel physicist Wolfgang Pauli was angry when his wife left him for a chemist: 'Had she taken a bullfighter then I would have understood, but an ordinary chemist . . . ' The second photograph shows a more mature Rutherford with his colleague John Ratcliffe at the Cavendish Laboratory. The TALK SOFTLY PLEASE sign above their heads was aimed at Rutherford, who had a predilection for singing 'Onward Christian Soldiers' at the top of his voice, disturbing the laboratory's sensitive equipment.

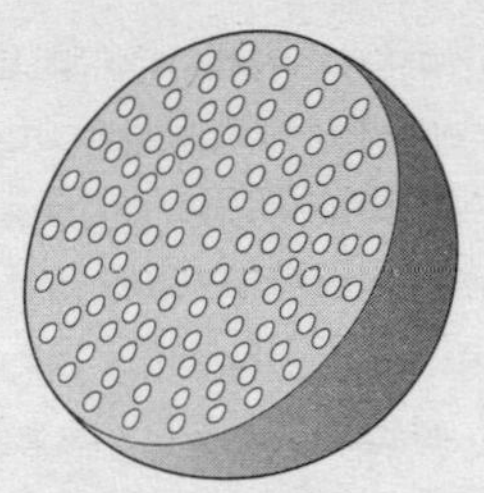

Figure 69 This cross-section shows J.J. Thomson's plum pudding model of the atom, whereby each atom consisted of a number of negative particles (the plums) embedded within a positively charged dough (the pudding). A light hydrogen atom would have one negative particle embedded within a small amount of positive dough, whereas a heavy gold atom would have many negative particles embedded within a larger amount of positive dough.

By the time Rutherford embarked on his research in the early 1900s, the picture of the atom was slightly more sophisticated than the simple, structureless sphere envisioned in the nineteenth century. Atoms were now regarded as containing two ingredients, a positively charged material and a negatively charged one. Opposite charges attract, which was why these materials remained bound within the atom. Then, in 1904, the eminent Cambridge physicist J.J. Thomson offered a refinement that became known as the plum pudding model, in which the atom consisted of a number of negative particles embedded within a positively charged dough-like material, as shown in Figure 69.

One form of radioactivity involved the emission of alpha radiation, which seemed to consist of positively charged particles, known as *alpha particles*. Presumably this could be explained in terms of atoms spitting out bits of positive dough. To test this hypothesis and the whole plum pudding model, Rutherford decided to see what would happen if he took the alpha particles emitted from one set of atoms and fired them into another set of atoms. In other words, he wanted to use alpha particles to probe the atom.

In 1909, Rutherford asked two young physicists, Hans Geiger and Ernest Marsden, to conduct the experiment. Geiger would later become famous for his invention of a radiation detector, the Geiger counter, but for the time being the duo had to make do with only the most primitive equipment. The only way to detect the presence of alpha particles was to place a screen made of zinc sulphide where the alpha particles were supposed to arrive. The alpha particles would emit a tiny flash of light as they struck the zinc sulphide, but seeing the flash would require Geiger and Marsden to have spent thirty minutes adapting their eyes to the absolute darkness. Even then, they still had to view the zinc sulphide screen through a microscope.

A key part of the experiment was a radium sample, which sprayed out alpha particles in all directions. Geiger and Marsden surrounded the radium with a lead shield containing a narrow slit, which turned the spray into a controlled beam of alpha particles. Next they placed a sheet of gold foil in the line of fire to see what would happen to the alpha particles as they hit the gold atoms, as shown in Figure 70.

Alpha particles are positively charged, and atoms are a mixture of negative and positive charges; like charges repel, while unlike charges attract. Therefore, Geiger and Marsden hoped that the interaction between the alpha particles and the gold atoms would reveal something about the charge distribution within the gold atoms. For example, if gold atoms really did consist of negative particles spread through a positive dough, then alpha particles should be deflected only slightly, because they would be encountering a mix of evenly distributed charges. Sure enough, when Geiger and Marsden placed their zinc sulphide screen on the other side of the foil, directly opposite the radium sample, they noticed only a minimal deflection in the path of the alpha particles.

Rutherford then asked for the detector to be moved round to

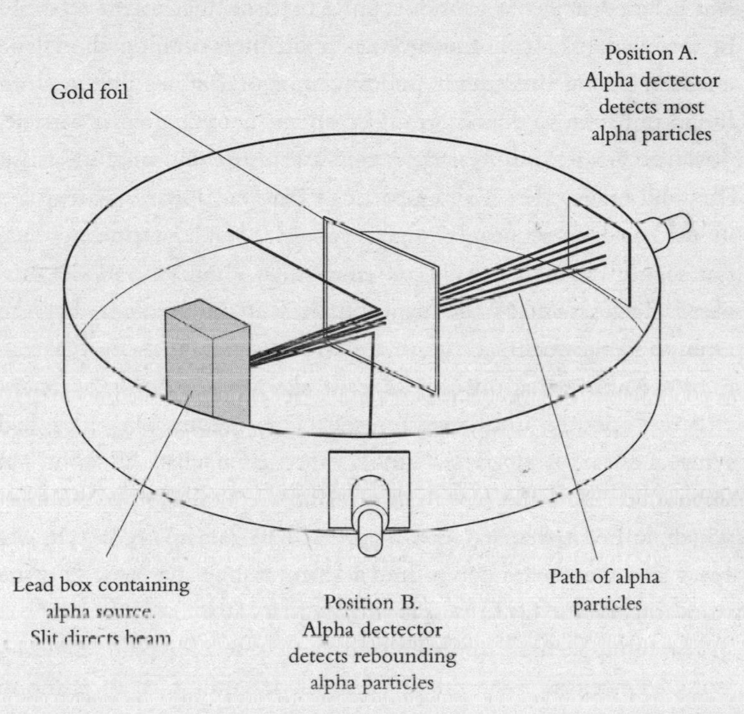

Figure 70 Ernest Rutherford asked his colleagues, Hans Geiger and Ernest Marsden, to study the structure of the atom using alpha particles. Their experiment used a radium sample to provide a source of alpha particles. A slit in a lead shield round the sample directed a beam of alpha particles onto a gold foil, and an alpha detector could be moved to different positions around the gold foil to monitor the deflection of alpha particles.

The vast majority of particles punched their way through the foil with little or no deflection and hit the detector in position A. This is what would be expected if Thomson's plum pudding model were correct, because it envisioned negative particles spread evenly in a positive dough.

However, in some cases the particles bounced back in a most surprising manner, and were picked up by the detector when it was moved to position B. This inspired Rutherford to build a new model of the atom.

the same side of the foil as the radium source 'for the sheer hell of it'. The idea was to look for alpha particles that might rebound off the gold foil. If Thomson was right, then nothing should be detected, because his plum pudding mix of charges in the atom should not have so drastic an effect on an incoming alpha particle. However, Geiger and Marsden were astonished by what they saw. They did indeed detect alpha particles that had apparently recoiled off the gold atoms. Only 1 in every 8,000 alpha particles was bouncing back, but this was one more than Thomson's model predicted. The results of the experiment seemed to contradict the plum pudding model.

To the uninitiated this might seem like just another experiment with a curiously unexpected result. For Rutherford, who had acquired a deep and visceral understanding of what the atom was supposed to look like, it was an utter shock: 'It was quite the most incredible event that has ever happened to me in my life. It was almost as incredible as if you fired a 15-inch shell at a piece of tissue paper and it came back and hit you.'

The result seemed impossible in the context of plum pudding atoms. Hence, the experiment compelled Rutherford to abandon Thomson's model and construct an entirely new model of the atom, one that would account for the rebounding alpha particles. He wrestled with the problem and eventually came up with an atomic structure that seemed to make sense. Rutherford offered a representation of the atom that is still largely valid today.

Rutherford's model concentrated all the positive charge in particles called *protons*, which were positioned at the centre of the atom, in a region dubbed the *nucleus*. The negatively charged particles, called *electrons*, orbited the nucleus, and were bound to the atom by the force of attraction between their negative charges and the positive charges within the nucleus, as shown in Figure 71. This model was sometimes called the planetary model of the atom, because the electrons orbited the nucleus just as the planets orbit the Sun. Electrons and protons have equal and opposite charges, and each atom contains the same number of electrons and protons, so Rutherford's atom had an overall charge of zero, which is to say that it was neutral.

The number of protons and electrons is crucial, because it defines the type of atom, and it is this number that appears next to each atom in the periodic table (Figure 67, p. 287). Hydrogen is labelled with the atomic number 1, because its atoms have one electron and one proton; helium has the atomic number 2, because its atoms have two electrons and two protons; and so on.

Rutherford suspected that the nucleus also contained a type of chargeless particle, and he would later be proved right; the *neutron* has almost the same mass as the proton, but it has no charge. As explained in Figure 71, the number of neutrons in the nucleus can vary, but as long as the number of protons in an atom stays the same, then it is still an atom of the same type of element. For example, most hydrogen atoms have no neutrons, but some have either one or two neutrons, and are called deuterium and tritium respectively. Plain hydrogen, deuterium and tritium are all forms of hydrogen because they all contain one proton and one electron; they are known as *isotopes* of hydrogen.

Although atoms vary in size depending on the number of protons, neutrons and electrons they possess, they are generally slightly smaller than one-billionth of a metre in diameter. However, Rutherford's scattering experiment suggested that the atomic nucleus has a diameter that is 100,000 times smaller still. In terms of volume, the atomic nucleus represents just $(1/100{,}000)^3$ or 0.0000000000001% of the entire atom.

This is extraordinary: atoms, which make up everything that is solid and tangible in the world around us, consist almost entirely of empty space. If a single hydrogen atom were enlarged to

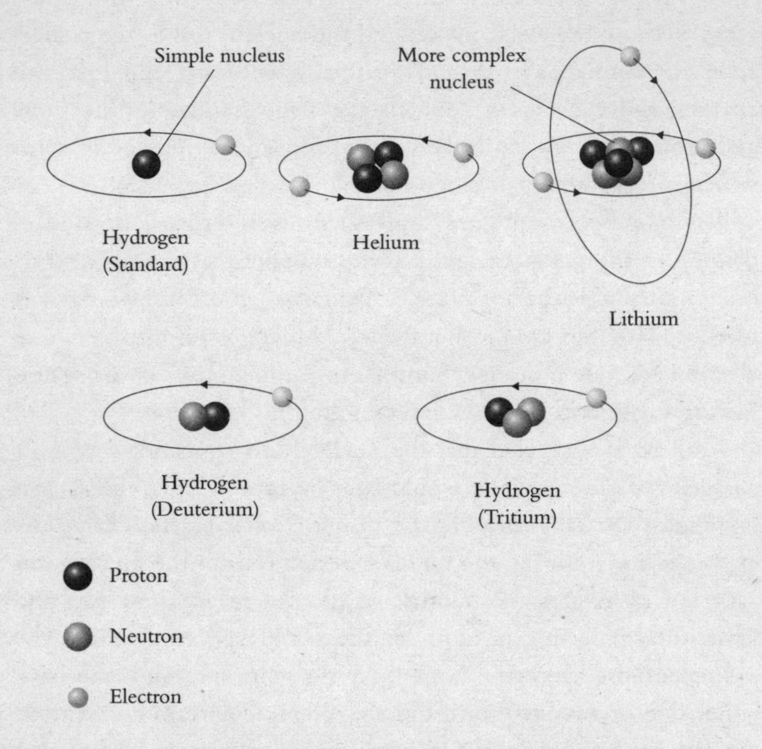

Figure 71 Rutherford's model of the atom had the positively charged protons concentrated in a central nucleus, surrounded by the orbiting, negatively charged electrons. These diagrams are not drawn to scale, because the diameter of a nucleus is roughly 100,000 times smaller than the diameter of the atom. The number of protons equals the number of electrons, and this *atomic number* is the same for all atoms of a particular element and determines its position in the periodic table (Figure 67). Hydrogen atoms have one electron and one proton, helium atoms have two electrons and two protons, lithium atoms have three electrons and three protons, and so on.

The number of neutrons in the nucleus can vary, but as long as the number of protons stays the same it is still considered to be an atom of the same chemical element. For example, most hydrogen atoms have no neutrons, but some have one neutron and are called deuterium, and others have two neutrons and are called tritium. Plain hydrogen, deuterium and tritium are said to be isotopes of hydrogen.

completely fill a concert hall, such as London's Royal Albert Hall, the nucleus would be the size of a flea, in the midst of the hall's vast emptiness, yet it would dwarf the even smaller electron hovering somewhere in the hall. Also, the proton and the neutron each weigh almost 2,000 times more than the electron, and the protons and neutrons reside in the infinitesimally small nucleus, so at least 99.95% of an atom's mass is squeezed into just 0.0000000000001% of its volume.

This revised atomic model provided a perfect explanation for the results of Rutherford's experiment. Because the bulk of an atom is empty space, then the vast majority of alpha particles would pass through the gold foil with only a minor deflection. However, a small fraction of positively charged alpha particles would have a head-on collision with the concentration of positive charge in an atomic nucleus, and this would cause a drastic rebound. These two forms of interaction are illustrated in Figure 72. Initially, the results of Rutherford's experiment had seemed shockingly impossible, but with a revised model everything seemed obvious. Rutherford once said: 'All of physics is either impossible or trivial. It is impossible until you understand it, and then it becomes trivial.'

Only one problem remained: there was still no evidence for the existence of Rutherford's neutrons, which were supposed to sit with the protons in the atomic nucleus. This missing piece of the atomic jigsaw was hard to pin down because the neutron was electrically neutral, unlike the positively charged proton and the negatively charged electron. James Chadwick, one of Rutherford's protégés, set out to prove its existence. He became so obsessed with the brand-new science of *nuclear physics* that he even continued research during his four years as a prisoner of war in Germany during the First World War. He knew that a certain brand of toothpaste contained radioactive thorium – it was supposed to give teeth a brilliant glow – and he managed to scrounge some from the guards

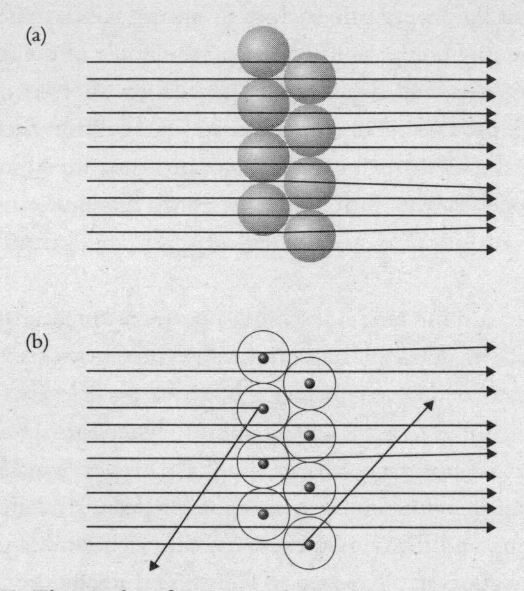

Figure 72 The results of Geiger and Marsden's experiment showed that a small fraction of alpha particles rebounded back when striking a gold foil. This makes no sense in the context of the Thomson plum pudding model. Diagram (a) shows a gold foil made of plum pudding atoms. The positive dough sprinkled with negative plum particles has a very even distribution of charge, so the positively charged alpha particles are hardly deflected.

Diagram (b) shows a gold foil made of Rutherford's atoms, which does explain the rebounding of alpha particles. In this model, the positive charge was concentrated in a central nucleus. Most alpha particles remain undeflected, because most of the atom is empty. However, if an alpha particle strikes the concentrated positive charge of a nucleus, it is deflected quite markedly.

---

so that he could experiment with it. Chadwick did not make much progress with his toothpaste experiments, but he returned to his laboratory after the war, toiled for another decade, and eventually discovered the atom's missing ingredient in 1932. In fact, the open door seen on the left in Figure 68 (p. 288) led to the laboratory in which James Chadwick discovered the neutron.

Armed with a proper understanding of the atom's structure and components, physicists could at last explain the underlying cause of the radioactivity that had been studied by Pierre and Marie Curie. Every atomic nucleus was made up of individual protons and neutrons, and these ingredients could be swapped around to transform one nucleus into another nucleus, thereby transforming one atom into another atom. This was the mechanism behind radioactivity.

For example, the nuclei of heavy atoms, such as radium, are very large. Indeed, the radium nuclei studied by the Curies contained 88 protons and 138 neutrons, and such large nuclei are often unstable and therefore liable to transform into smaller nuclei. In the case of radium, the nucleus spits out a pair of protons and a pair of neutrons in the shape of an alpha particle (which also happens to be the nucleus of a helium atom), thus transforming itself into a radon nucleus consisting of 86 protons and 136 neutrons, as shown in Figure 73. The process whereby a large nucleus is split into smaller nuclei is called *fission*.

Although we normally associate nuclear reactions with very heavy nuclei, they are also possible with very light nuclei such as hydrogen. It is possible to transform hydrogen nuclei and neutrons into helium by merging them in a process called *fusion*. Hydrogen is relatively stable, so this process does not occur spontaneously, but given the right conditions of high temperature and pressure then hydrogen will fuse into helium. The incentive for hydrogen to fuse into helium is that helium is even more stable than hydrogen, and there is always a tendency for nuclei to seek the greatest possible stability.

In general, the most stable atoms are the ones found in the middle of the periodic table, such as iron, and these are also the ones with middling numbers of protons and neutrons in their nuclei. Therefore, while the very largest of nuclei might undergo

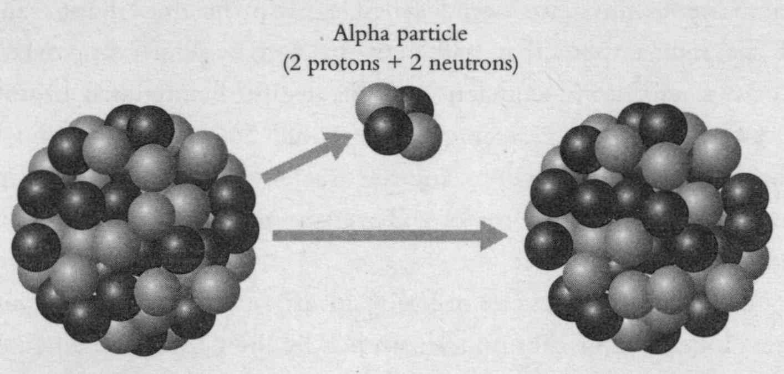

Figure 73 There are various isotopes of radium, but this particular nucleus is the most common and it is called radium-226, because it consists of 88 protons and 138 neutrons, making a total of 226 particles. The radium nucleus is large and therefore highly unstable, so it undergoes fission and ejects two neutrons and two protons in the shape of an alpha particle, transforming itself into a smaller radon nucleus, which is itself rather unstable.

---

fission and the very smallest of nuclei might undergo fusion, the vast majority of the medium-sized nuclei virtually never undergo any kind of nuclear reaction.

Although this explains how nuclear reactions work, and why radium is radioactive (and iron is not), it does not explain why the Curies detected such huge amounts of energy when radium underwent fission. Nuclear reactions are notorious for the amount of energy they release, but where does it come from?

The answer lies in Einstein's special theory of relativity, and in one particular aspect which was not covered in Chapter 2. When Einstein analysed the speed of the light and realised its implications for space and time, he also derived the most famous equation in physics, namely $E = mc^2$. In essence, this says that energy (E) and mass (m) are equivalent and can be transformed into each other with a conversion factor of c^2, where c is the speed of light. The speed of light is 3×10^8 m/s, so c^2 is 9×10^{16} (m/s)^2, which means that a tiny amount of mass can be converted into a huge amount of energy.

And, indeed, the energy released during nuclear reactions comes directly from converting tiny amounts of mass into energy. When a radium nucleus is transformed into a radon nucleus and an alpha particle, the combined mass of the products is less than the mass of the radium nucleus. The loss in mass is only 0.0023%, so 1 kg of radium would be converted into 0.999977 kg of radon and alpha particles. Although the mass loss is tiny, the conversion factor (c^2) is huge, so the missing 0.000023 kg is converted into more than 2×10^{12} joules of energy, which is equivalent to the energy from over 400 tonnes of TNT. Energy is released in fusion in exactly the same way, except that the amount of energy released is generally even greater. A hydrogen fusion bomb is far more devastating than a plutonium fission bomb.

It has been a while since astronomy or cosmology was mentioned in this chapter, but it has been important to introduce the breakthroughs that were made in atomic and nuclear physics, because they were destined to play a crucial role in testing the Big Bang model. Rutherford's nuclear model of the atom and the understanding of nuclear reactions (fission and fusion) that emerged from it set the stage for a new way of studying the heavens. Before returning to our main subject, here is a recap of the key points that emerged out of nuclear physics:

1. Atoms consist of electrons, protons and neutrons.
2. Protons and neutrons occupy the atom's centre, i.e. the nucleus.
3. Electrons orbit the atomic nucleus.
4. Large nuclei are often unstable and can split (fission).

5. Small nuclei are more stable, but can be made to merge (fusion).
6. The nuclei after fission/fusion weigh less than the initial nuclei.
7. Thanks to $E = mc^2$, this mass reduction leads to an energy release.
8. Medium nuclei are the most stable, rarely undergoing reactions.
9. Even very light or very heavy nuclei sometimes need high energies and pressures before they will undergo fusion or fission.

One of the first scientists to link these rules of nuclear physics with astronomy was a courageous and principled physicist named Fritz Houtermans, well known for his charm and wit. He is possibly the only physicist whose jokes have been collated and published in a forty-page booklet. Houtermans' mother was half-Jewish, and he sometimes countered anti-Semitic remarks by retorting: 'When your ancestors were still living in the trees, mine were already forging cheques!'

Houtermans was born in 1903 in Zoppot, near what was then the German Baltic port of Danzig, known today as Gdansk, in Poland. His parents moved to Vienna, where Houtermans spent his childhood, and from there he moved back to Germany to study physics at Göttingen in the 1920s, where he went on to obtain a post as a researcher. Working alongside the British scientist Robert d'Escourt Atkinson, he became fascinated with the notion that nuclear physics could be used to explain how the Sun and other stars were fuelled.

It was known that the Sun consisted mainly of hydrogen and partly of helium, so it seemed natural to assume that the energy generated by the Sun was the result of nuclear reactions whereby hydrogen was fusing into helium. Nobody had observed nuclear fusion on Earth, so the details of the mechanism were uncertain. But it was known that if hydrogen could somehow be transformed into helium, there would be a 0.7% loss in mass: 1 kg of hydrogen would somehow be fused into 0.993 kg of helium, resulting in a mass loss of 0.007 kg. Again, this may seem a small loss in mass, but Einstein's formula $E = mc^2$ explains how even a seemingly small loss of mass can result in an immense amount of energy:

$$\text{Energy} = mc^2 = \text{mass} \times (\text{speed of light})^2$$
$$= 0.007 \times (3 \times 10^8)^2 = 6.3 \times 10^{14} \text{ joules}$$

So, in theory, 1 kg of hydrogen could be fused into just 0.993 kg of helium and generate 6.3×10^{14} joules of energy, which is equal to the energy generated by burning 100,000 tonnes of coal.

The main question that bothered Houtermans was whether or not the conditions in the Sun were extreme enough to trigger fusion. It was mentioned earlier that fusion reactions cannot happen spontaneously, and require high tempertures and pressures to occur. This is because they need an initial input of energy to trigger the reaction. In the case of fusing two hydrogen nuclei, this energy is necessary to overcome an initial repulsion. A hydrogen nucleus is a proton with a positive charge, so it will repel another hydrogen nucleus with its positive charge, because like charges repel. However, if the protons can get sufficiently close to each other, then there is an attractive force, known as the *strong nuclear force*, which will overpower the repulsion and securely bind them together to form helium.

Houtermans calculated that the critical distance was 10^{-15} metres, which is one-trillionth of a millimetre. If two approaching hydrogen nuclei could get this close to each other then fusion would take place. Houtermans and Atkinson were convinced that the pressure and temperature in the deep interior of the Sun were great enough to force the hydrogen nuclei to within this critical distance of 10^{-15} metres, which would result in fusion, thereby releasing energy to maintain the temperature and encourage further fusion. They published their ideas on stellar fusion in 1929 in the journal *Zeitschrift für Physik*.

Houtermans was convinced that he and Atkinson were on the right track to explaining why the stars shine, and was so proud of his research that he could not help boasting about it to a girl that he was dating. He later recounted the exchange that took place the night after he had completed his research paper on stellar fusion:

> That evening, after we had finished our paper, I went for a walk with a pretty girl. As soon as it grew dark the stars came out, one after another, in all their splendour. 'Don't they shine beautifully?' cried my companion. But I simply stuck out my chest and said proudly: 'I've known since yesterday why it is that they shine.'

Charlotte Riefenstahl was clearly impressed. She later married him. Houtermans, however, had developed only a partial theory of stellar fusion. Even if it were possible for the Sun to fuse two hydrogen nuclei into a helium nucleus, it would only be a very light and unstable isotope of helium – stable helium requires two more neutrons to be added to the nucleus. Houtermans was confident that the neutron existed, and indeed was present in the Sun, but it had yet to be discovered when he published his 1929 paper with Atkinson. Houtermans was therefore largely ignorant of the neutron's various properties and was unable to complete his calculations.

When the neutron was eventually discovered by Chadwick in 1932, Houtermans was in an ideal position to fill in the details of his theory, but politics soon intervened. He had been a member of the Communist Party and feared that he would become a victim of Nazi persecution. In 1933 he fled Germany for Britain, where neither the culture nor the food was to his taste. He said he could not tolerate the ever-present odour of boiled mutton and called England 'the domain of the salted potatoes'. At the end of 1934 he left for the Soviet Union. According to his biographer Iosif Khriplovich, his emigration was driven by 'idealism and English cooking'.

Houtermans' work progressed well at the Ukrainian Physico-Technical Institute until Stalin instigated a purge of the scientific community. Having fled the Nazis, Houtermans was now under the absurd suspicion of being a Nazi spy and was arrested by the NKVD, the Soviet secret police, in 1937. For the next three years he was either locked in a cramped cell along with more than a hundred other prisoners, or he was being questioned and pressured into an admission of guilt. Houtermans was interrogated for up to eleven days continuously, during which time he was deprived of sleep and forced to stand throughout. The Nazi–Soviet pact led to his release in 1940, but he was arrested immediately by the Gestapo and grilled once again. He was in the uniquely unpleasant position of being able to compare and contrast the NKVD and the Gestapo: 'The NKVD is the more serious organisation. When I was being interrogated by the Gestapo, the examiner kept my file open in front of him. But I can read upside down. The NKVD would never make such a blunder.'

During Houtermans' detention in the late 1930s, other physicists picked up on his ideas about stellar fusion and calculated the exact details of the processes that were taking place in the Sun. The man most responsible for completing Houtermans' research was Hans Bethe, who had been dismissed from his job at the University of Tübingen in 1933 because his mother was Jewish. He found sanctuary, first in Britain and then in America, eventually becoming head of the theoretical division at Los Alamos, home of the nuclear bomb project.

Bethe identified two nuclear routes for turning hydrogen into helium that were feasible given the temperatures and pressures then thought to prevail in the Sun. In one route, standard hydrogen (one proton) reacted with deuterium, a rarer and heavier isotope of hydrogen (one proton and one neutron). This formed a relatively stable isotope of helium containing two protons and one neutron.

Next, two of these light helium nuclei would fuse together to form a standard and stable helium nucleus, releasing two hydrogen nuclei as a by-product. This process is shown in Figure 74.

Bethe's other proposed route for turning hydrogen into helium employed a carbon nucleus as a way of trapping hydrogen nuclei. If the Sun contained a small amount of carbon, then each carbon nucleus could capture and swallow hydrogen nuclei one at a time,

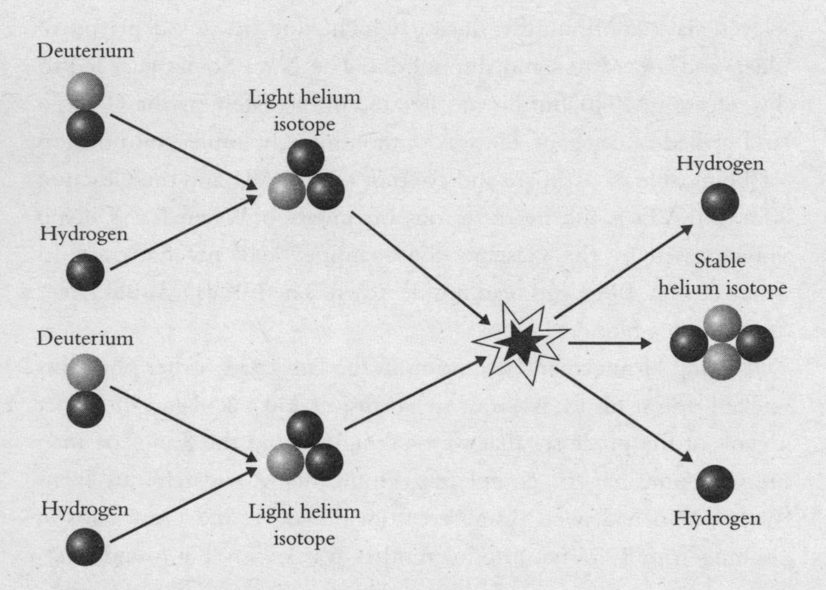

Figure 74 This diagram shows one of the ways in which hydrogen can be converted into helium in the Sun. The dark spheres represent protons and the pale spheres represent neutrons. In the first stage of the reaction, standard hydrogen and deuterium fuse to form helium. Helium usually has two protons and two neutrons, but this isotope has two protons and only one neutron. In the second stage, two of the light helium nuclei fuse to form the stable isotope of helium, releasing two hydrogen nuclei (protons) in the process. These hydrogen nuclei can go on to form further helium atoms. In theory, two deuterium nuclei (one proton and one neutron each) could fuse directly to form a stable helium nucleus (two protons and two neutrons). However, deuterium nuclei rarely interact with each other, so the indirect route is more productive.

transforming itself into increasingly heavy nuclei. Eventually, the transformed carbon nucleus would become unstable, causing it to spit out a helium nucleus and convert itself back into a stable carbon nucleus, whereupon the process would start all over again. In other words, the carbon nucleus acts as a factory, using hydrogen nuclei as its raw material and churning out helium nuclei.

These two nuclear reaction routes were initially speculative, but other physicists checked the equations and confirmed that the reactions were viable. At the same time, astronomers became more certain that the Sun's internal environment was intense enough to initiate the nuclear reactions. By the 1940s it became clear that both of Bethe's proposed nuclear reactions were taking place in the Sun and were responsible for generating its energy. Astrophysicists could envisage exactly how the Sun converted 584 million tonnes of hydrogen into 580 million tonnes of helium each second, transforming the missing mass into sunshine energy. Despite this massive rate of consumption, the Sun will continue to generate energy for billions of years to come, as it currently contains roughly 2×10^{27} tonnes of hydrogen.

This was a milestone in the relationship between the atomic and the cosmic. Nuclear physicists had proved that they could make a concrete contribution to astronomy by explaining how the stars shone. Now, Big Bang cosmologists hoped that nuclear physics could help them tackle an even bigger question: how did the universe evolve into its current state? It was now clear that stars could turn simple atoms such as hydrogen into slightly heavier atoms such as helium, so perhaps nuclear physics could show how the Big Bang created the various abundances of the atoms we see today.

The stage was set for the arrival of a new pioneer in cosmology. He would be a scientist capable of applying the rigorous rules of nuclear physics to the speculative realm of the Big Bang. By straddling the disciplines of nuclear physics and cosmology, he would establish a make-or-break test for the Big Bang model of the universe.

The First Five Minutes

George Gamow was a gregarious Ukrainian-born maverick with a penchant for hard drinking and card tricks. Born in Odessa in 1904, he showed an interest in science from an early age. He became fascinated by a microscope given to him by his father and used it to analyse the process of transubstantiation. Having attended Communion at the local Russian Orthodox church, he dashed home with a piece of bread and a few drops of wine secreted in his cheeks. He put them under the microscope and compared what he saw with everyday bread and wine. He could find no evidence that the structure of the bread had transformed into the body of Christ, and he later wrote: 'I think this was the experiment that made me a scientist.'

Gamow made a name for himself as an ambitious young physicist at Odessa's Novorossia University, and then in 1923 he went to study in Leningrad with Alexander Friedmann, who at the time was still developing his nascent Big Bang theory. Gamow's interests diverged from those of Friedmann, and he rapidly made world-class discoveries in nuclear physics. His research prompted the state-owned newspaper *Pravda* to dedicate a poem to him when he was just twenty-seven years old. Another newspaper proclaimed: 'A Soviet fellow has shown the West that Russian soil can produce her own Platos and sharp-witted Newtons.'

Gamow, however, was becoming disaffected with life as a Soviet academic. The state would use the Marxist-Leninist philosophy of dialectical materialism to dictate whether scientific theories were valid or invalid, leading to periods when Soviet scientists were supposed to acknowledge the existence of the discredited ether and deny the tried and tested theory of relativity. Using politics to determine scientific truth was absurd to a freethinker like Gamow, and he grew to despise the Soviet attitude to science and indeed the whole of Communist ideology.

Consequently, in 1932, Gamow attempted to escape the Soviet Union by fleeing across the Black Sea to Turkey. It turned out to be a thoroughly amateurish escape bid. He and his wife, Lyubov Vokhminzeva, attempted to paddle their way to freedom across the 250 kilometres of water in a tiny kayak. He told the story in his autobiography:

> An important item was the food supply for the trip, which, we figured, would last five or six days . . . We hardboiled [some eggs] and saved them for the trip. We also managed to get several bricks of hard cooking chocolate, and two bottles of brandy, which turned out to be very handy when we were wet and cold at sea . . . One thing we found out was that it was rational to take turns in paddling, rather than paddling together, since in the latter case the speed of the boat did not increase by a factor of two . . . The first day was a complete success . . . I'll never forget the sight of a porpoise seen through a wave illuminated by the sun sinking below the horizon.

But after thirty-six hours their luck changed. The weather turned against them, and they were forced to paddle back to the bosom of the Soviet Union.

Gamow made another failed attempt, this time across Arctic waters from Murmansk to Norway. Then, in 1933, he adopted a new strategy. Having been invited to the Solvay Conference for physicists in Brussels, Gamow managed to arrange a meeting with senior politburo member Vyacheslav Molotov to seek special permission for his wife, also a physicist, to accompany him. He obtained the necessary papers, but only after a lengthy bureaucratic

Figure 75 Snapshots of George Gamow and his wife, Lyubov Vokhminzeva, and a picture of the Gamows as they prepared for their failed bid to flee the Soviet Union by paddling across the Black Sea in a kayak.

battle. The couple went off to the conference with no intention of ever returning to the Soviet Union. In due course they journeyed from Europe to America, and in 1934 Gamow joined George Washington University, where he spent the next two decades exploring, testing and defending the Big Bang hypothesis.

In particular, Gamow was interested in the Big Bang in relation to *nucleosynthesis* – the formation of atomic nuclei. Gamow wanted to see whether nuclear physics and the Big Bang could explain the observed atomic abundances. As we have seen, for every 10,000 atoms of hydrogen in the universe there are roughly 1,000 atoms of helium, 6 atoms of oxygen and 1 atom of carbon, and all the atoms of all the other elements put together are even less numerous than carbon atoms. Gamow wondered whether the early moments of the Big Bang could be responsible for our universe being dominated by hydrogen and helium. And he wondered whether the Big Bang could account for the various abundances of the heavier atoms, which are comparatively rare yet so vital for life.

Before looking at Gamow's research, let us recall Lemaître's view of nucleosynthesis. His universe started as a single, supermassive, primeval atom, the mother of all other atoms: 'The atom world broke up into fragments, each fragment into still smaller pieces. Assuming, for the sake of simplicity, that this fragmentation occurred in equal pieces, we find that two hundred and sixty successive fragmentations were needed in order to reach the present pulverization of matter into poor little atoms which are almost too small to be broken farther.' Based on the established principle that large nuclei are unstable, a supermassive atom would be highly unstable and would indeed split into lighter atoms. However, the debris would probably settle somewhere in the middle of the periodic table, which is where the most stable elements are found. This would lead to a universe dominated by elements such as iron. In Lemaître's model there seemed to be no way of creating the atoms of hydrogen and helium so abundant in today's universe. As far as Gamow was concerned, Lemaître was just plain wrong.

Spurning Lemaître's top-down approach, Gamow instead adopted a bottom-up strategy. What would happen if the universe started as a dense, compact soup of simple hydrogen atoms that expanded outwards? Could the Big Bang have created the right conditions for hydrogen to fuse into helium and the other heavier atoms? This seemed more likely than Lemaître's idea, because starting with 100% hydrogen was a more obvious way to explain why it still accounted for 90% of the atoms in today's universe.

But before he began to speculate on the nuclear physics of the

Figure 76 George Gamow discussing a calculation with John Cockcroft (left), who would win a Nobel prize for his contribution to nuclear physics. The pictures capture the intensity and joy of physicists at work.

Big Bang, Gamow studied the work of Houtermans and Bethe to find out exactly what stars were capable of in terms of fusing hydrogen into heavier atoms. He was struck by two key limitations of stellar fusion. First, the rate of stellar helium production was inordinately slow. Our Sun creates 5.8×10^8 tonnes of helium each second, which may sound a lot, but the Sun currently contains 5×10^{26} tonnes of helium. At the rate of stellar helium production, it would have taken over 27 billion years to make this amount of helium, yet the universe was supposed to be just 1.8 billion years old according to the Big Bang model. Gamow therefore concluded that the majority of helium must already have been present when the Sun was being formed, so perhaps it was created in the Big Bang.

The other limitation of stellar fusion was its apparent inability to create atoms of elements much heavier than helium. Physicists failed dismally to find any viable stellar nuclear route to elements such as iron or gold. Stars seemed to be a dead end in terms of creating anything but the lightest atoms.

Gamow took these two limitations as opportunities for the Big Bang model to prove itself by making up for stellar inadequacies. Where the stars failed to create enough helium or any heavier elements, perhaps the Big Bang could succeed. In particular, he hoped that the conditions in the early universe were sufficiently extreme to permit new types of nuclear reaction and open novel pathways that were not possible in the stars, which would then explain the creation of all the elements. If Gamow could link the Big Bang to the nucleosynthesis of heavy elements, it would be strong evidence in favour of the Big Bang model. If he could not, this ambitious theory of creation would be faced with a major embarrassment.

It was the early 1940s when Gamow embarked on his research project to explain the creation of elements in the wake of the Big Bang. He soon realised that he was just about the only physicist in America exploring the question of Big Bang nucleosynthesis, and he soon worked out why he had the privilege of having the entire field to himself. Working on the formation of nuclei required a deep understanding of nuclear physics, and almost everybody with this sort of background had been secretly recruited to work on the Manhattan Project at Los Alamos, designing and building the first atomic bombs. The only reason that Gamow had not been whisked away from George Washington University was that he failed to gain

Front Row { I. Joliot, A. Joffe, P. Langevin, E. Rutherford, M. De Broglie, L. Meitner
E. Schrödinger, N. Bohr, M. Curie, O. Richardson, T. De Donder, L. De Broglie, J. Chadwick

Figure 77 This group photo of the 1933 Solvay Conference in Brussels includes George Gamow (back row, centre), who engineered his escape from the Soviet Union by attending this conference. The conference was devoted to discussing the structure of atoms, so the photo includes many other notable figures. Ernest Rutherford and James Chadwick are seated in the front row, along with Marie Curie and her daughter Irene Joliot, who like her mother won a Nobel prize.

Pierre Curie had been killed many years earlier when he was hit by a horse-drawn wagon in 1906. Marie then started a relationship with Paul Langevin, who is in the photograph next to her. Langevin was still married, which led to a public scandal. When Curie received notice of her second Nobel prize she was asked not to come to Stockholm to collect her prize in person, because of the embarrassment it might cause to the Nobel committee. She ignored the request, explaining that the prize was presumably a reward for her science and not her personal life.

the highest level of security clearance, because he had once been a commissioned officer in the Red Army. Those responsible for issuing clearance failed to appreciate that Gamow had been given officer status merely so he could teach science courses to soldiers. Neither did the American authorities pick up on more obvious signs of Gamow's true loyalty, such as the fact that the Soviets had sentenced him to death in absentia for fleeing the USSR.

Gamow's strategy for exploring Big Bang nucleosynthesis was superficially simple. He started with observations of the universe as it is now. Astronomers had examined the distribution of stars and galaxies, so they could estimate the density of matter throughout the cosmos, which is roughly one gram per thousand Earth volumes. Next, Gamow took Hubble's measurement of the expansion of the universe and ran the clock backwards so that the universe was contracting. Gamow's contracting universe would become increasingly dense as it approached the moment of creation, and he could use relatively simple mathematics to work out the average density at any moment in the past. Compressing material usually generates heat, which is why a bicycle pump compressing air feels warm after just a few strokes. Therefore, Gamow could also use relatively simple physics to show that the younger, compressed universe would have been much hotter than today's universe. In short, Gamow found that he could easily work out the temperature and density of the universe at any point in time from soon after its creation (hot and dense) right up to the present day (cool and spread out).

Establishing the conditions that prevailed in the early universe was critical, because the outcome of any nuclear reaction depends almost entirely on density and temperature. The density dictates the number of atoms in a given volume, and the higher the density, the greater likelihood of two atoms colliding and fusing. And as the temperature increases, there is more energy available and the atoms move faster, which also means that their nuclei are more likely to fuse. It was only because astrophysicists knew the temperature and density inside the Sun that they could work out which nuclear reactions occurred inside stars. Gamow, with similar information about the early universe, hoped that he could work out which nuclear reactions took place soon after the Big Bang.

Gamow's first step in his research into modelling Big Bang nucleosynthesis was to assume that the extreme heat of the very early universe would have broken all matter down into its most elementary form. So he assumed that the initial components of the universe would have been separate protons, neutrons and electrons, the most fundamental particles known to physicists at the time. He called this mix *ylem* (pronounced 'eye-lem'), a word he stumbled upon in Webster's Dictionary. This obsolete Middle English word means 'the primordial substance from which the elements were formed' – a perfect description of Gamow's hot soup of neutrons, protons and electrons. A single proton is equivalent to a hydrogen nucleus, and with the addition of an electron it becomes a complete hydrogen atom. However, the early universe was so hot and so full of energy that the electrons were all moving far too fast to attach themselves to any nucleus. In addition to the particles of matter, the early universe contained a turbulent sea of light.

Starting from this hot, dense soup, Gamow wanted to run the clock forwards and, tick by tick, work out how the fundamental particles might begin to stick together and form the nuclei of the familiar atoms that exist today. Ultimately, his ambition was to show how these atoms would coalesce and form stars and galaxies, evolving into the universe we see around us. In short, Gamow wanted to prove that the Big Bang model could explain how we had arrived at where we are today.

Unfortunately, as soon as he started to calculate the nuclear reactions that might have taken place, Gamow was struck by the sheer magnitude of the gargantuan task that lay ahead of him. He could have coped with calculating the nuclear reactions that would have taken place under a specific set of conditions, but the problem with the Big Bang scenario was that it was constantly evolving. At one moment in time there would be a specific temperature, density and mix of particles, but a second later the universe would have expanded, resulting in a cooler temperature, a lower density and a slightly different mix of particles, depending on the nuclear reactions that might already have taken place. Gamow struggled with the nuclear calculations, making very little progress. He was a great physicist but a weak mathematician, and the nuclear calculations were beyond him. This was also an era when computers were effectively non-existent, so they could not come to his rescue.

Eventually, in 1945, Gamow received some much-needed support when he took on a young student by the name of Ralph Alpher, who was struggling to establish himself in the scientific community. Alpher's academic career had started promisingly in 1937, when, as a sixteen-year-old prodigy, he received a scholarship to the Massachusetts Institute of Technology. Unfortunately, while chatting to one of the institute's alumni, he casually mentioned that his family was Jewish – and the scholarship was promptly withdrawn. It was a terrible shock for an aspiring teenager: 'My brother had told me not to get my hopes up and he was damn right. It was a searing experience. He said it was unrealistic to think that a Jew could go anywhere back then.'

The only way that Alpher could get back on the academic track was by holding down a day job and attending evening classes at George Washington University, where he eventually completed his bachelor's degree. It was during this period that Gamow met Alpher and took a shine to him, possibly because Alpher's father was from Odessa, his own birthplace. Gamow recognised that Alpher was mathematically talented and had a good eye for detail, which contrasted with his own mathematical failings and rather slapdash attitude. He immediately took Alpher on as his doctoral student.

Gamow set Alpher to work on the problem of nucleosynthesis in the early universe, presenting his student with a starting point and an outline of the key issues based on what he had gleaned so far. For example, Gamow pointed out that Big Bang nucleosynthesis could be confined to a relatively short window of time and temperature. The very early universe was so hot and so energetic that the protons and neutrons were travelling too fast to stick to one another. A little later, the universe was cool enough for nucleosynthesis to commence. However, after a little more time had elapsed the universe's temperature would have dropped to the point where protons and neutrons no longer had enough energy or speed to initiate nuclear reactions. In short, nucleosynthesis could take place only when the universe was cooler than trillions of degrees and hotter than millions of degrees.

Another restriction on the window for nucleosynthesis was the fact that neutrons are unstable and decay into protons, unless they are trapped within a nucleus such as helium. Hence the free neutrons in the early universe had to form nuclei before they disappeared. Free neutrons have a so-called half-life of roughly 10 minutes, which means that half of them disappear within 10 minutes, half of those remaining disappear in another 10 minutes, and so on. Therefore, less than 2% of the original neutrons would be left one hour after the moment of creation, unless the neutrons had already reacted with protons to form stable nuclei. On the other hand, there is a temperature-dependent nuclear reaction that can create neutrons, which further complicates the situation. Because neutrons are a vital ingredient in nucleosynthesis, both the neutron half-life and the rate of neutron creation were critical factors in determining the amount of time during which nucleosynthesis could take place after the Big Bang.

Concentrating on this complex time window for nucleosynthesis, Gamow and Alpher began to estimate the likelihood of protons and neutrons interacting. One of the inputs into their calculations, and another complicating factor, was the *cross-section* for neutrons and protons. A particle's cross-section is an indication of how big a target it presents to other particles. If two people stand on opposite sides of a room and throw tiny marbles at each other, it is unlikely that the marbles will collide in mid-air. If, instead, they throw footballs at each other, there will be a much greater likelihood of two footballs colliding, or at least glancing off each other. So footballs have a bigger cross-section than marbles. The critical question in terms of nucleosynthesis was this – how big a cross-section or target do neutrons and protons present to each other?

Nuclear particle cross-sections are measured in *barns*, and 1 barn equals 10^{-28} square metres. The name is an ironic coinage from expressions such as 'couldn't hit a barn door'; some etymologists suggest that the term was first used as a code by physicists working on the Manhattan Project, so that spies overhearing mentions of barns would not be able to tell what was meant. Understanding cross-sections had been crucial to the bomb-makers, who were trying to work out how much uranium they would have to amass in order to create a nuclear explosion. The higher the cross-section for interactions in uranium, the greater the likelihood of nuclear interactions and the less uranium would be required to guarantee a nuclear explosion.

Importantly for Alpher, the secrecy surrounding the atomic bomb project was declining in the years immediately after the war. This meant that valuable cross-section measurements were in the process of being declassified just as Alpher was embarking on his research into Big Bang nucleosynthesis. Another boost came from scientists at the Argonne National Laboratory, who had been exploring the possibility of building a nuclear power station.

Alpher was delighted when they too released their latest data on nuclear cross-sections.

Gamow and Alpher spent three years working through their calculations, questioning their assumptions, updating their cross-sections and refining their estimates. Some of their deepest conversations took place in Little Vienna, a bar on Pennsylvania Avenue, where one or two drinks would help them to make sense of the early universe. This was an extraordinary adventure. They were applying concrete physics to a previously vague Big Bang theory, attempting to mathematically model the conditions and events of the early universe. They were estimating initial conditions and applying the laws of nuclear physics to see how the universe evolved with time and how the processes of nucleosynthesis progressed.

As each month passed, Alpher became increasingly convinced that he could accurately model the formation of helium in the few minutes after the Big Bang. His confidence increased when he found that his calculations agreed closely with reality. Alpher estimated that there should be roughly one helium nucleus for every ten hydrogen nuclei at the end of the Big Bang nucleosynthesis phase, which is exactly what astronomers observed in the modern universe. In other words, the Big Bang could explain the ratio of hydrogen to helium that we see today. Alpher had not yet seriously attempted to model the formation of other elements, but even predicting the formation of hydrogen and helium in the observed proportions was in itself a highly significant achievement. After all, these two elements accounted for 99.99% of all the atoms in the universe.

Several years earlier, astrophysicists had been able to show that the stars fuelled themselves by turning hydrogen into helium, but the rate of stellar nuclear reaction was so slow that stellar nucleosynthesis could account for only a tiny fraction of the helium known to exist. Alpher, however, could explain the abundance of helium by assuming that there had been a Big Bang. This result was the first major triumph for the Big Bang model since Hubble had observed and measured the redshifts of galaxies.

Keen to announce their breakthrough, Gamow and Alpher set out their calculations and conclusions in a formal paper entitled 'The Origin of Chemical Elements', and submitted it to the journal *Physical Review*. It was due for publication on 1 April 1948, and perhaps this was what spurred Gamow to do something he had been secretly considering for many months. Gamow was a close friend of Hans Bethe, who was famous for his work on stellar nuclear reactions, and he wanted to add Bethe's name to the list of authors, even though he had contributed nothing to this particular research paper. His motivation for adding the extra name was that readers could enjoy the sight of a paper authored by Alpher, Bethe and Gamow, a pun on the Greek letters alpha (α), beta (β) and gamma (γ).

Not surprisingly, Alpher took exception. He feared that crediting Bethe would diminish how the rest of the world perceived his own contribution to the research. Alpher's name was already overshadowed by Gamow's co-authorship, because Alpher was the young Ph.D. student and Gamow the famous physicist, and adding Bethe's even more eminent name would only make things worse for him. Alpher had done more than his fair share of the work, and now it seemed that he was going to receive only a tiny fraction of the credit. Throughout this authorship tussle between Gamow and Alpher, Bethe remained unaware of Alpher's strength of feeling and had no idea that this would be one of the most important scientific papers in the history of cosmology. He was simply happy to be part of one of Gamow's little japes.

As soon as the paper was sent off for publication, with Bethe's name still in place, Gamow tried to patch up the quarrel with his student by arranging a small celebration to mark their great achievement. Gamow brought a bottle of Cointreau into the office, its label doctored to read 'Ylem', his word for the primordial soup of particles that first filled the universe. Pouring the orange liqueur out of the bottle and into a couple of glasses became a playful recreation of the Big Bang.

Although Gamow could now relax a little, Alpher still had plenty of work to do. This research was Alpher's Ph.D. project, so he had to write it up independently and explain it in excruciating detail to demonstrate that he was truly worthy of a doctorate. Unfortunately, he was struck by a severe case of mumps soon after he started to write his thesis. Aching and swollen, Alpher had to complete his thesis from his bed, dictating it to his wife, Louise. The couple had met while they were both attending evening classes at George Washington University, but Louise was studying psychology, not physics, so she was largely baffled by Alpher's research. Nevertheless, she dutifully and accurately typed up the abstruse equations that formed the core of his thesis.

Alpher's work was still not complete. Next he had to undergo the ordeal of defending his thesis, the final hurdle on the journey to earning his doctorate. He would have to sit alone in front of a panel of experts and convince them that hydrogen and helium could have been created in the correct proportions in the moments after the Big Bang. He also wanted to argue that there was a reasonable chance that other atoms could have been created during this phase. Essentially, he was going to defend the results of his collaboration with Gamow, but relying solely on his own wits, unable to turn to his mentor for advice. If he succeeded, then he would be awarded his Ph.D. If he failed, then he would have wasted three years. His thesis defence was scheduled for the spring of 1948.

Such thesis defences are often public occasions, but they are not generally considered to be a spectator sport with mass appeal, so the audience tends to be just friends, close family and a few academics with a particular interest in the subject. In this case, however, news

Figure 78 The famous cartoonist Herbert L. Block ('Herblock') showed an interest in Alpher's research. This cartoon, which appeared in the *Washington Post* on 16 April 1948, shows an atomic bomb musing over the news that the world was created in five minutes. The bomb seems to be having the mischievous thought that it could destroy the world in just five minutes.

that a twenty-seven-year-old novice had made a major breakthrough had spread across Washington, and Alpher found himself arguing his case before a packed audience of three hundred people, including newspaper reporters. They listened intently to the baffling series of questions and Alpher's even more arcane answers. At the end of his defence, the examiners were sufficiently convinced to award Alpher his doctorate.

Meanwhile, reporters had taken special note of one of Alpher's comments – that the primordial nucleosynthesis of hydrogen and helium had taken only 300 seconds. And that was what made the headlines in newspapers all across America over the next few days. On 14 April 1948, the *Washington Post* announced WORLD BEGAN IN 5 MINUTES, which then inspired a cartoon in the same paper two days later, shown in Figure 78. On 26 April *Newsweek* ran the same story, but stretched the timescale to account for the creation of other varieties of atoms: 'According to this theory, all the elements were created out of a primordial fluid in a single hour, and have been reshuffling themselves into the material of stars, planets and life ever since.' In fact, Alpher had said very little about elements heavier than hydrogen and helium.

For the next few weeks, Alpher enjoyed a degree of celebrity. Academics showed interest in his work, a curious public sent him fan mail and religious fundamentalists prayed for his soul. However, the spotlight soon faded and, as he anticipated, he became lost in the shadow of his illustrious co-authors, Gamow and Bethe. When physicists read the paper they assumed that Gamow and Bethe were responsible for the breakthrough, and Alpher's name was overlooked. The spurious addition of Bethe's name for comic effect had extinguished any possibility that Alpher would receive proper recognition for his crucial role in the development of the Big Bang model.

Divine Curves of Creation

The Alpha–Beta–Gamma paper, as it became known, was a milestone in the Big Bang versus eternal universe debate. It showed that it was possible to do real calculations relating to the nuclear processes that might have occurred after a hypothetical Big Bang, and thus test this theory of creation. Big Bang supporters could now point to two pieces of observational evidence, the expansion of the universe and the abundances of hydrogen and helium, and show that they were entirely consistent with the Big Bang model of the universe.

Critics of the Big Bang theory fought back by trying to undermine the supposed success of Big Bang nucleosynthesis. Their first reaction was to dismiss the agreement between Gamow and Alpher's calculations and the observed helium abundance as mere coincidence. A second and more substantial criticism was aimed at Gamow and Alpher's failure to explain the creation of nuclei heavier than hydrogen and helium.

Gamow and Alpher had largely put this problem to one side in their published paper, intending to address it later, but in fact they soon realised that their research had reached a dead end: trying to synthesise any nuclei that were heavier than helium in the heat of the Big Bang looked to be almost impossible.

Their greatest difficulty was the so-called 5-nucleon crevasse. A *nucleon* is the generic term for any component of the nucleus, which means that it covers both protons and neutrons. Thus:

common hydrogen contains 1 proton + 0 neutrons	= 1 nucleon
deuterium hydrogen contains 1 proton + 1 neutron	= 2 nucleons
tritium hydrogen contains 1 proton + 2 neutrons	= 3 nucleons
common helium contains 2 protons + 2 neutrons	= 4 nucleons

The next heaviest nucleus would contain five nucleons, but such a nucleus cannot exist because it is inherently unstable, a result of the complicated way that nuclear forces interact. However, beyond the unstable 5-nucleon nucleus is a whole range of stable nuclei, such as carbon (usually 12 nucleons), oxygen (usually 16 nucleons) and potassium (39 nucleons).

To get a feel for why the number of nucleons determines the stability and existence of certain nuclei (and the instability and non-existence of others), we can consider the situation of vehicles and their stability in relation to how many wheels they have. One-wheeled unicycles exist, as do two-wheeled bicycles, three-wheeled tricycles and four-wheeled cars. Five-wheeled vehicles, however, are virtually non-existent, because the fifth wheel would be pointless and, if anything, it might be detrimental to the vehicle's stability and performance. However, one more wheel improves balance and spreads the vehicle load, and many lorries do indeed have six or more wheels. Similarly, but for different reasons, 1-nucleon, 2-nucleon, 3-nucleon, 4-nucleon and 6-nucleon nuclei are all stable, but a 5-nucleon nucleus is effectively forbidden.

But why was the lack of a 5-nucleon nucleus so disastrous for Gamow and Alpher? It turned out to be an apparently unbridgeable crevasse across the road of nucleosynthesis that led to heavier nuclei such as carbon and beyond. The path of transformation that turns a light nucleus into a heavier one contains one or more intermediate steps, and if one of the intermediate steps is not allowed then the entire path is blocked. The obvious path to heavier nuclei would start by adding a proton or a neutron to a helium nucleus (4 nucleons) to create a 5-nucleon nucleus – but this was exactly the type of nucleus that was not allowed. Therefore the path to heavier nuclei was blocked.

One solution would be for a helium nucleus to simultaneously absorb both a neutron and a proton, thereby skipping the unstable

5-nucleon nucleus and transforming directly into a stable 6-nucleon lithium nucleus (three protons and three neutrons). However, the chances of a proton and a neutron simultaneously hitting a helium nucleus in exactly the right way were vanishingly small. Even one nuclear reaction caused by one collision is hard to induce, so it was too much to expect a reaction caused by two collisions happening at almost exactly the same moment.

Another way of skipping the 5-nucleon step would be for two 4-nucleon helium nuclei to merge and create an 8-nucleon nucleus, but this nucleus is also inherently unstable for the same sort of reasons that the 5-nucleon nucleus is unstable. Nature had annoyingly contrived to block the two most obvious paths by which light nuclei might transform into heavier ones.

Figure 79 The Hungarian-born physicist Eugene Wigner tried unsuccessfully to find alternative routes to get from helium across the 5-nucleon crevasse to carbon and beyond. George Gamow drew a cartoon to illustrate one of Wigner's failed pathways. Gamow's caption explained: 'Another ingenious method of crossing the mass 5 crevasse was proposed by E. Wigner. It is known as the method of the nuclear chain bridge.'

Gamow and Alpher persevered. They refined their calculations with the latest neutron lifetime and cross-section data. Also, the calculations in their original paper had relied on nothing more than an electrified Marchant & Friden desk calculator, but now they brought the latest developments in computing to bear on the problem. They obtained a Reeves analogue computer, which they then upgraded to a magnetic drum storage computer. Then they invested in an IBM programmable punchcard calculator and finally a SEAC, an early digital computer.

The good news was that their estimate of the hydrogen and helium abundances remained accurate. Even independent calculations by academic rivals, as shown in Figure 80, confirmed that the relative amounts of hydrogen and helium created in the early universe were in rough agreement with the ratio observed in the current universe. The bad news was that the refined calculations still showed no hint of a mechanism for resolving the problem of creating nuclei heavier than helium.

While the nucleosynthesis of heavy atoms was running into problems, Alpher began to work on another aspect of the Big Bang theory, alongside a colleague by the name of Robert Herman. Alpher and Herman had much in common. Both were sons of Russian Jewish émigrés who had settled in New York, and both were still young researchers trying to make a name for themselves. When Herman overheard snippets of cosmological discussions between Alpher and Gamow, he could not resist becoming involved in their research. The idea of making calculations that related to the earliest moments of the universe was simply too much of a temptation.

Alpher and Herman started their new collaboration by revisiting the early history of the universe according to the Big Bang model. The earliest phase was pure chaos, with too much energy around for any significant evolution of matter. The next few minutes were

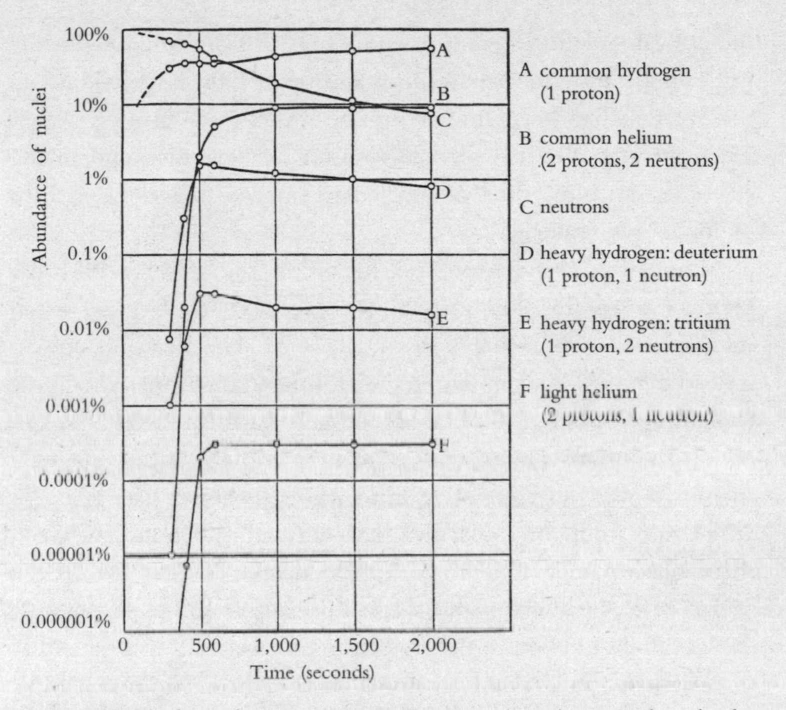

Figure 80 Nuclear physicists Enrico Fermi and Anthony Turkevich also calculated the abundances of the elements in the early universe. Their results agreed with Gamow and Alpher and are shown in this graph, which illustrates the chemical evolution of the universe during its first 2,000 seconds.

The number of neutrons is continually falling as they decay into protons, which is why the number of protons (equivalent to hydrogen nuclei) is increasing. Another reason for the decline in neutrons is that they are incorporated in helium nuclei, and the abundance of helium is continually increasing, making it the second most abundant nucleus in the universe. The other nuclei represented on the graph are other hydrogen and helium isotopes created on the path from common hydrogen to common helium.

Astronomers measured the present-day abundances of deuterium and tritium (heavy hydrogen isotopes), and these measurements were consistent with the predictions made by Gamow, Alpher, Fermi and Turkevich. This was a further endorsement of the Big Bang model, which could now explain the abundances of the lightest nuclei in the universe as a result of nuclear reactions that took place during the hot, dense period that followed the Big Bang. Gamow called the lines in this graph the 'divine curves of creation'.

the critical Goldilocks era – not too hot and not too cool, just the right temperature to form helium and other light nuclei. This was the era that had been studied in the Alpha–Beta–Gamma paper. Thereafter, the universe was too cool for further fusion and, in any case, the unstable 5-nucleon nucleus seemed to block the path to building heavier nuclei.

Although it was now too cool for fusion, the universe still had a temperature of roughly a million degrees, which resulted in all matter existing in a state known as *plasma*. The first and coolest state of matter is solid, in which the atoms and molecules are tightly locked together, as in ice. The second and warmer state is liquid, in which the atoms or molecules are only loosely linked, allowing them to flow, as in water. The third and even hotter state is gas, in which the atoms or molecules have virtually no bonds between them, allowing them to move independently, as in steam. In the fourth state of matter, plasma, the temperature is so hot that atomic nuclei cannot hold on to their electrons, so that matter is a mixture of unattached nuclei and electrons, as shown in Figure 81. Most people are unaware of the plasma state, even though many of us create plasmas every day by switching on a fluorescent light tube, which turns the gas inside into a plasma.

So, an hour after its creation the universe was still a plasma soup of simple nuclei and free electrons. The negatively charged electrons would try to latch on to the positively charged nuclei because opposite charges attract, but they were simply moving too fast to settle into orbits around the nuclei. Instead, the nuclei and electrons bounced off one another over and over again, and the plasma state persisted.

The universe also contained one more ingredient, namely an overwhelming sea of light. Surprisingly, however, being present at the birth of the universe would not have been a very illuminating experience, because it would have been impossible to see anything.

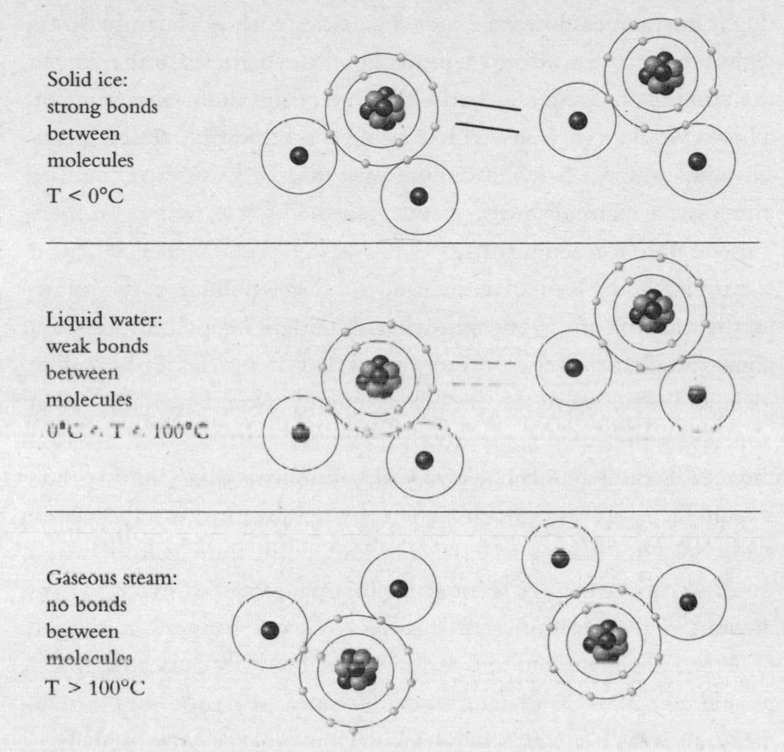

Figure 81 These four diagrams represent the four states of matter, using water as an example. Water is H_2O, each molecule consisting of two hydrogen atoms connected to an oxygen atom. These molecules can be bonded to each other to form a solid, but heat energy can weaken these bonds to create a liquid, or it can break them to form a gas. Further heat energy can strip the electrons from the nuclei to create a plasma.

Light interacts easily with charged particles, such as electrons, so the light would have scattered repeatedly off the particles in the plasma, resulting in an opaque universe. Because of this multi-scattering, the plasma would have behaved like a fog. It is impossible to see the car ahead of you in a fog, because the light from it is scattered countless times by the fine droplets of water, so the light is redirected many times before it reaches you.

Alpher and Herman continued to develop their early history of the universe and wondered what else might happen to this sea of light and plasma as the universe expanded with time. They realised that as the universe expanded, its energy would become spread through a greater volume, so the universe and the plasma within it would steadily cool. The two young physicists deduced that there would be a critical moment when the temperature would become too cool for a plasma to exist, at which point the electrons would latch on to nuclei and form stable, neutral atoms of hydrogen and helium. The transition from plasma to atoms happens at roughly 3,000°C for hydrogen and helium, and the duo estimated that it would take 300,000 years or so for the universe to cool to this temperature. This event is generally known as *recombination* (which is a little confusing because it implies that the electrons and nuclei had previously been combined, which was not the case).

After recombination, the universe became full of gaseous neutral particles, because the negatively charged electrons had combined with the positively charged nuclei. This dramatically changed the behaviour of the light that filled the universe. Light interacts easily with charged particles in a plasma, but not with neutral particles in a gas, as shown in Figure 82. Hence, according to the Big Bang model, the moment of recombination was the first time in the history of the universe that rays of light could start to sail through space unhindered. It was as though the cosmic fog had suddenly lifted.

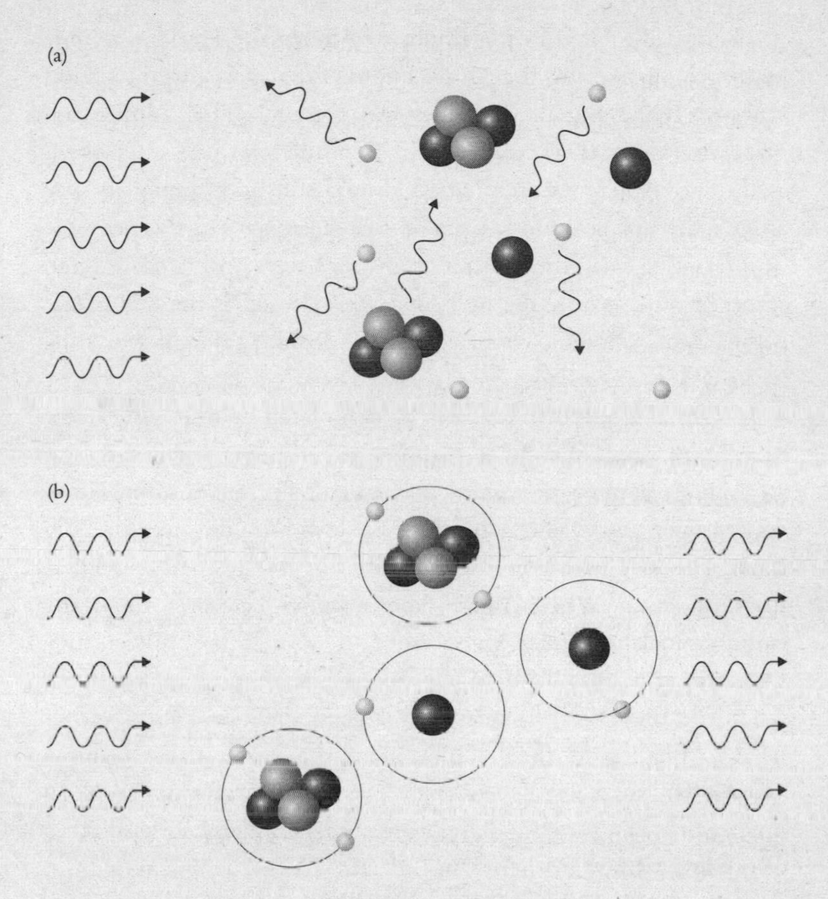

Figure 82 The moment of recombination is a critical milestone in the history of the early universe, according to the Big Bang model. Diagram (a) illustrates conditions in the universe during the first 300,000 years after the Big Bang, when everything was plasma. The light rays would be continually scattered by the particles they encountered, because many of the particles were charged, and this enabled the scattering process. Diagram (b) illustrates conditions during the period after recombination, when the universe had cooled sufficiently to allow hydrogen and helium nuclei to capture electrons and form stable atoms. Because atoms are neutral, there were no unattached charges to enable the scattering of light. The universe was therefore transparent to light, and the rays passed through the cosmos unhindered.

The fog also lifted in the minds of Alpher and Herman as they began to appreciate the implications of a post-recombination universe. If the Big Bang model was correct, and if Alpher and Herman had got their physics right, then the light that was present at the moment of recombination should still be beaming its way around the universe today, because that light was largely incapable of interacting with the neutral atoms that were sprinkled through space. In other words, the light that was released at the end of the plasma epoch should currently exist as a fossil. This light would be a legacy of the Big Bang.

Alpher and Herman's research, completed within just a few months of the Alpha–Beta–Gamma paper being published, was arguably even more important than calculating the transformation of hydrogen into helium in the first few minutes after the Big Bang. The original paper was brilliant, but it was open to accusations of fixing. When Alpher and Gamow had performed the earlier calculation, they knew from the outset the answer they were trying to find, namely the observed helium abundance. So, when the theoretical calculation matched the observation, critics tried to undermine their achievement by claiming that Gamow and Alpher had steered their calculation in the right direction. In other words, the anti-Big-Bang campaigners unfairly accused them of fiddling with their theory in order to get the desired result, just as Ptolemy had fiddled with the epicycles to match the retrograde motion of Mars.

In contrast, the remnant light from 300,000 years after creation could in no way be interpreted as an ad hoc postdiction. There could be no accusations of fiddling. This luminous echo was a clear prediction based solely on the Big Bang model, so Alpher and Herman had provided a make-or-break test. Detecting this light would provide powerful evidence that the universe really did start with a Big Bang. Conversely, if the light did not exist, then the Big Bang could not have happened, and the entire model would collapse.

Alpher and Herman estimated that the sea of light released at the moment of recombination had a wavelength of roughly one-thousandth of a millimetre. This wavelength was a direct consequence of the temperature of the universe when the plasma fog cleared, which was 3,000°C. However, all these light waves would have been stretched because the universe has been expanding ever since recombination. This was similar to the stretching and redshift of light from the apparently receding galaxies, which had already been measured by astronomers such as Hubble. Alpher and Herman confidently predicted that the stretched Big Bang light should now have a wavelength of roughly a millimetre. This wavelength is invisible to the human eye, and is located in the so-called microwave region of the spectrum.

Alpher and Herman were making a specific prediction. The universe should be full of a feeble microwave light with a wavelength of one millimetre, and it should be coming from all directions because it had existed everywhere in the universe at the moment of recombination. Anybody who could detect this so-called *cosmic microwave background radiation* (CMB radiation) would prove that the Big Bang really happened. Immortality was waiting for whoever could make the measurement.

Unfortunately, Alpher and Herman were completely ignored. Nobody made any serious effort to search for their proposed CMB radiation.

There were various reasons why the academic community shunned the prediction of CMB radiation, but first and foremost was the interdisciplinary nature of the research. Gamow's team had been applying theoretical nuclear physics to cosmology to provide a prediction that required the detection of microwaves in order to test it. The ideal person to test the prediction of CMB radiation was therefore someone with an interest and expertise in astronomy, nuclear physics and microwave detection, but there were very few people with such a breadth of knowledge.

Even if a scientist did have the necessary range of skills, he would be unlikely to believe that it was technically possible to detect the CMB radiation, because microwave technology was still relatively primitive. And if by chance he was optimistic about the technical challenge, then he was probably sceptical about the premise behind the project. The majority of astronomers had not accepted the Big Bang model of the universe, and clung to their conservative view of an eternal universe. Hence, why should they bother to look for a CMB radiation that apparently emerged from a Big Bang that might never have happened? Alpher later recalled how he, Herman and Gamow spent the next five years trying hard to persuade astronomers that their work was worth taking seriously: 'We expended a hell of a lot of energy giving talks about the work. Nobody bit; nobody said it could be measured.'

To compound their problems, Alpher, Herman and Gamow suffered from an image problem. They were often looked down upon as two young upstarts led by a joker. Gamow was infamous for his limericks and his sometimes offbeat application of physics. On one occasion, he argued that God lived 9.5 light years from the Earth. This estimate relied on the fact that in 1904, at the outbreak of the Russo-Japanese War, churches across Russia had offered prayers requesting the destruction of Japan, but it was not until 1923 that Japan was struck by the Kanto earthquake. Presumably prayers and God's wrath were limited by the speed of light, and the time delay indicated the distance to the Lord's abode. Gamow also became famous for *Mr Tompkins in Wonderland*, a book in which he described a world where the speed of light was just a few kilometres per hour, so that a bicycle ride would reveal the weird effects of relativity, such as time dilation and length contraction.

Figure 83 Robert Herman (left) and Ralph Alpher (right) created a montage of themselves with Gamow, along with the bottle of ylem used to celebrate the submission of the Alpha–Beta–Gamma paper. Alpher smuggled the image into a set of slides, which meant that Gamow was as surprised as the audience when it suddenly appeared on the screen during a lecture that he was giving at Los Alamos in 1949. Gamow is shown as a genie escaping from the bottle, along with the primordial ylem soup.

Unfortunately, some rivals viewed this approach to popularisation as childish and trivial. Alpher summarised their predicament: 'Because he wrote on physics and cosmology at a popular level and because he injected a considerable amount of humour into his presentations, he was frequently not taken seriously by too many of his fellow scientists. His not being taken seriously is something that rubbed off on the two of us as his colleagues, particularly because we were working in such a speculative area as cosmology.'

Faced with the overwhelming apathy that greeted their work, the three men reluctantly brought their research programme to a close in 1953, when they published a final paper summarising their work and latest calculations. Gamow moved into other areas of research, including a dalliance with the chemistry of DNA. Alpher left academia and became a researcher at General Electric, while Herman joined General Motors Research Laboratories.

The departure of Gamow, Alpher and Herman was symptomatic of the sorry state of Big Bang cosmology. After a few encouraging years, the Big Bang model faced a pair of awkward problems. First, based on the redshifts of the galaxies, the age of the Big Bang universe was less than the age of the stars it contained, which was clearly nonsensical. Second, attempts to build atoms out of the Big Bang had hiccupped at helium, which was embarrassing because this implied that the universe should not contain any oxygen, carbon, nitrogen or any other heavy elements. But although the outlook was grim, the Big Bang was not yet a lost cause. The model could be salvaged and its credibility boosted if somebody could detect the cosmic microwave background radiation predicted by Alpher and Herman. Unfortunately, nobody could be bothered to look for it.

Meanwhile, the situation for those who supported the idea of an eternal universe was looking more positive. They were about to fight back with their own revamped model. A team of cosmologists based in Britain were developing a theory that not only gave rise to an eternal universe, but was also capable of explaining Hubble's observations of redshifts. This new model of an eternal universe was to become the greatest rival to the Big Bang model of creation.

Plus ça change, plus c'est la même chose

Fred Hoyle was born in Bingley on 24 June 1915. He was a Yorkshireman, a cosmologist, a rebel and a creative genius. He would prove to be the most formidable and aggressive critic of the Big Bang model, and would make a huge contribution towards our understanding of the universe.

Hoyle showed his talent for observation and deduction at an early age. When he was just four years old, he worked out for himself how to tell the time through a process of detailed analysis. Fred noticed that when one of his parents asked the time, the other would look at the grandfather clock before answering. So Fred began to ask the time over and over again to find out what was going on. One evening he was sent to bed having been told that it was 'twenty past seven', and in the moments before falling asleep he solved the mystery:

> An idea suddenly occurred to me. Could it be that the 'time', instead of being a mysterious number unknown to me called 'twenty past seven', was really two separate numbers, twenty and seven? . . . There were two hands on the clock. Perhaps one number belonged to one hand and the other number belonged to the other hand. A few more repetitions of the question 'What's the time?' the following day showed that this was indeed so. Because the numbers on the clock face were big and clear, it was easy now to see there were two sets of them. One hand went with one set and the other hand went with the other set. Refinements remained, like the meaning of 'past' and 'to', but, to all intents and purposes, the problem was solved and I could turn to other puzzling things, like what made the wind blow.

Fred preferred teaching himself about the world, so he was a regular truant from school and occasionally abandoned the classroom for several weeks at a time. In his autobiography he recalled the day that a teacher attempted to teach him about Roman numerals, a lesson that seemed utterly pointless when Arabic numerals were so much more sensible and ubiquitous: 'This was more than I could reasonably stomach, and the day this outrage to the intelligence was perpetrated became my last at that particular school.' At another school, Fred brought a flower into the classroom to prove that it had more petals than the teacher had stated the previous day. The teacher responded by slapping him for being insolent. Not surprisingly, Fred walked out again and never returned.

Young Fred seemed to spend more time in his local fleapit cinema than in the classroom. He made up for some of the lessons that he was missing by studying the captions on the silent movies: 'I learned to read while patronising the bug hole in the Hippodrome cinema . . . a superior educational establishment . . . and, at 1d per admission, a good deal cheaper than school.'

When he was a few years older, Fred developed an interest in astronomy. His father, an uneducated cloth merchant, would often walk with him to a neighbouring town to visit a friend who had a telescope. There they would spend the night studying the stars, returning home early in the morning. Fred's early fascination with astronomy was reinforced at the age of twelve when he read Arthur Eddington's *Stars and Atoms*.

Eventually Hoyle was persuaded to give the British education system a chance. He settled down at Bingley Grammar School and then embarked on a traditional academic path. In 1933 he won a scholarship to Emmanuel College, Cambridge, where he studied mathematics. He excelled and won the Mayhew Prize, which is given to the best student in applied mathematics. After graduation he earned a place as a Ph.D. student at Cambridge, working alongside such greats as Rudolf Peierls, Paul Dirac, Max Born and his hero, Arthur Eddington. After earning his doctorate in 1939 he

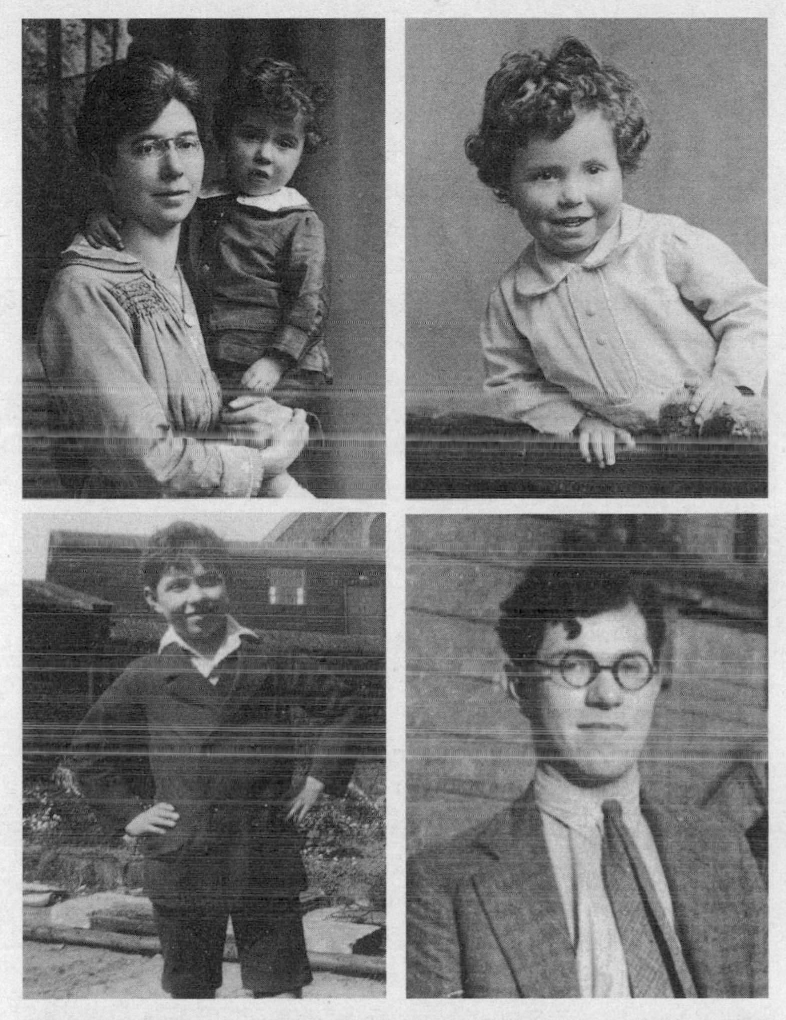

Figure 84 The picture of Fred Hoyle as a baby with his mother was kept by his father as he fought in the trenches of the First World War. Commenting on the picture that showed him as a toddler with his teddy bear, Hoyle later described himself as 'evidently persuaded in the mistaken belief that the world is a better place than I ever subsequently found it to be'. The photograph of Hoyle aged about ten shows him at the height of his truanting phase, while the final photograph shows him as a young student at Cambridge.

was elected a fellow of St John's College, and his research began to focus on the evolution of stars.

Hoyle's academic progress was then suddenly interrupted: 'War would change everything. It would destroy my comparative affluence, it would swallow my best creative period, just as I was finding my feet in research.' He was initially sent to work in the Admiralty Radar Group near Chichester, and in 1942 he was promoted to section leader in the Admiralty Signal Establishment at Witley in Surrey, where he continued to conduct radar research. It was here that he met Thomas Gold and Hermann Bondi, who shared his interest in astronomy. In the years that lay ahead, the collaboration of Hoyle, Bondi and Gold would become as famous as that of their great American rivals, Gamow, Alpher and Herman.

Bondi and Gold, who had both grown up in Vienna and then studied at Cambridge together, shared a house close to the Admiralty's research laboratory. Hoyle would often spend several nights a week with them, as his own home was 80 kilometres away and he hated having to commute. After a day of intense research into building better radar systems, the three men would often relax at home by holding mini-seminars on the subjects that had interested them before the outbreak of war.

In particular, they became fascinated by Hubble's observations of an expanding universe and its implications. Whenever they tackled the subject of cosmology, each of them would take a particular role. Bondi, who had a talent for mathematics, provided a logical foundation for their discussions and would manipulate the equations that emerged. Gold, who was more scientifically inclined, would provide a physical interpretation for Bondi's equations. Hoyle, who was the senior figure, guided the speculation. According to Gold:

> Fred Hoyle kept urging us – what could the Hubble expansion mean? That was always Hoyle's challenge to us. Fred would have Bondi sit cross-legged on the floor, then sit behind him in an arm-chair and kick him every five minutes to make him scribble faster, just as you might whip a horse. He would sit there and say, 'Now come on, do this, do that,' and Bondi would calculate with furious speeds, though *what* he was calculating was not always clear to him – as on the occasion he asked Fred, 'Now at this point do I multiply or divide by 10^{46}?'

After the war, Hoyle, Bondi and Gold pursued separate careers in astronomy, mathematics and engineering, respectively, but they all lived in Cambridge and continued with their part-time cosmological brainstorming. Hoyle and Gold would regularly convene at Bondi's house and discuss the pros and cons of the two competing theories of the universe: the Big Bang model and the eternal static model. Their discussions were heavily biased against the Big Bang, partly because it indicated that the universe was younger than the stars that were in it, and partly because nobody had any idea about what came before the Big Bang. At the same time, all three of them had to admit that Hubble's observations did imply an expanding universe.

Then, in 1946, the Cambridge trio suddenly made a break-through. They concocted a radically new model of the universe. Their model was extraordinary because it seemed to make an impossible compromise: it described a universe that was expanding but which was still truly eternal and essentially unchanging. Until this point, cosmic expansion had been synonymous with a Big Bang moment of creation, but the new model suggested that Hubble's redshifts and the receding galaxies could also be allied with the traditional view of a universe that had existed for ever.

The inspiration for this new model seems to have come from a film called *Dead of Night*, released in September 1945. Although it

Figure 85 Fred Hoyle made contributions to many areas of physics and astronomy, but he is most famous for his Steady State model of the universe.

was made by Ealing Studios, it was a far cry from their usual output of genteel English comedies. In fact, it was the first horror film to be made in Britain after the repeal of wartime censorship, which had prohibited any form of entertainment that might damage morale.

Dead of Night, starring Mervyn Johns, Michael Redgrave and Googie Withers, is the tale of an architect called Walter Craig who wakes up one day, journeys into the countryside and visits a farmhouse to discuss a new design project. Upon arrival he tells the various house guests that they are already familiar to him from a recurring and disturbing dream. The guests react with a mixture of suspicion and curiosity, and one by one they reveal their own strange experiences, treating Craig to a series of five horror stories. They range from a tale of sibling murder to a psychiatrist's account of a psychotic ventriloquist. Craig becomes increasingly agitated by each story until the film reaches its peak in a flurry of nightmarish terror. Suddenly he wakes up and realises that the sequence of events has all been a nasty dream. He scrambles out of bed, gets dressed, journeys into the countryside and visits a farmhouse to discuss a new design project. Upon arrival he tells the various house guests that they are already familiar to him from a recurring and disturbing dream . . .

The film has a strange property, because the story evolves with time, with new characters appearing and the plot developing throughout, yet it finishes exactly where it started. Lots of things happen, but at the end of the film nothing has changed. Because of this circular structure, the film could have continued for ever.

The three men watched the film in a Guildford cinema in 1946, and soon afterwards it prompted Gold to come up with a remarkable idea. Hoyle later described Gold's reaction to *Dead of Night*:

> Tommy Gold was much taken with it and later that evening he remarked, 'How if the universe is constructed like that?' One tends

> to think of unchanging situations as being necessarily static. What the ghost-story film did sharply for all three of us was to remove this wrong notion. One can have unchanging situations that are dynamic, as for instance a smoothly flowing river.

The film inspired Gold to develop a completely new model of the universe. In this model the universe was still expanding, but it contradicted the Big Bang model in every other aspect. Remember, the Big Bang supporters had assumed that an expanding universe implied that the universe had been smaller, denser and hotter in the past, which pointed logically to a moment of creation a few billion years ago. In contrast, Gold could now conceive of an expanding universe that might have existed for ever in a largely unchanged state. Just as in *Dead of Night*, Gold imagined a universe that developed with time, yet remained largely unchanged.

Before explaining Gold's apparently paradoxical idea in more detail, it is worth noting that this notion of continual change coupled with immutability is all around us. Hoyle gave the example of a river, which is continually flowing but remains largely unchanged. Also, there is a type of cloud, lenticular altocumulus, that loiters at the peak of a mountain even during fierce winds. Moist air is blown up towards one side of the cloud, where it cools, condenses, forms new droplets and adds to the cloud. Simultaneously, the wind is blowing away some of the water droplets on the other side of the cloud, at which point the droplets descend the mountain, warm up and evaporate. Droplets are added to the cloud and droplets are lost, but overall the cloud seems unchanged. Even our own bodies demonstrate this principle of change in harmony with constancy, because our cells die, only to be replaced by fresh cells, which also die, only to be replaced by fresh cells, and so on. In fact, we change almost all our cells over the course of a few years, but we still remain the same person.

So, how did Gold apply this principle – continual development resulting in no change – to the entire universe? The continual development was obvious, because the universe seemed to be continually expanding. If there was nothing more than expansion, then the universe would change and become less dense with time, which is exactly what the Big Bang model suggested. However, Gold introduced a second aspect to the universe's development, one that counteracted the thinning effect of the expansion and resulted in no overall change. This was the idea that the universe compensated for its expansion by creating new matter in the growing gaps between the receding galaxies, so that the overall density of the universe would remain the same. Such a universe would apparently be developing and expanding, yet it would be largely unchanging, constant and eternal. A universe depleted by expansion would be replenished.

The notion of an evolving yet unchanging universe would become known as the *Steady State model*. When Gold first introduced the idea, Hoyle and Bondi called it a crazy theory. It was early evening in the Bondi household, and Hoyle was of the opinion that Gold's theory could be torn apart and disproved before dinner. As they grew hungrier and hungrier, it became clearer and clearer that Gold's cosmology was self-consistent and compatible with a wide range of astronomical observations. It was a perfectly sensible theory of the universe. In short, if the universe was infinite, then it could double in size and remain infinite and unchanged, as long as matter was created in between the galaxies, as shown in Figure 86.

All cosmological thinking had previously been guided by the cosmological principle, which stated that our bit of the universe, the Milky Way and its environs, is essentially the same as anywhere else in the universe. In other words, we do not inhabit a special place in the universe. Einstein exploited this principle when he first

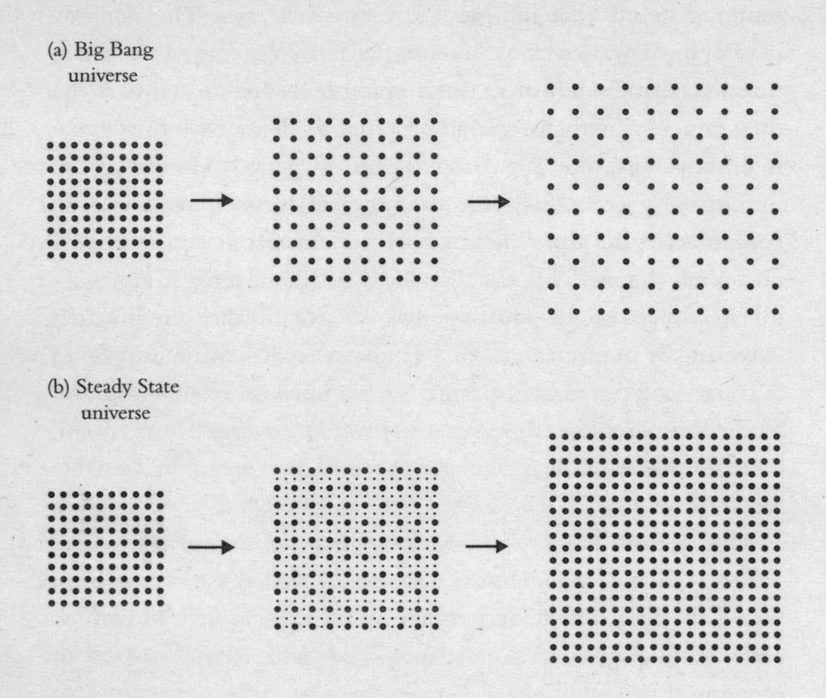

Figure 86 Diagram (a) shows expansion in a Big Bang universe. A small patch of the universe doubles its area and then doubles its area again. The dots representing galaxies become more thinned out, so as time passes the universe becomes less dense.

Diagram (b) shows expansion in a Steady State universe. Again, a small patch of the universe doubles in area twice over, but this time new galaxies appear in between the old ones, as shown in the intermediate stage of evolution. These seed galaxies grow to become fully fledged galaxies, so the third phase of the universe looks the same as the first. Critics might complain that while the density of the universe is the same, the universe has changed because it is now four times larger. But if the universe is infinite, then four times infinity is still infinity. Therefore an infinite universe can indeed expand while remaining unchanged, as long as the gaps created by the expansion are filled with new galaxies.

applied general relativity to the whole universe. Gold, however, was going one step further and postulated the *perfect cosmological principle:* not only is our patch of the universe the same as any other patch, but our era in the universe is the same as any other era. In other words, we live neither in a special place in the universe, nor at a special time. Not only is the universe broadly the same everywhere, but also everywhen. Gold believed that the Steady State model of the universe was a natural consequence of his perfect cosmological principle.

The Cambridge trio developed Gold's idea further, culminating in two papers published in 1949. The first paper, authored by Gold and Bondi, described the Steady State model in broad philosophical terms. Hoyle wanted to express it in more mathematical detail, which is why he published separately. This stylistic split was only superficial, and Hoyle, Gold and Bondi continued to work together to promote the Steady State model to the rest of the world.

There were two immediate questions levelled at the Steady State model. Where was all this matter that was being created, and where was it coming from? Hoyle replied that nobody should expect to see stars and galaxies appearing from nowhere. Compensating for the universe's expansion required a creation rate of only 'one atom every century in a volume equal to the Empire State Building', which observers on Earth would be unable to detect. To explain the creation of these atoms, Hoyle proposed the *creation field*, also known as the C-field. This entirely hypothetical entity was supposed to permeate the entire universe, spontaneously generating atoms and maintaining the status quo. Hoyle had to admit that he had no idea of the physics behind his notional C-field, but as far as he was concerned his model of continuous creation was far more sensible than creation in one almighty Big Bang.

There was now a clear choice for cosmologists. They could opt for a Big Bang universe, with a moment of creation, a finite history and a future that would be very different from the present. Or they could choose a Steady State universe, with continuous creation, an eternal history and a future that would be largely the same as the present.

Hoyle was anxious to prove that the Steady State model represented the true universe, and he suggested a definitive test that would show that he was right. According to the Steady State model, new matter was being created everywhere, which over the course of time would give rise to new galaxies everywhere. These baby galaxies should exist in our neighbourhood, and on the far side of the universe, and everywhere in between. If the Steady State model was right, then astronomers should be able to see these baby galaxies throughout the universe. The Big Bang model, however, predicted a very different situation. It claimed that the entire universe was born simultaneously, and everything should have evolved in a vaguely similar manner, so there would have been a time when all galaxies were babies, and then a time when they were mostly adolescents, and now they should all be fairly mature. Therefore the only way to see a baby galaxy today would be with an exceedingly powerful telescope that could see into the far reaches of the universe. This is because the light emitted by a very distant galaxy would have taken such a long time to reach us that we would effectively be seeing it as it was in the distant past, when it would have been a baby galaxy.

So the Steady State model predicted that baby galaxies were sprinkled evenly throughout the universe, while according to the Big Bang model we should be able to see baby galaxies only at huge cosmic distances. Unfortunately, when the debate between Steady Staters and Big Bangers began in the late 1940s, even the world's best telescopes were not powerful enough to allow astronomers to distinguish between baby galaxies and more mature galaxies. The distribution of baby galaxies remained unknown, and the Big Bang versus Steady State debate remained unresolved.

Figure 87 Thomas Gold, Hermann Bondi and Fred Hoyle, who invented the Steady State model of the universe.

Without precise observations or hard data that could differentiate between the Big Bang and Steady State models, the two rival camps resorted to sprinkling their scientific arguments with barbed remarks. For example, George Gamow pointed out that most of the Steady State supporters were based in England, and used this to tease them: 'It is not surprising that the Steady State theory is so popular in England, not only because it was proposed by its three (native-born and imported) sons H. Bondi, T. Gold, and F. Hoyle, but also because it has ever been the policy of Great Britain to maintain the status quo in Europe.'

Hoyle and Gold, and to some extent Bondi, were thoroughbred rebels, so Gamow's jibe that the Steady State model was born out of typical British conservatism was rather an unfair joke. In fact, Hoyle was almost obsessive in his questioning of orthodoxy. Sometimes he turned out to be right, but on many occasions he showed himself as a scientist out of his depth. Most notoriously, Hoyle denounced an archaeopteryx fossil as a forgery, and he also expressed serious doubts about Darwin's theory of evolution by natural selection. He wrote in the journal *Nature*: 'The likelihood of the formation of life from inanimate matter is one to a number with 40,000 noughts after it . . . It is big enough to bury Darwin and the whole theory of Evolution.'

Hoyle later came up with a dramatic analogy for illustrating the apparent impossibility of complex evolution: 'Imagine a tornado sweeping through a junk yard, and as it passes on its way, there in its wake is a brand new Boeing 747 jumbo jet, which, of course, has been fashioned and assembled by random chance out of the junk in the yard.'

Comments like this damaged Hoyle's standing, and, by association, they also slightly tarnished the Steady State's reputation among cosmologists. The three Steady Statesmen were also criticised for having no connection with observational astronomy. The Canadian astronomer Ralph Williamson said of Hoyle that 'he has had no real experience with handling the large telescopes which make modern astronomy possible'. In other words, Williamson was claiming that only those who actively explored the cosmos should theorise about it.

Bondi defended Hoyle by directly attacking Williamson's glib comment: 'It is on the same plane as the statement that only plumbers and milkmen have the right to pronounce on questions of hydrodynamics.'

Williamson also attacked Hoyle for being too speculative and not basing his cosmology on concrete astronomical observations, so-called hard facts. Again, Bondi was quick to speak up for Hoyle: 'But what is an astronomical fact? At most it is a smudge on a photographic plate!' Both sides of the debate had descended to the level of petty squabbling and backbiting.

Fed up with petty politics and personal attacks, Hoyle went through periods when he preferred to explain his ideas about the universe to the public rather than address fellow academics. He wrote several articles and published a series of popular books, which were written with a lively and lucid style. He once wrote: 'Space isn't remote at all. It's only an hour's drive away if your car could go straight upwards.' Indeed, he was such an accomplished wordsmith that he would eventually write a BBC TV drama series called *A for Andromeda*, a West End play for children called *Rockets in Ursa Major* and a series of science-fiction novels, including *The Black Cloud*.

In his first major work of popular science, *The Nature of the Universe*, Hoyle presented a detailed defence of the Steady State model: 'This may seem a strange idea and I agree that it is, but in science it does not matter how strange an idea may seem so long as it works – that is to say, so long as an idea can be expressed in a precise form and so long as its consequences are found to be in agreement with observation.'

It is interesting that George Gamow, Hoyle's main opponent in the Big Bang versus Steady State debate, also set out his theories in popular texts. Both men had a significant impact on the public understanding of science, which is why they both won the prestigious UNESCO Kalinga Prize for the Popularisation of Science, Gamow in 1956 and Hoyle in 1967.

The battle for popular support is well illustrated by a bizarre opera scene from *Mr Tompkins in Wonderland*, Gamow's science adventure fantasy. Gamow included Hoyle in the opera and made him sing a song that parodied his own Steady State theory. To make his point, Gamow introduced Hoyle into the story by having him materialise 'from nothing in the space between the brightly shining galaxies'.

The most significant incident in the populist battle for control of the cosmos occurred on the airwaves of the British Broadcasting Corporation in 1950. The BBC kept files on prospective contributors, and Hoyle's file was marked with the words 'Do not use this man', probably because he was considered to be a troublemaker who continually kicked against the establishment. Nevertheless, producer and fellow Cambridge academic Peter Laslett disregarded the warning label and invited Hoyle to broadcast a series of five lectures on the Third Programme radio network. The series was aired at eight o'clock on Saturday evenings, and transcripts were published in the *Listener* magazine. The entire project was a huge success, turning Hoyle into a celebrity.

The radio series is still remembered today because of a historic moment in the final lecture. Although the term 'Big Bang' has appeared in previous chapters of this book, its use has actually been anachronistic, because the term was originated by Hoyle during this radio broadcast. Up until the moment that Hoyle coined this catchy title, the theory had generally been known as the *dynamic evolving model*.

The term 'Big Bang' emerged while Hoyle was explaining that there were two rival theories of the cosmos. There was, of course, his own Steady State model, and then there was the model which involved a moment of creation:

> One of them is distinguished by the assumption that the universe started its life a finite time ago in a single huge explosion. On this supposition, the present expansion is a relic of the violence of this explosion. Now this Big Bang idea seemed to me to be unsatisfactory . . . On scientific grounds this Big Bang assumption is much the less palatable of the two. For it is an irrational process that cannot be described in scientific terms . . . On philosophical grounds, too, I cannot see any good reason for preferring the Big Bang idea.

When Hoyle used the term 'Big Bang', his voice took on a rather disdainful tone, and it seems that he intended the phrase as a derisory comment on the rival theory. Nevertheless, both fans and critics of the Big Bang model gradually adopted it. The greatest critic of the Big Bang model had inadvertently christened it.

CHAPTER 4 – MAVERICKS OF THE COSMOS

SUMMARY NOTES

① LEMAÎTRE TOOK HUBBLE'S OBSERVATIONS OF AN EXPANDING UNIVERSE AS EVIDENCE THAT HIS BIG BANG MODEL OF THE UNIVERSE (CREATION AND EVOLUTION) WAS CORRECT.

② EINSTEIN CHANGED HIS VIEW AND SUPPORTED THE BIG BANG MODEL.

- BUT THE MAJORITY OF SCIENTISTS CONTINUED TO BELIEVE THE TRADITIONAL MODEL OF AN ETERNAL STATIC UNIVERSE.
- THEY CRITICISED THE BIG BANG MODEL BECAUSE IT IMPLIED A UNIVERSE THAT WAS YOUNGER THAN THE STARS IT CONTAINED.

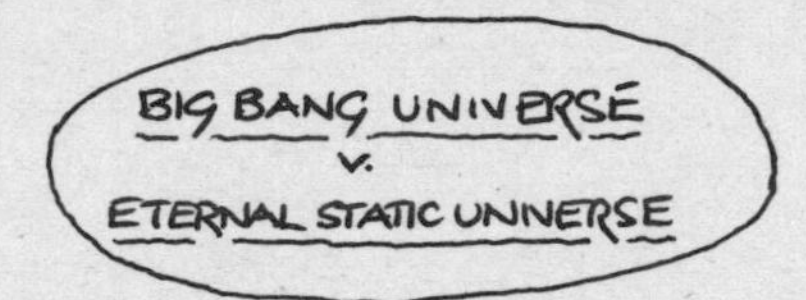

THE ONUS WAS ON THE BIG BANG SUPPORTERS TO FIND SOME EVIDENCE THAT THEIR THEORY WAS CORRECT. OTHERWISE THE ETERNAL STATIC UNIVERSE WOULD REMAIN THE DOMINANT THEORY.

ATOMIC PHYSICS WAS A VITAL TESTING GROUND: COULD THE BIG BANG MODEL EXPLAIN WHY LIGHT ATOMS (eg HYDROGEN AND HELIUM) ARE MORE COMMON THAN HEAVY ATOMS (eg IRON AND GOLD) IN TODAY'S UNIVERSE?

③ RUTHERFORD DEDUCED THE STRUCTURE OF THE ATOM. THE CENTRAL NUCLEUS CONTAINS PROTONS ⊕ AND NEUTRONS AND IT IS ORBITED BY ELECTRONS ⊖.

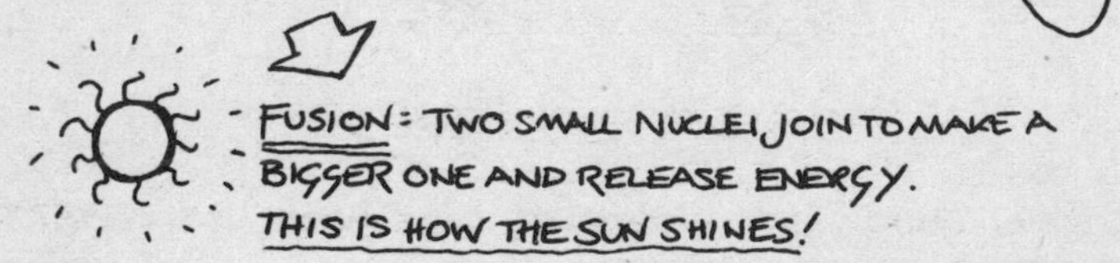

④ 1940s – GAMOW, ALPHER AND HERMAN PICTURED THE EARLY UNIVERSE AS A SIMPLE, DENSE SOUP OF PROTONS, NEUTRONS AND ELECTRONS. THEY HOPED THEY COULD BUILD BIGGER AND BIGGER ATOMS BY FUSION IN THE HEAT OF A BIG BANG.

SUCCESS: THE BIG BANG WAS ABLE TO EXPLAIN WHY TODAY'S UNIVERSE IS COMPOSED OF 90% HYDROGEN ATOMS AND 9% HELIUM ATOMS.

FAILURE: THE BIG BANG COULD NOT EXPLAIN THE FORMATION OF ATOMS THAT WERE HEAVIER THAN HELIUM.

⑤ MEANWHILE, GAMOW, ALPHER AND HERMAN PREDICTED THAT A LUMINOUS ECHO OF THE BIG BANG WAS RELEASED 300,000 YEARS OR SO AFTER THE MOMENT OF CREATION AND MIGHT STILL BE DETECTABLE TODAY.

DISCOVERING THIS ECHO WOULD PROVE THAT THERE WAS A BIG BANG BUT NOBODY SEARCHED FOR THIS SO-CALLED COSMIC MICROWAVE BACKGROUND (CMB) RADIATION.

⑥ ALSO IN THE 1940s HOYLE, GOLD AND BONDI PROPOSED THE STEADY STATE MODEL OF THE UNIVERSE. THIS SAID THAT THE UNIVERSE WAS EXPANDING. BUT NEW MATTER WAS CREATED AND FORMED INTO NEW GALAXIES IN THE INCREASING GAPS BETWEEN OLD GALAXIES.

THEY ARGUED THAT THE UNIVERSE EVOLVES, BUT OVERALL REMAINS UNCHANGED, AND HAS LASTED FOR EVER. THIS VIEW WAS COMPATIBLE WITH HUBBLE'S REDSHIFT OBSERVATIONS AND REPLACED THE TRADITIONAL ETERNAL STATIC MODEL OF THE UNIVERSE.

THE COSMOLOGICAL DEBATE NOW CENTRED ON THESE MODELS:

BIG BANG UNIVERSE
V.
STEADY STATE UNIVERSE

COSMOLOGISTS WERE DIVIDED OVER WHICH MODEL WAS CORRECT.

Chapter 5

THE PARADIGM SHIFT

You see, wire telegraph is a kind of a very, very long cat. You pull his tail in New York and his head is meowing in Los Angeles. Do you understand this? And radio operates exactly the same way: you send signals here, they receive them there. The only difference is that there is no cat. **ALBERT EINSTEIN**

The most exciting phrase to hear in science, the one that heralds new discoveries, is not 'Eureka!' (I found it) but 'That's funny . . .' **ISAAC ASIMOV**

In general we look for a new law by the following process. First you guess. Don't laugh, this is the most important step. Then you compute the consequences. Compare the consequences to experience. If it disagrees with experience, the guess is wrong. In that simple statement is the key to science. It doesn't matter how beautiful your guess is or how smart you are or what your name is. If it disagrees with experience, it's wrong. That's all there is to it.

RICHARD FEYNMAN

There were now two dominant theories fighting for control of the universe. In one corner was the Big Bang model, which had evolved out of Einstein's theory of general relativity, thanks to Lemaître and Friedmann. It proposed a unique moment of creation followed by a rapid expansion, and sure enough Hubble had observed that the universe was expanding and the galaxies were receding. Also, Gamow and Alpher had shown that the Big Bang could explain the abundances of hydrogen and helium. In the other corner was the Steady State model, invented by Hoyle, Gold and Bondi, which harked back to the conservative view of an eternal universe, except that it included an element of continuous creation and expansion. This creation and expansion made the model compatible with all the astronomical observations, including Hubble's observed red-shifts from the receding galaxies.

Scientific debates over the strengths of competing theories usually take place in university coffee-rooms or at the elite conferences where great minds convene. However, when it came to the question of whether the universe was eternal or created – the ultimate cosmological question – the discussion spilled over into the public arena, partly encouraged by the various popular books and radio broadcasts by Hoyle, Gamow and other cosmologists.

Not surprisingly, the Catholic Church was keen to make known its view on the cosmological debate. Pope Pius XII, who had already proclaimed that evolutionary biology was not in conflict with the Church's teaching, appeared at the Pontifical Academy of Sciences on 22 November 1951 to deliver an address entitled 'The Proofs for the Existence of God in the Light of Modern Natural Science'. In particular, the Pope strongly endorsed the Big Bang model, which he perceived as a scientific interpretation of Genesis and evidence for the existence of God:

> Thus everything seems to indicate that the material universe had a mighty beginning in time, endowed as it was with vast reserves of energy, in virtue of which, at first rapidly and then ever more slowly, it evolved into its present state . . . In fact, it would seem that present-day science, with one sweeping step back across millions of centuries, has succeeded in bearing witness to that primordial *Fiat lux* uttered at the moment when, along with matter, there burst forth from nothing a sea of light and radiation, while the particles of chemical elements split and formed into millions of galaxies . . . Therefore, there is a Creator. Therefore, God exists! Although it is neither explicit nor complete, this is the reply we were awaiting from science, and which the present human generation is awaiting from it.

The Pope's address, which also included a specific mention of Hubble and his observations, made headlines in newspapers around the world. One of Hubble's friends, Elmer Davis, read the address and could not resist writing to Hubble and joking: 'I am used to seeing you earn new and ever higher distinctions; but till I read this morning's paper I had not dreamed that the Pope would have to fall back on you for proof of the existence of God. This ought to qualify you, in due course, for sainthood.'

Surprisingly, the atheist George Gamow enjoyed the Papal

attention given to his field of research. He wrote to Pius XII after the address, sending him a popular article on cosmology and a copy of his book *The Creation of the Universe*. He even went as far as mischievously quoting the Pope in a research paper he published in 1952 in the prestigious journal *Physical Review*, knowing full well that this would annoy many of his colleagues, who were anxious to avoid any overlap between science and religion.

The overwhelming majority of scientists felt strongly that deciding the validity of the Big Bang model had nothing whatsoever to do with the Pope and that his endorsement should not be used in any serious scientific debate. In fact, it was not long before the Papal endorsement backfired and became an embarrassment for the Big Bang proponents. Supporters of the rival Steady State model began to use the Papal address as a way of mocking the Big Bang. The British physicist William Bonner, for example, suggested that the Big Bang theory was part of a conspiracy aimed at shoring up Christianity: 'The underlying motive is, of course, to bring in God as creator. It seems like the opportunity Christian theology has been waiting for ever since science began to depose religion from the minds of rational men in the seventeenth century.'

Fred Hoyle was equally scathing when it came to the Big Bang's association with religion, condemning it as a model built on Judeo-Christian foundations. His views were shared by his Steady State collaborator, Thomas Gold. When Gold heard that Pius XII had backed the Big Bang, his response was short and to the point: 'Well, the Pope also endorsed the stationary Earth.'

Scientists had been wary of the Vatican's attempts to influence the course of science ever since Urban VIII had forced Galileo to recant in 1633. However, this wariness sometimes bordered on paranoia, as noted by the English Nobel Laureate George Thomson: 'Probably every physicist would believe in a creation if the Bible had not unfortunately said something about it many years ago and made it seem old-fashioned.'

Perhaps the most important voice in the debate over the role of theology in cosmology was Monsignor Georges Lemaître, co-inventor of the Big Bang model and a member of the Pontifical Academy of Sciences. It was Lemaître's firm belief that scientific endeavour should stand isolated from the religious realm. With specific regard to his Big Bang theory, he commented: 'As far as I can see, such a theory remains entirely outside any metaphysical or religious question.' Lemaître had always been careful to keep his parallel careers in cosmology and theology on separate tracks, in the belief that one led him to a clearer comprehension of the material world, while the other led to a greater understanding of the spiritual realm: 'To search thoroughly for the truth involves a searching of souls as well as of spectra.' Not surprisingly, he was frustrated and annoyed by the Pope's deliberate mixing of theology and cosmology. One student who saw Lemaître upon his return from hearing the Pope's address to the Academy recalled him 'storming into class . . . his usual jocularity entirely missing'.

Lemaître was determined to discourage the Pope from making proclamations about cosmology, partly to halt the embarrassment that was being caused to supporters of the Big Bang, but also to avoid any potential difficulties for the Church. If the Pope – caught up as he was by his enthusiasm for the Big Bang model – were to endorse the scientific method and utilise it to support the Catholic Church, then this policy might rebound if new scientific discoveries contradicted Biblical teachings. Lemaître contacted Daniel O'Connell, director of the Vatican Observatory and the Pope's science advisor, and suggested that together they try to persuade the Pope to keep quiet on cosmology. The Pope was surprisingly compliant and agreed to the request – the Big Bang would no longer be a matter suitable for Papal addresses.

While cosmologists in the West were beginning to have some success in divorcing themselves from religious influence, those in the East were still having to deal with non-scientists trying to influence the scientific debate. In the Soviet Union, the influence was not theological but political, and it was not pro-Big-Bang but anti-Big-Bang. Soviet ideologues were antagonistic towards the Big Bang model because it failed to comply with the tenets of Marxist-Leninist ideology. In particular, they could not accept any model that posited a moment of creation, because creation was synonymous with a Creator. Also, they perceived the Big Bang as a Western theory, even though it was Alexander Friedmann in St Petersburg who had laid its foundations.

Andrei Zhdanov, who helped to coordinate the Stalinist purges of the 1930s and 1940s, encapsulated the Soviet position on the Big Bang: 'Falsifiers of science want to revive the fairy tale of the origin of the world from nothing.' He sought out and persecuted those he called 'Lemaître's agents'. His victims included the astrophysicist Nikolai Kozyrev, who was sent to a labour camp in 1937 and sentenced to be executed for continuing to discuss his belief in the Big Bang model. Fortunately his death sentence was commuted to ten years' incarceration when officials were unable to drum up a firing squad. After appeals by his colleagues, Kozyrev was eventually released and allowed to return to work at the Pulkovo Observatory.

Vsevolod Frederiks and Matvei Bronstein, who were also supporters of the Big Bang model, received the harshest punishments of all. Frederiks was imprisoned in a series of camps and died after six years of hard labour, while Bronstein was shot after being arrested on trumped-up charges of being a spy. By making examples of these and other scientists, the Soviets effectively gagged serious cosmological research and delivered a message that echoed on through the decades of Communism. The Russian astronomer

V.E. Lov followed the party line by stating that the Big Bang model is a 'cancerous tumour that corrodes modern astronomical theory and is the main ideological enemy of materialist science'. And Boris Vorontsov-Vel'iaminov, one of Lov's colleagues, maintained solidarity by calling Gamow an 'Americanised apostate' because of his defection to the West, stating that he 'advances new theories only for the sake of sensation'.

If the Big Bang theory was considered bourgeois science, then the Steady State theory hardly fared any better in the great scheme of communist ideology, because it too involved creation, albeit on a more gradual and continual basis. In 1958 Fred Hoyle attended a meeting of the International Astronomical Union in Moscow and recorded his reaction to the political undercurrent that dominated Soviet science: 'Judge my astonishment on my first visit to the Soviet Union when I was told in all seriousness by Russian scientists that my ideas would have been more acceptable in Russia if a different form of words had been used. The words "origin" or "matter-forming" would be OK, but creation in the Soviet Union was definitely out.'

The fact that politicians and theologians alike were using cosmology to shore up their beliefs struck Hoyle as ridiculous. As he wrote in 1956: 'Both Catholics and Communists argue by dogma. An argument is judged "right" by these people because they judge it to be based on "right" premises, not because it leads to results that accord with the facts. Indeed, if the facts should disagree with the dogma then so much worse for the facts.'

But regardless of the Pope's point of view or the Kremlin's stance, how did the cosmologists line up in the Big Bang versus Steady State debate? Throughout the 1950s, the scientific community was divided. In 1959 the *Science News-Letter* conducted a survey and asked thirty-three prominent astronomers to declare their position. The results showed that eleven experts backed the

Big Bang model, eight stood by the Steady State model, and the remaining fourteen were either undecided or thought that both models were wrong. Both models had established themselves as serious contenders for representing the reality of the universe, but neither had yet won a majority of support among scientists.

The reason for the lack of consensus was that the evidence for and against both models was inconclusive and contradictory. Astronomers were making observations that were at the very limit of their technology and understanding, so the 'facts' deduced from these observations had to be treated with a high degree of caution. For example, each measurement of a galaxy's recessional velocity might be called a fact, but it was open to criticism because of the convoluted chain of logic and observation that underpinned it. First, measuring the recessional velocity relied on detecting faint rays of galactic light and making assumptions about how they were or were not affected during their passage through the intervening space and the Earth's atmosphere. Second, the wavelengths of the light had to be measured and the galactic atoms that had emitted the light identified. Third, it was necessary to determine the spectral shift and then relate this shift to a recessional velocity via the cosmological Doppler effect. Finally, astronomers had to take into account the errors inherent in all the equipment and processes used, such as the telescope, the spectroscope, the photographic plate and even the developing process. This was a highly intricate set of connections, so astronomers had to be absolutely confident of every single step. Actually, measuring galactic recessional velocities was one of the more certain facts within cosmology; the chain of logic in other areas of the subject was even more convoluted and more open to criticism.

In the absence of conclusive evidence for or against either the Big Bang or the Steady State, many scientists based their cosmological preference on gut instinct or on the personalities of those who championed the rival models. This was certainly the case for Dennis Sciama, who would become one of the foremost cosmologists of the twentieth century, and whose supervision would inspire Stephen Hawking, Roger Penrose and Martin Rees. Sciama himself had been inspired by Hoyle, Gold and Bondi, whom he called 'an exciting influence for a younger person like myself.'

Sciama also found himself drawn to various philosophical aspects of their theory: 'The Steady State theory opens up the exciting possibility that the laws of physics may indeed determine the contents of the universe through the requirement that all features of the universe be self-propagating . . . The requirement of self-propagation is thus a powerful new principle with whose aid we see for the first time the possibility of answering the question why things are as they are without merely saying: it is because they were as they were.'

And he would later find another reason for preferring the Steady State over the Big Bang: 'It's the only model in which it seems evident that life will continue somewhere . . . even if the galaxy ages and dies out, there will always be new, young galaxies where life will presumably develop. And therefore the torch keeps being carried forward. I think that was probably the most important item for me.'

Sciama's largely subjective reasons for opting for the Steady State model were symptomatic of the uncertainty and tumult within cosmology. At the start of the twentieth century, cosmology was a comfortable subject with a well-established view of an eternal, unchanging, static universe, but measurements and new theories in the 1920s showed that this position was clearly unsatisfactory. Unfortunately, neither of the two emerging alternatives was entirely convincing. The Steady State cosmology was a revised version of the original eternal, static world-view, but there was little observational evidence to either support it or undermine it. The

Big Bang cosmology was a more radical and catastrophic view of the universe, with some evidence in its favour and some against. In short, cosmology was in limbo. Or, more technically, cosmology was in the middle of a *paradigm shift*.

The traditional view of the history of science was that scientific understanding developed gradually through a series of minor changes, with established theories being refined over the decades and new theories emerging from old ones. This was science developing by Darwinian evolution and natural selection. Theories mutated, and then it was a case of survival of the fittest, in the sense that the theory that best fitted observation would be adopted.

However, the philosopher of science Thomas S. Kuhn felt that this was only part of the story. In 1962 he wrote *The Structure of Scientific Revolutions*, in which he described scientific progress as a 'series of peaceful interludes punctuated by intellectually violent revolutions'. The peaceful interludes were periods during which theories would gradually evolve as described already, but every so often there would need to be a major shift in thinking, known as a paradigm shift.

For example, astronomers had for centuries tinkered with the paradigm of an Earth-centred model of the universe, adding epicycles and deferents to make the model a better fit to the observed paths of the Sun, stars and planets. Gradually there arose a series of problems to do with predicting planetary orbits, which most astronomers ignored through natural conservatism and ingrained respect for the existing paradigm. Eventually, when the problems had mounted and reached an intolerable level, rebels such as Copernicus, Kepler and Galileo offered a new Sun-centred paradigm. Within a couple of generations, the entire community of astronomers abandoned the old paradigm and shifted to the new one. Thereafter, a new era of scientific stability began, with a fresh programme of research built on new foundations and a new paradigm. The Earth-centred model did not evolve into the Sun-centred model, rather it was replaced by it.

The shift from the plum pudding atomic model to Rutherford's nuclear atomic model is another example of a paradigm shift, as is the shift from an ether-filled universe to one devoid of any ether. In each case, the shift from one paradigm to another could happen only once the new paradigm was properly fleshed out and the old paradigm had been fully discredited. The speed of the transition depends on numerous factors, including the weight of evidence in favour of the new paradigm and the extent to which the old guard resists change. Older scientists, having invested so much time and effort in the old paradigm, are generally the last to accept the change, whereas younger scientists are generally more adventurous and open-minded. The paradigm shift might therefore be completed only when the older generation has retired from scientific life, and the younger generation has become the new establishment. The old paradigm might have prevailed for centuries, so a transition period that lasts a couple of decades is still comparatively short.

The situation in cosmology was slightly unusual, inasmuch as the old paradigm of a static, eternal universe had already been discredited (because the galaxies were clearly not static) and there were two new paradigms vying for superiority, the Steady State and Big Bang models. Cosmologists hoped that this period of uncertainty and conflict would be ended by finding indisputable evidence that would prove which one of the two new models was correct.

In order to resolve whether we lived in the aftermath of a Big Bang or in the middle of a Steady State, astronomers would have to focus on a series of key criteria that were critical to the two competing models. These are summarised in Table 4, in which each criterion is briefly evaluated to indicate which model was more successful according to the data available in 1950.

While this table does not include every potential criterion for distinguishing between the two models, it does contain the main ones, such as each model's ability to explain the abundances of the various elements. Judged against this second criterion, the Big Bang model could accurately explain the abundances of hydrogen and helium in the universe, but not the abundances of heavier atoms. The Big Bang model earns a question mark on this point because of its partial success. The Steady State model also warrants a question mark here because it was unclear how the matter created in between the receding galaxies developed into the atomic abundances that we observe.

Not only did the two models have to explain the formation and abundances of the various atoms, they also had to explain how these atoms gathered together to form stars and galaxies, the third criterion in the table. This issue, which has not been discussed in any detail in previous chapters, posed a major problem for the Big Bang model. The universe would have expanded rapidly after the moment of creation, which would have tended to pull apart any baby galaxies that were trying to form. Also, because a Big Bang universe has only a finite history, there would have been only a billion or so years for the galaxies to evolve – a relatively short timescale. In other words, nobody could explain how the galaxies formed in the context of the Big Bang model. The Steady State theory was more confident on this issue, because in an eternal universe there would be more time for galaxies to develop.

The two columns that reflect the specific successes and failures of the two rival models contain a mixture of ticks, crosses and question marks, neither theory being completely satisfactory. One can therefore imagine cosmologists settling their differences by accepting that the Big Bang model could explain some features of the universe and the Steady State model could explain the others. However, cosmology is not a sport in which competing models

Table 4

This table lists various criteria against which the Big Bang and Steady State models could be judged. It shows how the two models fared on the basis of data available in 1950. The ticks and crosses give a crude indication of how

Criterion	Big Bang Model	Success
1. Redshift and the expanding universe	Expected from a universe that is created in a dense state and then expands	✓
2. Abundances of the atoms	Gamow and colleagues showed that the Big Bang predicts the observed ratio of hydrogen to helium, but fails to explain the other atomic abundances	?
3. Formation of galaxies	The Big Bang expansion would perhaps have pulled apart baby galaxies before they could grow; nevertheless, galaxies did evolve, but nobody could explain how	✗
4. Distribution of galaxies	Young galaxies existed in the early universe and should therefore be observable only at great distances, which effectively provides a window onto the early universe	?
5. Cosmic microwave background (CMB) radiation	This echo of the Big Bang should still be detectable with sufficiently sensitive equipment	?
6. Age of the universe	The universe is apparently younger than the stars it contains	✗
7. Creation	There is no explanation of what caused the creation of the universe	?

well each model fared in relation to each criterion, and a question mark indicates a lack of data or a mixture of agreement and disagreement. Criteria 4 and 5 warrant question marks because of a lack of observations.

Criterion	Steady State Model	Success
1. Redshift and the expanding universe	Expected from an eternal universe that expands, with new matter being created in the gaps	✓
2. Abundances of the atoms	Matter is created in between the galaxies moving apart, so somehow this material has to be transformed into the atomic abundances that we observe	?
3. Formation of galaxies	There is more time and no initial violent expansion; this allows galaxies to develop and die, to be replaced by new galaxies built from created matter	✓
4. Distribution of galaxies	Young galaxies should appear to be evenly distributed, because they can be born anywhere and at any time out of the matter created in between old galaxies	?
5. Cosmic microwave background (CMB) radiation	There was no Big Bang so there was no echo, which is why we cannot detect it	?
6. Age of the universe	The universe is eternal, so the age of the stars is not a difficulty	✓
7. Creation	There is no explanation of the continuous creation of matter in the universe	?

can share the glory. The Big Bang and Steady State models were contradictory and incompatible at a fundamental level. One model claimed that the universe was eternal, while the other claimed that the universe was created, and they could not both be correct. Assuming that one of the two models was correct, then whichever model would be victorious ultimately would have to crush its rival.

The Timescale Difficulty

The most pressing problem for the Big Bang supporters was the sixth criterion in Table 4 – the age of the universe. The cross highlights an absurdity in the Big Bang model: it implied a universe that was younger than the stars within it. This is as ridiculous as a mother being younger than her daughter – surely the stars could not be older than the universe itself? Chapter 3 described how Hubble had measured the distance to the galaxies and their apparent velocity; Big Bang cosmologists had then divided the distance by the velocity to deduce that roughly 1.8 billion years ago the entire mass of the universe had been concentrated into a single point of creation. But measurements of radioactive rocks had shown that the Earth was at least 3 billion years old, and it was logical to assume that the stars were even older.

Even Einstein, who supported the Big Bang, admitted that this problem would demolish the model, unless someone could find a drastic solution: 'The age of the universe . . . must certainly exceed that of the firm crust of the Earth as found from the radioactive minerals. Since determination of age by these minerals is reliable in every respect, the [Big Bang model] would be disproved if it were found to contradict any such results. In this case I see no reasonable solution.'

The age discrepancy became known as the *timescale difficulty*, a phrase that did not truly reflect the huge embarrassment it caused the Big Bang model. The only real prospect for resolving the age paradox was for an error to be discovered in previous measurements of either the distances to the galaxies or their velocities. For example, if the distances to the galaxies were greater than Hubble's estimates, then it would have taken the galaxies longer to reach their current distances, which would mean an older universe. Alternatively, if the speeds of recession of the galaxies were slower than Hubble's estimates, then again it would have taken the galaxies longer to reach their current distances, which again would mean an older universe. Hubble, though, was the most highly respected observational astronomer in the world, famous for his precision and diligence, so nobody seriously doubted the accuracy of his observations. Furthermore, his measurements had been independently checked by others.

When America entered the Second World War, observational astronomy and the activities of the major observatories largely ground to a halt. Any plans to try to resolve the Big Bang versus Steady State debate were postponed as astronomers dedicated themselves to serving their country. Even Hubble, who was in his fifties, left Mount Wilson to become head of ballistics at the Aberdeen Proving Ground, Maryland, the highest civilian post outside Washington, DC.

The only senior figure to remain at Mount Wilson was Walter Baade, a German émigré who had joined the observatory's staff in 1931. Despite a decade of living and working in America, he was still under suspicion and was forbidden to join any military research group. From Baade's point of view, the situation was quite satisfactory because it looked as though he would have sole use of the prestigious 100-inch Hooker Telescope. Moreover, wartime blackouts stopped the annoying light pollution from the Los Angeles suburbs, improving viewing conditions to a level unknown since the telescope was built in 1917. The only problem was that Baade's enemy alien status meant that he was confined to his house from sundown to sunrise, which was not ideal for an astronomer. Baade pointed out to the authorities that he was already in the process of applying for American naturalisation and eventually convinced them that he was not a security risk. He was still blocked from conducting military research, but within a few months his curfew was lifted and Baade had complete control of the world's best telescope under ideal viewing conditions. He also made the most of the increasingly sensitive photographic plates that were becoming available, creating images of unparalleled sharpness.

Baade spent the war years studying a particular type of star known as an *RR Lyrae star*, a type of variable star similar to a Cepheid variable star. Williamina Fleming, who worked alongside Henrietta Leavitt at the Harvard Observatory, had shown that the variability of RR Lyrae stars could be used, like Cepheids, to measure distances. So far her technique had been used only within the Milky Way, because RR Lyrae stars are less luminous than Cepheids. However, Baade's ambition was to use the ideal viewing conditions to find RR Lyrae stars in the Andromeda Galaxy, our nearest large galaxy. In this way he could use the variability of RR Lyrae stars to measure the distance to Andromeda and cross-check earlier distance measurements based on Cepheid variables.

In fact, Baade soon realised that the RR Lyrae stars in Andromeda were beyond the reach of the 100-inch Hooker Telescope, so he had to satisfy himself with using the 100-inch instrument to do background work on these stars in the Milky Way in preparation for observations to be made with the 200-inch telescope, which would be completed as soon as the war was over. He was optimistic that the new giant telescope would bring Andromeda's RR Lyrae stars clearly into view.

The 200-inch telescope, George Hale's greatest astronomical

engineering project, was being constructed on Mount Palomar, 200 kilometres south-east of Mount Wilson. Hale died in 1938, just two years after building started, so he never got to see what would be the most spectacular view of the universe ever achieved. When the instrument was eventually completed, it was named the Hale Telescope in his honour.

On 3 June 1948 the Los Angeles glitterati attended the telescope's inauguration. They marvelled at the 1,000-tonne revolving dome that housed the giant instrument, its concave mirror polished to an accuracy of fifty millionths of a millimetre. When Charles Laughton, star of *Mutiny on the Bounty*, was asked if the Hale Telescope was inspiring, he replied: 'Inspiring, my eye! It's damned frightening. What are they going to do with it? Start a war with Mars?'

By the time the Hale Telescope was fully operational, both Mount Wilson and Mount Palomar had a full complement of research staff. Nevertheless, Baade still had a head start in the search for RR Lyrae stars in the Andromeda Galaxy, thanks to his hard work during the Second World War when he had the 100 inch telescope to himself. He immediately aimed the new 200-inch telescope at the Andromeda Galaxy and scanned it for faint stars with a rapid variation in brightness, which would be indicative of RR Lyrae stars.

After a month of meticulous surveying, Baade found absolutely no sign of the RR Lyrae stars that he had expected to see. He persevered, yet still could not find what should have been discernible with the powerful Hale Telescope. He was baffled. He knew that his ability to see the RR Lyrae stars in the Andromeda Galaxy depended on only three things – the brightness of the stars, the power of the 200-inch telescope and the distance to the galaxy – and his calculations showed that the stars should definitely be visible. Unsure of what was behind his failure to detect any RR Lyrae stars, he revisited the three factors that determined his ability to see: he was confident of the brightness of the RR Lyrae stars from his wartime research, and he was sure that he understood the power of his telescope . . . so was it possible that the distance to Andromeda was farther than everybody had assumed?

Baade convinced himself that an error in the accepted distance to the Andromeda Galaxy was the only logical and possible explanation. His colleagues were initially sceptical, but they were persuaded that Baade was right when he was able to pinpoint exactly how and why the Andromeda Galaxy had been previously mismeasured.

As explained in Chapter 3, the original measurement to the Andromeda Galaxy had been performed using Cepheid variable stars, which had become the basic yardstick for measuring intergalactic distances. Henrietta Leavitt had shown that Cepheids have the useful property that the time period between two peaks in brightness is an excellent indication of their inherent luminosity, which can be compared with their apparent luminosity to ascertain their distance from the Earth. Hubble had been the first to find Cepheids outside the Milky Way and thereby measure the distance to another galaxy, namely the Andromeda Galaxy.

However, by the 1940s it was becoming evident that most stars could be grouped into two broad types, called *populations*. Older stars belong to Population II, and after these stars have expired their debris becomes an ingredient of newer, younger, Population I stars, which are generally hotter, brighter and bluer than their counterparts in Population II. Baade assumed that Cepheids were also split into these two categories, and suggested that this was what lay behind the contradictions over the distance to the Andromeda Galaxy.

Baade's argument that Andromeda was farther away was based on two simple steps. First, Population I Cepheids are intrinsically brighter than Population II Cepheids that have the same period of variation. And second, astronomers tended to see only the brighter Population I Cepheids in the Andromeda Galaxy, but they had inadvertently built their Cepheid distance scale by using the dimmer Population II Cepheids in the Milky Way.

Unaware that there were two types of Cepheid, Hubble had made the mistake of comparing dim, local Population II Cepheids with Andromeda's relatively bright Population I Cepheids. The consequence was that he had erroneously estimated the Andromeda Galaxy to be closer than it really is.

To set matters straight, Baade set about assiduously recalibrating the Cepheid yardstick according to the two types of Cepheid. In this way he could properly estimate the distance to the Cepheids in the Andromeda Galaxy, and therefore the distance to Andromeda itself. He worked out that Population I Cepheids are on average four times as luminous as Population II Cepheids that have the same period of variation. Conveniently, if a star is moved twice as far away from an observer, then it appears to be four times fainter. Therefore the Andromeda Galaxy had to be moved twice as far away – to a distance of roughly 2 million light years – to compensate for the fact that the Population I Cepheids visible in Andromeda were, on average, four times brighter than the Population II Cepheids traditionally used for measuring distance. The distance to Andromeda had now been corrected. At a distance of 2 million light years, it was no longer a surprise that the RR Lyrae stars were too faint to be seen.

If adjusting the distance to the Andromeda Galaxy had been the only consequence of Baade's work, it would not have merited a mention in this book. However, the distance to Andromeda had been used to estimate the distances to the other galaxies, using a method that will be discussed shortly. Hence, doubling the distance to Andromeda meant doubling the distances to all the other galaxies too.

Yet, the estimated recession speeds of these galaxies remained the same, as they were derived from spectroscopy and redshifts, which were unaffected by Baade's research. This had a massive positive impact on the Big Bang model. If the distances doubled and the speeds remained the same, then the time taken for all the galaxies to have reached their current distances from a moment of creation would also have to be doubled. In other words, the age of the universe in the Big Bang model could now be revised upwards to 3.6 billion years, a figure that was no longer incompatible with the age of the Earth.

Critics of the Big Bang model pointed out that the stars and galaxies were older than the Earth and therefore probably more ancient than 3.6 billion years, which meant that the universe still seemed to contain objects that were older than the universe itself. So, these critics claimed that the so-called timescale difficulty was still a problem. But the Big Bang supporters were not rattled by this perfectly valid point, because Baade's research had demonstrated that there was still a lot to learn about measuring galactic distances and the age of the universe. He had found one error and doubled the universe's age, so it was quite possible that another error would be found, and perhaps the age would be doubled again.

Baade's breakthrough had gone a long way towards fixing a major fault in the Big Bang model, but more importantly it had highlighted a weakness in astronomy more generally – the habit of blind obedience. Because of Hubble's reputation, astronomers had for too long unhesitatingly accepted his proclamation on the distances to Andromeda and the other galaxies. A failure to question and challenge such fundamental statements, even when they are made by eminent authorities, is one of the key features of poor science.

Many years later, and inspired by the Andromeda distance blunder, the Canadian astronomer Donald Fernie would acerbically highlight the undesirable quality of compliance in science: 'The definitive study of the herd instincts of astronomers has yet to be written, but there are times when we resemble nothing so much as a herd of antelope, heads down in tight formation, thundering with firm determination in a particular direction across the plain. At a given signal from the leader we whirl about, and, with equally firm determination, thunder off in a quite different direction, still in tight parallel formation.'

Baade formally announced that the universe was twice as old as previously believed when he attended the 1952 meeting of the International Astronomical Union in Rome. Those in the room who backed the Big Bang model immediately saw that this new measurement supported their belief in a moment of creation – or at least it removed a stumbling block. As luck would have it, the official recorder for that particular session was the Big Bang's fiercest critic, Fred Hoyle. He dutifully noted the result, but his deeply held belief in an eternal universe obliged him to choose his words in such a way as to carefully avoid any reference to the Big Bang or creation. He wrote: 'Hubble's characteristic time scale for the Universe must now be increased from about 1.8 billion years to about 3.6 billion years.'

The only person more disappointed than Hoyle by the result was Edwin Hubble. His frustration had nothing to do with whether or not the Big Bang was true, for he had never bothered himself with such cosmological questions. Hubble cared only about the accuracy of his measurements, not the interpretations and theories that were based upon them. Consequently, he was devastated because Baade had found a major flaw in his distance measurements.

As Hubble took on board the significance of Baade's new measurements, he felt a twinge of bitterness. Despite the numerous national and international prizes and awards he had won, he always regretted that he had never been honoured with a Nobel prize, which had always been his ultimate goal. Now that Baade had highlighted an error in his work, it seemed as though a Nobel prize would remain out of his reach.

In fact, the Nobel physics committee had no doubt that Hubble was the greatest astronomer of his generation, and Baade's research had hardly tarnished the great man's reputation in their eyes. After all, Hubble had settled the Great Debate in 1923 by proving the existence of galaxies beyond the Milky Way, and he had laid the foundation for the Big Bang versus Steady State debate with his law of galactic redshifts in 1929. The only reason why the Nobel Foundation had ignored him was that they had never considered astronomy to be part of physics. Hubble had lost out on a technicality.

Hubble had to be satisfied with the acclaim he received from the press and the public, who adored their cosmic hero and who rightly praised his achievements. As one journalist put it: 'While Columbus sailed three thousand miles and discovered one continent and some islands, Hubble has roved through infinite space and discovered hundreds of vast new worlds, islands, sub-continents, and constellations not just a few thousand miles away, but trillions of miles out yonder.'

Hubble died of a cerebral thrombosis on 28 September 1953. Tragically, he was completely unaware that the Nobel physics committee had secretly decided to change their rules and recognise his achievements with a Nobel prize. In fact, the committee was preparing to make the announcement of his nomination when Hubble passed away.

The prize cannot be awarded posthumously, and protocol dictated that committee discussions should remain confidential. Hubble's nomination would thus have been a secret for ever had it not been for two committee members, Enrico Fermi and Subrahmanyan Chandrasekhar, who decided to contact Grace Hubble. They were anxious to let Grace know that her husband's unparalleled contribution to our understanding of the universe had not been overlooked.

Dimmer, Further, Older

By challenging and then correcting the accepted distance to the Andromeda Galaxy, Walter Baade was reminding his colleagues that past measurements should be challenged and reviewed, and discarded if found wanting. This was an essential part of a healthy scientific climate. Only when a measurement had been checked, double-checked, triple-checked and cross-checked might it just about earn the title of 'fact'; even then, the occasional rebellious remeasurement would never do any harm.

This culture of doubt and criticism was even applied to Baade's distance measurements. In fact, it was Baade's own student, Allan Sandage, who would revise his master's measurements, thereby increasing the age of the universe again.

Sandage, like so many of his colleagues, had been hooked on astronomy ever since he had first peered through the eyepiece of a telescope. He never forgot that childhood moment when 'a firestorm took place in my brain'. He went on to earn a doctoral position at the Mount Wilson Observatory, working alongside Baade, who asked him to take fresh images of the most distant galaxies that had been observed. Baade simply wanted Sandage to check that his distance estimates were correct.

Astronomers could not use the Cepheid yardstick technique for measuring the distance to the farthest galaxies because it had been impossible to detect Cepheid variable stars so far away. Instead, they were forced to adopt a completely different measuring technique, which relied on the reasonable assumption that the brightest star in the Andromeda Galaxy was intrinsically as bright as the brightest star in any other galaxy. Therefore, if the brightest star in a distant galaxy was apparently $\frac{1}{100}$ ($\frac{1}{10^2}$) as bright as the brightest star in Andromeda, then the distant galaxy was assumed to be 10 times farther away, because brightness falls off with the square of the distance.

Although the brightness of stars varies enormously, this approach to measuring distance was not unreasonable. Human height, for example, also varies enormously. However, in a randomly selected group of fifty adults, it would be reasonable to assume that the height of the tallest person would be roughly 190 centimetres. So if there are two large groups of people in the distance and the tallest person in one group is apparently one-third the height of the tallest person in the other group, then it would be reasonable to guess that the first group is three times farther away than the second group. This is because the tallest person in both groups should be of roughly equal height, and apparent height falls away in proportion with distance. This method is not perfect, as one group might be on their way to a basketball tournament, and the other might be going to a demonstration to campaign for more rights for jockeys. In most cases, however, the distance estimation should be accurate to within a few per cent.

The technique would be even more accurate if one assessed the average height of people or the average brightness of stars, but astronomers were studying such distant objects that they were forced to apply the technique to the brightest star in each galaxy, which they were more likely to be able to see. Astronomers had used this technique for measuring distances to galaxies since the 1940s and were confident that it was basically reliable, although they were prepared to accept that distances might occasionally need to be tweaked, which is why Baade had asked Sandage to check his estimates. In fact, Sandage would reveal that the brightest-star approach had been plagued by a fundamental flaw.

Thanks to improved photography, Sandage could see that what had been repeatedly perceived as the brightest star in a distant galaxy was in fact something else altogether. Much of the hydrogen in the universe has coalesced into familiar compact stars, but there is also a significant amount of it in the form of vast clouds known as HII regions. An HII region absorbs light from surrounding stars, which heats it to over 10,000°C. Because of its temperature and size, an HII region can outshine almost any star.

Before Sandage, astronomers had been accidentally and incorrectly comparing the brightest star visible in the Andromeda Galaxy with the brightest HII region in more distant, newly discovered galaxies. Thinking that the HII regions were stars, astronomers had assumed that these new galaxies were relatively close because their brightest 'stars' appeared to be comparatively bright. When Sandage obtained images that were sharp enough to distinguish these HII regions from genuine stars, he concluded that the brightest genuine stars in distant galaxies were actually much fainter than the misinterpreted HII regions, so the galaxies had to be farther away than previously estimated.

The distances to these far-off galaxies were absolutely critical in terms of estimating the age of the universe according to the Big Bang model. In 1952 Baade had doubled galactic distances, and in so doing he had doubled the age of the universe to 3.6 billion years. Then, two years later, Sandage pushed back the galaxies even farther, increasing the age of the universe to 5.5 billion years.

Despite these increases, the measurements were still underestimated. Sandage continued working on his distance measurements throughout the 1950s, and both the distances and resulting age continued to grow. Indeed, Sandage would become the dominant figure in measuring distances and the age of the universe, and largely thanks to his observations it eventually became clear that the universe was between 10 and 20 billion years old. This broad range was certainly compatible with other objects in the universe. No longer could the Steady Staters mock the Big Bangers for positing a universe that was younger than the stars contained within it.

F. Hoyle   H.C. van de Hulst   A.R. Sandage   J.A. Wheeler   H. Zanstra   L. Ledoux

O.S. Klein   W.W. Morgan   B.V. Kukarkin   M. Fierz   W. Baade   H. Bondi   T. Gold   L. Rosenfeld   A.C.B. Lovell   J. Géhéniau

V.A. Ambartsumian   E. Schatzman

W.H. McCrea   J.H. Oort   G. Lemaître   C.J. Gorter   W. Pauli   W.L. Bragg   J.R. Oppenheimer   C. Møller   H. Shapley   O. Heckman

Figure 88 This group photograph from the 1958 Solvay Conference shows Allan Sandage and Walter Baade, whose revised distance measurements to the galaxies increased the age of the universe in the context of the Big Bang model. Many of the main figures in the Big Bang versus Steady State debate are pictured here, including Hoyle, Gold, Bondi and Lemaître.

Despite the bitter academic rivalry, there were some personal friendships between the two camps. For instance, Hoyle was very fond of Lemaître, whom he described as 'a round, solid man, full of jokes and laughter'. Hoyle affectionately recalled a car trip they made across Italy after a conference in Rome: 'Only in one respect did Georges's presence raise an issue, and this was over lunch. I always wanted a light lunch, so that I could continue driving in the afternoon, whereas Georges wanted a heavy lunch with a bottle of wine, so that he could sleep in the afternoon. We compromised by allowing Georges to sleep in the back of the car, which, unfortunately, led to his awakening almost always with a shocking headache.'

Cosmic Alchemy

Although the timescale difficulty had now been resolved, the Big Bang model still suffered from other problems. The foremost puzzle concerned nucleosynthesis, specifically the creation of the heavier elements. George Gamow had once boasted: 'The elements were cooked in less time than it takes to cook a dish of duck and roast potatoes.' In short, he believed that all the various atomic nuclei were created in the hour immediately following the Big Bang. However, despite the best efforts of Gamow, Alpher and Herman, it had been impossible to find a mechanism that would create anything but the lightest atoms, such as hydrogen and helium, even though the aftermath of the Big Bang was a period of intense heat. If the heavier elements were not created in the moments immediately after the Big Bang, then the problem was clear: where and when were they created?

Arthur Eddington had already put forward one possible theory about nucleosynthesis: 'I think the stars are the crucibles in which lighter atoms are compounded into more complex elements.' However, the temperature of stars was estimated to be just a few thousand degrees at the surface and just a few million degrees at the core. This temperature was certainly sufficient to turn hydrogen into helium slowly, but was wholly inadequate for fusing these helium nuclei into truly heavy nuclei, which required a temperature of a few billion degrees.

For example, creating neon atoms would require a temperature of 3 billion degrees, and creating heavier silicon atoms would require an even higher temperature of 13 billion degrees. And this leads to another problem. If there was an environment capable of creating neon, then it would not be hot enough to create silicon. Alternatively, if it was hot enough to create silicon, then all the neon would be converted into something heavier. It seemed as though every type of atom needed its own tailor-made crucible of creation, and that the universe would have to house a vast array of intense environments. Alas, nobody could work out where, or even if, these crucibles existed.

It was Fred Hoyle who would contribute most to solving this mystery. He did not see the problem of nucleosynthesis as a Big Bang versus Steady State issue, but rather as a matter of concern for both theories. The Big Bang model somehow had to explain how the fundamental particles at the start of the universe had been transformed into heavier atoms of varying abundances. Similarly, the Steady State model had to explain how the particles being continually created in between the receding galaxies were converted into heavier atoms. Hoyle had been thinking about the problem of nucleosynthesis ever since he was a junior researcher, but he did not take his first tentative steps towards a solution until the late 1940s. He began to make progress when he speculated about what would happen to a star as it passed through the various phases of its life.

A middle-aged star is generally stable, generating heat by fusing hydrogen into helium and losing heat by radiating light energy. At the same time, all of the mass of the star is being pulled inwards by its own gravitational attraction, but this is counteracted by the huge outward pressure caused by the high temperatures at the core of the star. As discussed in Chapter 3, this stellar equilibrium is similar to the balance of forces acting on a balloon, where the stretched rubber skin is trying to collapse the balloon inwards, and the air inside the balloon exerts a pressure that is pushing outwards. This analogy was used to explain why Cepheid stars are variable.

Hoyle was very familiar with the theoretical research that had been done on stars and the balance between the threat of gravitational collapse and the resistance of outward pressure, but he wanted to see what would happen when this balance was disrupted. In particular, Hoyle wanted to understand what would happen towards the end of a star's life, when it began to run out of hydrogen fuel. Not surprisingly, the fuel shortage would cause the star to begin to cool down. The fall in temperature would result in a fall in outward pressure, and the gravitational force would become overpowering and would initiate a stellar contraction. Crucially, however, Hoyle realised that this contraction was not the end of the story.

As the entire star falls inwards, the compression would cause the stellar core to heat up and generate an increased outward pressure, which would halt the collapse. The temperature rise associated with compression has several causes, but one of them is that compression encourages more nuclear reactions, resulting in the generation of more heat.

Although this extra heat re-establishes some level of stability in the star, it is only a temporary hiatus; the star's death has only been deferred. The star continues to consume more fuel, and eventually its dwindling fuel supply becomes critical. Lack of fuel means lack of energy production, so the core begins to cool again, which leads to another collapsing phase. Again, this heats the core, again halting the collapse until the next fuel shortage. This stop–start collapse means that many stars endure a slow, lingering death.

Hoyle set about analysing the various types of star (e.g. small, medium, large, Population I, Population II), and after several years of dedicated research he successfully completed his calculations of all the temperature and pressure changes that happened in different stars as they neared the end of their lives. Most importantly of all, he also worked out the nuclear reactions in each death spasm, and crucially showed how the various combinations of extreme temperatures and pressures could lead to a whole range of medium-weight and heavyweight atomic nuclei, as shown in Table 5.

Stage	Temp (°C)	Density (g/cm^3)	Duration of stage
Hydrogen → helium	4×10^7	5	10^7 years
Helium → carbon	2×10^8	7×10^2	10^6 years
Carbon → neon + magnesium	6×10^8	2×10^5	600 years
Neon → oxygen + magnesium	1.2×10^9	5×10^5	1 year
Oxygen → sulphur + silicon	1.5×10^9	1×10^7	6 months
Silicon → iron	2.7×10^9	3×10^7	1 day
Core collapse	5.4×10^9	3×10^{11}	0.25 seconds
Core bounce	23×10^9	4×10^{14}	0.001 seconds
Explosive	about 10^9	varies	10 seconds

Table 5
Fred Hoyle calculated the conditions in different stars at different stages of their life to see how nucleosynthesis might occur. This table shows the nucleosynthesis reactions that take place in a star with roughly twenty-five times the mass of our Sun. Such a heavy star has a remarkably short lifetime compared with typical stars. Initially the star spends several million years fusing hydrogen into helium. The temperature and pressure increase during the latter phases of its life, and allows for the nucleosynthesis of oxygen, magnesium, silicon, iron and other elements. A variety of even heavier atoms are generated during the final and most intense stages.

It became apparent that each type of star could act as a crucible for creating several different elements because stellar interiors changed dramatically during the course of a star's life and death. Hoyle's calculations could even account for the exact abundances of almost all the elements that we see today, explaining why oxygen and iron are common, while gold and platinum are rare.

In exceptional cases, the early collapsing phase of a very massive star becomes unstoppable and the star dies quite rapidly. This is a supernova, the most violent example of stellar death, which causes an implosion of unparalleled intensity. When it goes supernova, a single star can release enough energy to outshine more than 10 billion ordinary stars (which is why a supernova had confused astronomers involved in the Great Debate, as discussed earlier in Chapter 3). Hoyle showed that supernovae create the most extreme stellar environments and thus allow rare nuclear reactions to take place, thereby manufacturing the heaviest and most exotic atomic nuclei.

One of the most important outcomes of Hoyle's research was that the death of a star did not mark the end of the nucleosynthesis process. As a star implodes it sends out massive shock waves, which leads to an explosion, sending atoms flying out across the universe. Importantly, some of these atoms are the products of the nuclear reactions that took place in the final phases of the star's life. This stellar debris mixes with whatever else might be floating in the cosmos, including the atoms from other dead stars, eventually condensing to form completely new stars. These second-generation stars have a head start in terms of nucleosynthesis because they already contain some heavier atoms. This means that when they in turn die and implode they will build even heavier atoms. It is thought that our own Sun is probably a third generation star.

Marcus Chown, author of *The Magic Furnace*, described the significance of stellar alchemy as follows: 'In order that we might live, stars in their billions, tens of billions, hundreds of billions even, have died. The iron in our blood, the calcium in our bones, the oxygen that fills our lungs each time we take a breath – all were cooked in the furnaces of the stars which expired long before the Earth was born.' Romantics might like to think of themselves as being composed of stardust. Cynics might prefer to think of themselves as nuclear waste.

Hoyle had tackled one of the greatest puzzles in cosmology, and found a solution that was almost complete, except that there was one outstanding problem. Table 5 shows the chain of nucleosynthesis in one particular type of star: hydrogen is converted into helium, then helium into carbon, then carbon into all the heavier elements.

Although the table explicitly shows the helium to carbon phase, Hoyle could not actually work out how this step happened. As far as he could see, there was no viable nuclear pathway for transforming helium into carbon. This was a major problem, because unless he could explain the formation of carbon, he could not explain how all the other nuclear reactions took place because they all required carbon at some point in the chain that led to their creation. And this was a problem for all types of star – there was simply no way of turning helium into carbon.

Hoyle had run into exactly the same nuclear brick wall that had halted the progress of Gamow, Alpher and Herman towards an explanation of how helium was converted into heavier elements in the early moments of the Big Bang. If you recall, Gamow's team found that any nuclear reactions undergone by helium produced only unstable nuclei. Adding a hydrogen nucleus to a helium nucleus gave an unstable lithium-5 nucleus; merging two helium nuclei gave an unstable beryllium-8 nucleus. It seemed as though nature had conspired to block the only two paths that could turn helium nuclei into heavier nuclei, most notably carbon. Unless these two obstacles could be circumvented, the problem of building heavier nuclei would undermine the whole of Hoyle's vision of stellar nucleosynthesis. His hopes of explaining the rich variety of elements would dissolve.

Gamow's team could not solve this problem in the context of Big Bang nucleosynthesis, and Hoyle could not solve it in the context of stellar nucleosynthesis. Transforming helium into carbon seemed to be impossible. But Hoyle refused to give up hope of finding a viable pathway for carbon production. All the complex nuclear reactions he had predicted within dying stars relied on the existence of carbon, so he dedicated himself to solving the mystery of how the carbon itself was formed.

The most common form of carbon is known as carbon-12,

because its nucleus contains twelve particles, namely six protons and six neutrons. The most common form of helium is known as helium-4, because its nucleus contains four particles, namely two protons and two neutrons. Hoyle's problem could therefore be boiled down to one straightforward question: is there a viable mechanism for transforming three helium nuclei into a single carbon nucleus?

One possibility was for three helium nuclei to simultaneously collide and form a carbon nucleus. It was a nice idea, but impossible in practice. The chances of three helium nuclei being in exactly the same place at exactly the same time and travelling at exactly the right speeds to fuse together was effectively nil. The alternative pathway was for two helium nuclei to fuse to form a beryllium-8 nucleus, with four protons and four neutrons, and then for this beryllium-8 nucleus later to fuse with another helium nucleus to form carbon. This pathway and the three-way helium collision mechanism are illustrated in Figure 89.

However, beryllium-8 is very unstable, which is why it was already regarded by Gamow as a block on the path to building nuclei heavier than helium. In fact, a beryllium-8 nucleus is so unstable that (on the rare occasions on which one does form) it typically lasts for less than a millionth of a billionth of a second before spontaneously breaking up. It is just about conceivable that a helium nucleus might merge with a beryllium-8 nucleus during its fleeting existence to form carbon-12, but even if this did happen there was another hurdle to overcome.

The combined mass of a helium nucleus and a beryllium nucleus is very slightly greater than the mass of a carbon nucleus, so if helium and beryllium did fuse to form carbon then there would be the problem of getting rid of the excess mass. Normally, nuclear reactions can dissipate any excess mass by converting it into energy (via $E = mc^2$), but the greater the mass difference, the longer the

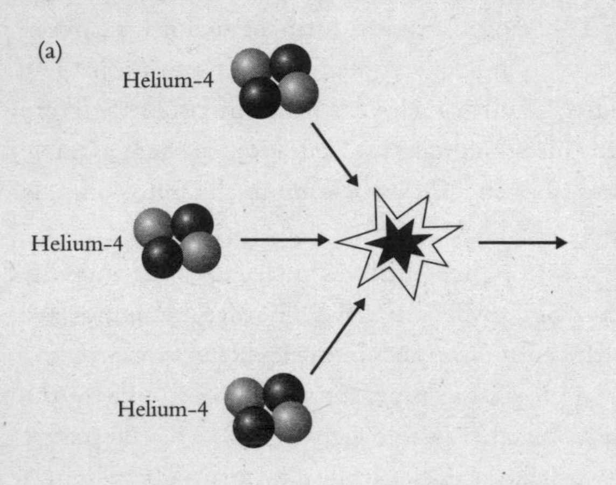

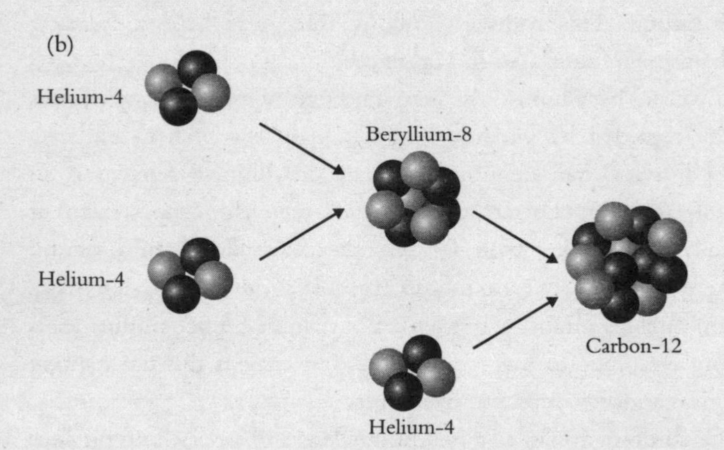

Figure 89 Diagram (a) illustrates a possible nuclear path from helium to carbon which requires three helium nuclei to collide simultaneously. This is very unlikely to happen. The second path, shown in diagram (b), requires two helium nuclei to collide and form beryllium. In turn, the beryllium nucleus collides and fuses with another helium nucleus to form carbon.

time required for the reaction to happen. And time is something that the beryllium-8 nucleus does not have. The formation of carbon has to happen almost instantaneously because beryllium-8 has such a short lifetime.

So, there were two barriers en route to carbon via beryllium-8. First, beryllium-8 was utterly unstable and did not last for more than the merest fraction of a second. Second, turning helium and beryllium into carbon required a significant time window because of the slight mass imbalance. The situation seemed impossible, because the two problems exacerbated each other. Hoyle could have given up at this point and turned his mind to something simpler. Instead, he made one of the greatest intuitive leaps in the history of science.

Although any given nucleus has a standard structure, Hoyle knew that alternative arrangements of the protons and neutrons were possible. We can think of the twelve particles that make up a carbon nucleus as twelve spheres; two possible arrangements of these spheres are illustrated in Figure 90. The first has two layers of six particles in a rectangular arrangement; the second has four layers of three particles in a triangular arrangement (this is a huge oversimplification, because things at the nuclear level are not so geometrically neat). Let us assume that the first arrangement is the one associated with the most common form of carbon, and the second is associated with the so-called excited form of carbon. It is possible to transform the common carbon nucleus into the excited form by injecting energy. Because energy and mass are equivalent ($E = mc^2$ again), the excited carbon nucleus has a slightly greater mass than the common carbon nucleus. Hoyle concluded that there must be an excited form of carbon-12 with exactly the right mass, one that perfectly matched the combined mass of beryllium-8 and helium-4. If there was such a carbon nucleus, then helium-4 could react more quickly with beryllium-8 to form carbon-12. Despite

Figure 90 The diagrams represent two possible forms of carbon, although in reality the protons (darker) and neutrons (lighter) do not arrange themselves so neatly, but tend to form a spherical cluster instead. The important point is that the carbon nucleus can exist in different arrangements with different masses.

---

the very short lifetime of beryllium-8, it would then be possible to create significant quantities of carbon-12.

Problem solved!

But scientists cannot just imagine a solution to a problem. Just because Hoyle knew that an excited state of carbon-12 with just the right mass would open the door to the creation of carbon and all the heavier elements, this did not necessarily mean that such a state existed. Excited nuclei can possess only very particular masses, and scientists cannot simply wish that they have a convenient value. Fortunately, Hoyle was more than just a wishful thinker. His confidence in the existence of just the right excited state of carbon was based on a strange but valid chain of logical reasoning.

Hoyle's premise was that he existed in the universe. Furthermore, he pointed out, he was a carbon-based life form. Therefore carbon existed in the universe, so there must have been a way of creating carbon. However, the only way to create carbon seemed to rely on the existence of a specific excited state of carbon. Consequently, such an excited state must exist. Hoyle was rigorously applying what would later become known as the *anthropic principle.* This principle can be defined and interpreted in various ways, but one version states:

> We are here to study the universe, so the laws of the universe must be compatible with our own existence.

In Hoyle's argument, he stated that he is partly made from the carbon-12 nucleus, so the correct excited state of carbon must exist, otherwise neither carbon-12 nor Fred Hoyle would exist.

Technically, Hoyle predicted that his proposed excited state of carbon would have 7.65 megaelectronvolts (MeV) more energy than the basic carbon nucleus. The megaelectronvolt is a minuscule unit of energy well suited to measuring the amounts of energy associated with minuscule objects such as atomic nuclei. Hoyle now wanted to know if this excited state actually existed.

In 1953, soon after he postulated this excited state of carbon, Hoyle was invited to spend a sabbatical at the California Institute of Technology (Caltech), where he had the chance to test out his theory. On the Caltech campus was the Kellogg Radiation Laboratory, where Willy Fowler had earned a reputation as one of the greatest experimental nuclear physicists in the world. One day Hoyle wandered into Fowler's office and told him about his prediction of an excited state of carbon, 7.65 MeV above the common state. Nobody had ever before made such a precise prediction about the excited state of a nucleus, because the physics and mathematics were far too complex. But Hoyle's prediction was based on pure logic, not mathematics or physics. Hoyle wanted

Fowler to look for his predicted state of carbon-12 and prove that he was right.

This was Fowler's first encounter with Hoyle, and he had no real idea what was going on in the Yorkshireman's mind. Fowler's initial response was that carbon-12 had already been measured in detail and there was no record of an excited state at 7.65 MeV. He later recalled that his reaction to Hoyle was wholly negative: 'I was very sceptical that this Steady State cosmologist, this theorist, should ask questions about the carbon-12 nucleus . . . Here was this funny little man who thought that we should stop all this important work that we were doing . . . and look for this state, and we gave him the brush off. Get away from us, young fellow, you bother us.'

Hoyle continued to press his argument, pointing out that Fowler could check this theory within a few days by specifically searching for the 7.65 MeV carbon-12 state. If Hoyle was wrong, then Fowler would have to work a few late nights to catch up on his schedule, but if Hoyle was right then Fowler would be rewarded with having made one of the biggest discoveries in nuclear physics. Fowler was convinced by this simple cost-benefit analysis. He asked his team to start searching for the excited state immediately, just in case it had been overlooked during earlier measurements.

After ten days of analysing the carbon-12 nucleus, Fowler's team found a new excited state. It was at 7.65 MeV, exactly where Hoyle said it should be. This was the first and only time that a scientist had made a prediction using the anthropic principle and had been proved right. It was an instance of extreme genius.

At last, Hoyle had proved and identified the mechanism by which helium could be transformed into beryllium and then into carbon. He had confirmed that carbon was synthesised at temperatures of roughly 200,000,000°C via the reaction shown in Figure 89(b). It was a slow process, but billions of stars over billions of years could create significant amounts of carbon.

And explaining the creation of carbon confirmed the starting point for the other nuclear reactions that created all the other elements in the universe. Hoyle had solved the problem of nucleosynthesis. This was a breakthrough for the Steady State model, because Hoyle could claim that the simple matter supposedly created in between receding galaxies would clump together to form stars and new galaxies, whereupon it would be forged in the various stellar furnaces into the heavier elements we see today. Hoyle's work was also a boost for the Big Bang model, which was otherwise incapable of explaining the creation of the heavy elements from all the hydrogen and helium that supposedly emerged in the period immediately after the creation of the universe.

At first sight, resolving the issue of nucleosynthesis could now be considered an honourable draw between the two rival cosmological camps. After all, both the Big Bang and the Steady State model could explain the synthesis of heavy elements by invoking the same stellar processes. In fact, the Big Bang had emerged as the stronger of the two models, because when it came to the creation of lighter elements, such as helium, only the Big Bang model could explain their abundances satisfactorily.

Helium is the second most abundant and second lightest element in the universe, after hydrogen. The stars do turn hydrogen into helium, but only very slowly, so from a Big Bang point of view the stars could not account for the large amounts of helium that actually exist in the universe today. However, Gamow, Alpher and Herman had shown that the helium in today's universe could be accounted for if hydrogen had been fused into helium in the moments after the Big Bang. The latest Big Bang calculations estimated that helium should make up 10% of all the atoms in the universe, which was very close to the latest estimates based on observations, so theory and observation were consistent.

In contrast, the Steady State model failed to explain the helium abundance. Therefore, the Big Bang and the Steady State were on a par in terms of heavy element nucleosynthesis, but only the Big Bang could really explain helium nucleosynthesis.

The case in favour of Big Bang nucleosynthesis was further strengthened by new calculations of the nucleosynthesis of the nuclei of elements such as lithium and boron, which are heavier than helium, but lighter than carbon. The calculations showed that these lithium and boron nuclei could not be synthesised within stars, but they could have emerged from the heat of the Big Bang at the same time that hydrogen was being converted into helium. Indeed, theoretical estimates of the abundances of lithium and boron created in the heat of the Big Bang matched exactly what was actually observed in the modern universe.

Ironically, although a complete explanation of nucleosynthesis was ultimately a victory for the Big Bang model, it would not have been possible without the immense contribution of Hoyle, who was from the opposing camp. George Gamow had huge respect for Hoyle and acknowledged his achievements in his light-hearted rewriting of Genesis, shown in Figure 91. Gamow's Genesis is actually an excellent summary of nucleosynthesis, from the creation of light nuclei in the heat of the Big Bang to the creation of heavy nuclei in supernovae.

The entire programme of research to explain nucleosynthesis in terms of processes inside stars involved dozens of steps and numerous refinements that took place over more than a decade. Hoyle remained at the heart of the effort throughout, but he was clearly supported by the experimental work of Willy Fowler, and he also collaborated with the husband-and-wife team of Margaret and Geoffrey Burbidge. The foursome collaborated on a definitive 104-page paper, entitled 'Synthesis of the Elements of Stars', which identified the role of each stellar phase and the consequences of each nuclear reaction. The paper contained an extraordinarily bold

Figure 91 Genesis according to George Gamow

In the beginning God created radiation and ylem. And ylem was without shape or number, and the nucleons were rushing madly over the face of the deep.

And God said: 'Let there be mass two.' And there was mass two. And God saw deuterium, and it was good.

And God said: 'Let there be mass three.' And there was mass three. And God saw tritium, and it was good.

And God continued to call numbers until He came to transuranium elements. But when He looked back on his work, He found that it was not good. In the excitement of counting, He missed calling for mass five and so, naturally, no heavier elements could have been formed.

God was very much disappointed, and wanted first to contract the Universe again, and to start all over from the beginning. But it would be much too simple. Thus, being almighty, God decided to correct His mistake in a most impossible way.

And God said: 'Let there be Hoyle.' And there was Hoyle. And God looked at Hoyle and told him to make heavy elements in any way he pleased.

And Hoyle decided to make heavy elements in stars, and to spread them around by supernova explosions. But in doing so, he had to obtain the same abundances which would have resulted from nucleosynthesis in ylem, if God would not have forgotten to call for mass five.

And so, with the help of God, Hoyle made heavy elements in this way, but it was so complicated that nowadays neither Hoyle, nor God, nor anybody else can figure out exactly how it was done.

Amen

statement: 'We have found it possible to explain, in a general way, the abundances of practically all the isotopes of the atoms from hydrogen through uranium by synthesis in stars and supernovae.'

The paper grew to be so famous that it became known simply by the initials of its authors (as the B^2FH paper), and was widely recognised as one of the greatest triumphs of twentieth-century science. Not surprisingly, it would earn a Nobel prize for one of its authors. What is surprising is that the 1983 Nobel Prize for Physics went to Willy Fowler, not Fred Hoyle.

The fact that Hoyle was ignored is one of the greatest injustices in Nobel history. The main reason that the Nobel committee snubbed Hoyle was that he had made numerous enemies over the years, thanks to his outspoken nature. For example, he had complained vociferously when the 1974 Nobel Prize for Physics was awarded for the discovery of pulsars. He agreed that the detection of these pulsating stars was a major breakthrough, but was outraged that the prize was not shared with the young astronomer Jocelyn Bell, who had made the crucial pulsar observations. The sensible strategy would have been to stay silent and keep out of the controversy, but Hoyle was incapable of putting decorum above honesty and integrity.

Similarly, instead of keeping his head down and getting on with his work in Cambridge, Hoyle battled against the absurd politics that governed the university. In 1972, after years of fighting the system, a frustrated Hoyle resigned his post:

> I do not see any sense in continuing to skirmish on a battlefield where I can never hope to win. The Cambridge system is effectively designed to prevent one ever establishing a directed policy – key decisions can be upset by ill-informed and politically motivated committees. To be effective in this system one must forever be watching one's colleagues, almost like a Robespierre spy system. If one does so, then of course little time is left for any real science.

Although Hoyle's forthright approach to physics and life made him unpopular in some circles, the majority of scientists were very fond of him, including the American astronomer George O. Abell:

> He is a brilliant lecturer and a wonderful teacher. He is also a warm human being who always found time to talk with students; his enthusiasm about almost everything is extremely contagious. And he did, indeed, turn out to be a man of ideas; he simply is the kind of person that things occur to, during almost any kind of conversation, under almost any circumstance . . . It is from such a wealth of ideas, some of which are wrong, others of which are wrong but brilliant, and still others of which are brilliant and right, that scientific progress is made.

After his resignation, Hoyle spent the next thirty years of his life as a vagabond astrophysicist, visiting various universities and spending time in the Lake District, before finally retiring to the coast at Bournemouth. As the Astronomer Royal Martin Rees points out, this was a sad end for such a great man: 'His consequent isolation from the broad academic community was probably detrimental to his own science; it was certainly a sad deprivation for the rest of us.'

Corporate Cosmology

Those who have contributed to the history of cosmology have financially supported their research in a variety of ways. Copernicus found time to study the Solar System in between his duties as physician to the Bishop of Ermland, while Kepler benefited from the patronage of Herr Wackher von Wackenfels. The rise of European universities provided ivory towers for the likes of Newton and Galileo, whereas some researchers, such as Lord Rosse, were independently wealthy and able to fund their own ivory towers, and ivory observatories to boot. Royal patronage was an important influence in Europe for many centuries, with monarchs such as King George III supporting the likes of Herschel. In contrast, American astronomers who wanted bigger telescopes at the start of the twentieth century turned to multi-millionaire philanthropists such as Andrew Carnegie, John Hooker and Charles Tyson Yerkes.

However, throughout the history of astronomy up to 1920, big business had invested nothing in the exploration of the heavens. This is not surprising, as probing the structure of the universe is not an obvious route to shareholder profits. Nevertheless, one American corporation did decide to become a major player in the development of cosmology, and made a significant contribution to the ongoing Big Bang versus Steady State debate.

The American Telephone and Telegraph (AT&T) Corporation established its reputation by constructing America's communications network and exploiting Alexander Graham Bell's telephone patents. Then, after merging with Western Electric in 1925, it established its research base at Bell Laboratories in New Jersey, which rapidly earned a reputation for world-class research. In addition to its applied communications research, Bell Labs also devoted major resources to pure and fundamental research. Its philosophy has always been that first-rate, arcane, pure research nurtures a culture of curiosity and builds bridges with universities, which ultimately leads to concrete commercial benefits. Those benefits aside, research discoveries at Bell Labs have netted six Nobel prizes in physics, shared among eleven scientists, a record that is matched only by the world's greatest universities. For example, in 1937, Clinton J. Davisson received the prize for his work on the wave nature of matter; in 1947, Bardeen, Brattain and Shockley were awarded the prize for inventing the transistor; and in 1998, Stormer, Laughlin and Tsui shared it for the discovery and explanation of the fractional quantum Hall effect.

The story of how Bell Labs came to be involved in cosmological research is rather convoluted and dates back to 1928, the year after AT&T began a transatlantic radio-based telephone service. The radio link could carry one call at a time at a rate of $75 for the first three minutes – equivalent to almost $1,000 at today's prices. AT&T was anxious to keep a grip on this lucrative market by offering a high-quality service, so it asked Bell Laboratories to undertake a survey of the natural sources of radio waves, which were interfering with long-distance radio communication by causing a background crackling noise. The task of surveying these annoying radio sources fell to Karl Jansky, a twenty-two-year-old junior researcher who had only just graduated in physics from the University of Wisconsin, where his father had been a lecturer in electrical engineering.

Radio waves, like waves of visible light, are part of the electromagnetic spectrum. However, radio waves are invisible and have wavelengths that are much longer than those of visible light. Whereas the wavelengths of visible light are less than a thousandth of a millimetre, radio wavelengths vary from a few millimetres (microwaves) to a few metres (FM radio waves) and a few hundred metres (AM radio waves). The wavelengths of concern in AT&T's radio-telephone system were of the order of a few metres, so Jansky built a giant, highly sensitive radio antenna at the Bell Labs Holmdel site, as shown in Figure 92, capable of detecting 14.6-metre radio waves. The antenna was mounted on a turntable that rotated three times each hour, allowing it to pick up radio waves from all directions. When Jansky's back was turned, local children would perch on the struts of the world's slowest carousel, which is why the antenna was nicknamed 'Jansky's merry-go-round'.

Having constructed the antenna by the autumn of 1930, Jansky spent several months laboriously measuring the strength of the radio interference from different directions and at different times of the day. He had hooked up the antenna to a loudspeaker, so he could actually hear the hiss, crackle and static of the natural radio interference. It slowly dawned on him that the interference fell into three categories. First, there was the occasional impact of local thunderstorms. Second, there was a weaker, more constant crackle from distant storms. Third, there was an even weaker category of interference, which Jansky described as 'composed of a very steady hiss type static the origin of which is not yet known'.

Most researchers would have ignored the unknown radio source, because it was insignificant compared with the other two sources and would have no serious impact on transatlantic communications. Jansky, however, was determined to get to the bottom of the mystery and spent several more months analysing the baffling interference. Gradually, it emerged that the hiss came from a particular region of the sky and that it peaked every 24 hours. Actually, when Jansky looked at his data more carefully, he found that the peak

Figure 92 Karl Jansky makes adjustments to the antenna that was designed to detect natural sources of radio waves. The Ford Model T wheels are part of the turntable that allowed the antenna to rotate.

came every 23 hours and 56 minutes. Almost a full day between peaks, but not quite.

Jansky mentioned the curious time interval to his colleague Melvin Skellet, who had a Ph.D. in astronomy and who was able to point out the significance of the missing four minutes. Each year the Earth spins on its axis 365¼ times, and each day lasts 24 hours, so one year consists of 365¼ × 24 = 8,766 hours. However, as well as spinning on its own axis 365¼ times, the Earth effectively makes one extra spin each year by going once around the Sun. Therefore, the Earth actually makes 366¼ rotations in 8,766 hours (one year), so each rotation takes 23 hours and 56 minutes, which is known as the *sidereal day*. The significance of the sidereal day is that it is the duration of our rotation with respect to the entire universe, as opposed to our provincial 24-hour day.

Skellet was very familiar with the duration of the sidereal day and its astronomical relevance, but it came as a surprise to Jansky, who immediately started to consider the implications for his radio interference. He realised that if the mysterious radio hiss peaked once each sidereal day, then its source had to be something far beyond the Earth and the Solar System. The sidereal day implied a cosmic radio source. Indeed, when Jansky tried to establish the direction of the radio signal, he discovered that it was coming from the centre of the Milky Way, our home galaxy. The only explanation was that our galaxy was generating radio waves.

At the age of just twenty-six, Karl Jansky had become the first person to detect and identify radio waves coming from outer space, a truly historic discovery. We now know that the centre of the Milky Way has intense magnetic fields which interact with fast-moving electrons, resulting in a constant output of radio waves. Jansky's research had opened a window onto this phenomenon. He announced his result in a paper entitled 'Electrical Disturbances Apparently of Extraterrestrial Origin'.

The story was picked up by the *New York Times*, which ran a front-page article on 5 May 1933, including this reassurance to readers: 'There is no indication of any kind . . . that these galactic radio waves constitute some kind of interstellar signaling, or that they are the result of some form of intelligence striving for intergalactic communication.' But this was not enough to stop a pile of letters landing on Jansky's desk claiming that he was receiving important messages from aliens which should not be ignored.

The true significance of Jansky's breakthrough surpasses even the momentous discovery that the Milky Way emits radio waves. His accomplishment was to establish the science of *radio astronomy* and to demonstrate that astronomers could learn a huge amount about the universe by looking beyond the narrow band of electromagnetic wavelengths that are visible to the human eye. As mentioned in Chapter 3, objects emit electromagnetic radiation at a vast range of wavelengths. These wavelengths, which are summarised in Figure 93, can be both shorter and longer than the familiar rainbow of visible wavelengths.

Even though we cannot see these extreme wavelengths with our eyes, they are real enough. The situation is the same with sound. Animals emit sound at a range of wavelengths, but we humans can hear only those within a very limited range. We can hear neither the infrasound (long wavelengths) generated by elephants, nor the ultrasound (short wavelengths) emitted by bats. We know that ultrasound and infrasound exist only because we can detect them with special equipment.

Jansky was ahead of his time, because the astronomers of his day were unfamiliar with radio technology and were reluctant to follow up his breakthrough. To make matters worse, it was the Great Depression, and Bell Labs could not justify diverting funds towards radio astronomy, so Jansky was forced to abandon his research. However, in time, Jansky's breakthrough would encourage

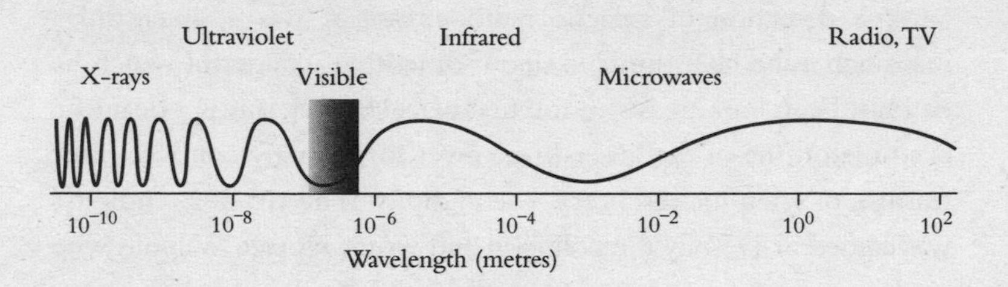

Figure 93 The spectrum of visible light is part of a much broader span of wavelengths known as the electromagnetic spectrum. All electromagnetic radiation, visible light included, consists of electric and magnetic vibrations. The range of visible light wavelengths is limited to a very narrow band of the electromagnetic spectrum. So, in order to study the universe as fully as possible, astronomers try to detect radiation across the full range of wavelengths, from billionths of a metre (X rays) to several metres (radio waves).

astronomers to broaden the scope of their observation beyond the visible spectrum.

Today's astronomers employ not only radio telescopes, but also infrared telescopes, X-ray telescopes and other equipment, giving them access to the entire electromagnetic spectrum of wavelengths. By exploring these different wavelengths, astronomers are able to study different aspects of the universe. For example, X-ray telescopes detect the shortest wavelengths, which is ideal for observing the most energetic events in the universe. And infrared telescopes are highly effective for peering through our own Milky Way, because infrared wavelengths punch through the galactic dust and gas that obscure visible light.

Exploiting every possible wavelength of light from celestial objects has become a central tenet of modern astronomy. Light, both visible and invisible, is the only avenue for studying the universe, so astronomers have to pick up every possible clue at every available wavelength.

On a slightly tangential point, it is interesting to note that

Jansky's detection of galactic radio emissions was a sheer fluke, inasmuch as he had stumbled upon something wonderful which he had not been looking for in the first place. In fact, this is a beautiful example of one of the lesser-known yet surprisingly commonplace features of scientific discovery – serendipity. The word 'serendipity' was coined in 1754 by the politician and writer Horace Walpole, who used it in a letter in which he recounted an accidental but fortunate discovery about an acquaintance:

> This discovery indeed is almost of that kind which I call serendipity, a very expressive word which, as I have nothing better to tell you, I shall endeavour to explain to you: you will understand it better by the derivation than by the definition. I once read a silly fairy tale, called *The Three Princes of Serendip*: as their highnesses travelled, they were always making discoveries, by accidents and sagacity, of things which they were not in quest of.

The history of science and technology is littered with serendipity. For example, in 1948 George de Mestral went for a stroll in the Swiss countryside, saw some prickly seeds on his trousers, noticed that their spiny hooks had got caught on the loops of the fabric and was inspired to invent Velcro. In another example of sticky serendipity, Art Fry was trying to develop superglue when he accidentally concocted a glue that was so weak that two objects that had been stuck together could easily be pulled apart. Fry, a keen member of his local church choir, coated bits of paper with his failed superglue and used them to mark pages in his hymnbook, at which point the Post-it note was born. An example of medical serendipity is Viagra, which was initially developed as a treatment for heart problems. Researchers became suspicious that it might have a positive side-effect only when the patients who had taken part in a clinical trial steadfastly refused to hand back their unused pills, even though the drug seemed to have had no significant impact on their heart problems.

It would be all too easy to label scientists who have exploited serendipity as merely lucky, but that would be unfair. All these serendipitous scientists and inventors were able to build upon their chance observations only once they had accumulated enough knowledge to put them into context. As Louis Pasteur, who himself benefited from serendipity, put it: 'Chance favours the prepared mind.' Walpole also highlighted this in his original letter when he described serendipity as the result of 'accidents and sagacity'.

Furthermore, those who want to be touched by serendipity must be ready to embrace an opportunity when it presents itself, rather than merely brushing down their seed-covered trousers, pouring their failed superglue down the sink or abandoning a failed medical trial. Alexander Fleming's discovery of penicillin depended on a speck of penicillium mould floating in through the window, landing in a petri dish and killing off a bacterial culture. It is highly likely that many microbiologists had previously had their bacterial cultures contaminated by penicillium mould, but they had all discarded their petri dishes in frustration instead of seeing the opportunity to discover an antibiotic that would save millions of lives. Winston Churchill once observed: 'Men occasionally stumble over the truth, but most of them pick themselves up and hurry off as if nothing had happened.'

Returning to radio astronomy, we shall see that serendipity would turn out to be responsible for more than just giving birth to this new observational technique. In the years to come it would play a central role in several discoveries in this field.

For example, during the Second World War, the schoolteacher Stanley Hey was seconded to the Army Operational Research Group to work on the British radar research programme. As well as looking into the transmission and reception of radio waves, which was the basis of radar, Hey was asked to address a particular problem that was confronting Allied radar. Operators monitoring their radar systems occasionally found their screens lighting up like Christmas trees, preventing them from identifying enemy bombers among the multitude of signals. The assumption was that German engineers had developed a new radar jamming technology based on blasting radio waves at British radar stations. Hey set himself the task of working out how the Germans were generating such powerful radio jamming signals, which in turn might help him to find a way of countering them. Then, in the spring of 1942, he figured out that the British problem had nothing to do with the Germans.

Hey noticed that the jamming appeared to be coming from the east in the morning, from the south around lunchtime and from the west in the afternoon, and then stopped at sunset. Clearly this was no Nazi secret weapon, but merely the result of radio emissions from the Sun. It so happened that the Sun was at a peak in its eleven-year sunspot cycle, and that the intensity of the radio emissions was linked with strong sunspot activity. By researching radar, Hey had accidentally discovered that the Sun – and presumably all stars – emit radio waves.

Hey seemed to have a knack for serendipity, because in 1944 he made another lucky discovery. Using a special radar system pointed up at a steep angle, which he had developed for spotting incoming V-2 rockets, Hey noticed that meteors also emitted radio signals as they sizzle through the atmosphere.

When the frenzy of wartime radar research ended in 1945, there was a large amount of redundant radio equipment and a large posse of equally redundant scientists who knew how to use it. It was for these reasons that radio astronomy now began to establish itself as a serious field of research. Two of the first full-time radio astronomers were Stanley Hey and fellow wartime radar researcher Bernard Lovell, who managed to obtain an ex-army mobile radar unit and

Figure 94 Stanley Hey's wartime discoveries were given new life when they were featured in a cartoon strip in the 'Frontiers of Science' section of the *Daily Herald* in April 1963.

embarked on a programme of radio astronomy observations. This was only the starting point for Lovell, who went on to set up a radio astronomy observatory in Manchester. Radio interference from passing trams forced him to move to Jodrell Bank, a botanical park some 30 kilometres south of the city, where he began to construct a world-class radio observatory. Meanwhile, Martin Ryle at Cambridge University tried to keep pace with Jodrell Bank, and it was he who would plunge radio astronomy into the heart of the Big Bang versus Steady State controversy.

Ryle, who graduated in physics in 1939, had also worked on radar during the war. He had been drafted into the Telecommunications Research Establishment to work on airborne radar, and then moved to the Air Ministry Research Department where he discovered how to jam the V-2 rocket guidance system. His greatest wartime achievement was as part of the top-secret Moonshine project, which could simulate a naval or airborne attack by generating fake signals on German radar. In the run-up to D-Day, he helped to distract and confuse the German military by simulating two massive naval assaults on the French coast far from where the actual landings took place.

After the war, Ryle scavenged ex-military equipment and set out to improve the accuracy of radio astronomy measurements. Compared with an optical telescope, radio telescopes are notoriously poor at resolving exactly where a signal is coming from, a consequence of the fact that radio waves are longer than waves of visible light. Ryle overcame this problem in 1946 by helping to pioneer a technique known as *interferometry*, whereby the signals from several radio telescopes can be combined to improve their overall accuracy.

Consequently, in 1948 Ryle was able to embark on a detailed survey of the sky to find out if there were objects that emitted very little visible light, but large amounts of radio waves instead. Such objects would have been invisible to optical telescopes, but might show up clearly with his radio telescope. Ryle's approach was similar to the way the police might go about searching for an escaped prisoner on a dark night. They could use a pair of optical binoculars to scan the horizon, but they would see nothing because the prisoner does not emit any light and the night is very dark. But if instead they use a thermal camera, designed to detect infrared radiation emitted by any warm body, then the prisoner would show up very clearly. Alternatively, if the prisoner were to use a mobile phone to contact his accomplices, the phone would be emitting radio waves and the police could use a radio detector to pin down his location. In other words, different objects emit energy at different wavelengths, and if you want to 'see' the objects then you have to use an appropriate detector tuned to the correct wavelength.

Ryle's first survey, known as the First Cambridge (or 1C) Survey, mapped fifty distinct radio sources. These celestial objects emitted strong radio signals, but were otherwise invisible. Immediately questions were raised over the interpretation of these objects. Ryle believed that they were a new type of star within our own Milky Way galaxy, but others, such as the Steady State supporter Thomas Gold, argued that they were independent galaxies. Gold had harboured ambitions to lead the Cambridge radio astronomy group but Ryle had beaten him to the job, so this scientific dispute was tainted with personal animosity.

Ryle did not take Gold's opinion seriously because Gold was a theorist and not an observational astronomer. Without specifically mentioning his name, Ryle publicly dismissed the views of Gold at a meeting at University College, London, in 1951: 'I think the theoreticians have misunderstood the experimental data.' In other words, theoreticians had no idea what they were talking about. Hoyle was present, and felt that Ryle's tone implied that theoreticians were 'some inferior and detestable species'.

The question of whether these celestial radio sources were stars or galaxies was settled over the course of the next year. The Cambridge group was able to specify the location of the radio source labelled Cygnus A with such precision that Walter Baade at the Mount Palomar Observatory was able to point the 200-inch telescope at the area in question in an attempt to detect an optical signal. For Baade, seeing was believing: 'I knew something was unusual the moment I examined the negatives. There were galaxies all over the plate, more than two hundred of them, and the brightest was at the centre . . . It was so much on my mind that while I was driving home for supper, I had to stop the car and think.'

Baade had shown that Ryle's radio source was in exactly the same position as a hitherto unseen galaxy. Therefore, he concluded that the galaxy was the source of the radio waves, not a star. Baade had proved that Ryle was wrong and that Gold was right. Having confidently associated one of Ryle's radio sources with a galaxy, astronomers would go on to link the majority of other radio sources in the 1C survey with galaxies. These galaxies, which predominantly emitted radio waves rather than visible light, became known as *radio galaxies*.

Gold always remembered the moment when Baade first approached him at a conference with the news that Cygnus A was a radio galaxy:

> In the large antechamber to the conference room one was milling around like one usually does, and Walter Baade was there. He said, 'Tommy! Come over here! Look what we've got!' . . . Then Ryle comes into the room. Baade shouts, 'Martin! Come over here! Have a look at what we've found!' Ryle comes and looks with a very stern face at the photographs, does not say a word, throws himself on a nearby couch – face down, buried in his hands – and weeps.

Ryle had staked his professional reputation on the fact that the radio sources in the 1C survey were stars, whereas his critics, mainly Hoyle and Gold, had relentlessly argued in favour of radio galaxies. This was a battle that had become increasingly antagonistic, so Ryle was devastated when he had to admit that Hoyle and Gold had been right all along.

Embarrassed and humiliated, Ryle decided that he would have his revenge on Hoyle and Gold if he could find new evidence against the Steady State model and in favour of the Big Bang. In particular, Ryle focused on trying to measure the distribution of young galaxies. The significance of this distribution was mentioned earlier as the fourth criterion in the table of decisive issues in the Steady State versus Big Bang debate (Table 4, pp. 370–1). Essentially, the two models predicted two distinctly different distributions for young galaxies:

(1) *The Big Bang model* says that young galaxies could have existed only in the early universe, because they would have matured as the universe grew older. Nevertheless, we should still be able to see young galaxies, but only in the far reaches of space because it would have taken billions of years for the light from distant galaxies to reach us, so we see them as they were in the early universe.

(2) *The Steady State model* says that young galaxies should be much more evenly distributed. In a Steady State universe young galaxies should be born all the time from the matter created throughout the universe in between the receding galaxies. Therefore we should see young galaxies in our own neighbourhood as well as far away.

Crucially, astronomers believed that radio galaxies were, in very general terms, younger than the average galaxy. Therefore, if the Big Bang model was right, radio galaxies should generally exist very far from our Milky Way. Alternatively, if the Steady State model was right, they should appear both near and far. Therefore, measuring the distribution of the radio galaxies would be a conclusive way to test which model was correct.

Ryle decided to apply this critical test, quietly hoping that it would go against the Steady State model and in favour of the Big Bang. Following on from his 1C survey, he embarked on a series of increasingly rigorous surveys, imaginatively titled the 2C, 3C and 4C surveys. Along the way he constructed the Mullard Observatory, thereby making Cambridge a world-class centre for radio astronomy. Radio astronomy is less vulnerable than optical astronomy when it comes to poor weather because radio waves are not blocked by clouds. Radio telescopes located in Cambridge could therefore compete with the rest of the world, even during a miserable British winter.

By 1961, Ryle had catalogued five thousand radio galaxies and analysed their distribution. He was unable to measure the exact distance to every radio galaxy, but he could apply a sophisticated statistical argument to deduce whether their distribution was compatible with the Steady State or the Big Bang model. The result was clear: the radio galaxies tended to be more common at greater distances, exactly as the Big Bang model predicted. Ryle checked his result with another radio astronomy group in Sydney, who had been conducting a similar survey in the southern hemisphere. They agreed that the distribution of radio galaxies favoured the Big Bang model.

Ten years earlier Baade had proved that most radio sources were galaxies, meaning that Ryle was wrong and that Gold and Hoyle were right. At last, Ryle could turn the tables and exact his revenge. He organised a press conference in London to present the results and invited Hoyle to attend. To maximise the impact of the announcement, Ryle did not warn Hoyle in advance of what he was going to say. This turned the press conference into a ritual humiliation for Hoyle, because he misinterpreted the invitation and expected a completely different set of results. Hoyle later recalled: 'Surely, if [the results] were adverse, I would hardly have been set up so blatantly. Surely, it must mean that Ryle was about to announce results in consonance with the Steady State theory . . . I sat there, hardly listening, becoming more and more convinced that, incredible as it might seem, I really had been set up.'

Ryle's observations clearly endorsed the Big Bang model, which described a universe with a finite history and a moment of creation. Within a few hours the evening newspaper hawkers were hollering 'The Bible was right!' Hoyle wanted to hide himself away and analyse Ryle's data, hoping to find a serious flaw, but neither the public nor the press would give him or his family any peace: 'For the next week my children were ragged about it at school. The telephone rang incessantly. I just let it ring, but my wife, fearing that something had happened to the children, always answered, fending off the callers.'

Gamow was cheered by the news of Ryle's measurements and marked the pro-Big-Bang breakthrough with one of his infamous pieces of doggerel, displayed in Figure 95. The poem paints a vivid picture of the ongoing tension between Ryle and Hoyle.

The Steady State brigade had put its neck on the line by making a firm prediction that the universe would be shown to be the same everywhere, with young galaxies distributed both near and far. Had Ryle's result supported that prediction, then Hoyle would have had no hesitation in embracing it as evidence in favour of his model. Hoyle should have had equal respect for Ryle's result even though it contradicted the Steady State model, but instead he tried to find fault with the observations, in terms of both how they had been gathered and how they were being interpreted.

Hoyle pointed out that Ryle's measurements varied significantly from the 2C to the 3C survey, and then from the 3C to the 4C survey, insinuating that a fifth survey might give a different result that was more in keeping with the Steady State model. Gold backed Hoyle, dubbing the constantly shifting results the 'Ryle effect'. Gold also promoted the idea that radio astronomy was a new discipline that could not yet be trusted, and said: 'I do not think that the kind of observations being referred to are capable of giving such a verdict.'

Ryle acknowledged that there had been errors in the past, but he was adamant that the 4C survey was reliable and reiterated that it had been independently confirmed by Australian astronomers. On one occasion, when Hermann Bondi was continuing the Steady State onslaught against the 4C survey, Ryle eventually snapped. According to Martin Harwit, Ryle 'flew into a rage, which resulted in the nastiest public display of tempers between scientists that I have seen in more than 30 years as a professional astrophysicist'.

Although Hoyle, Gold and Bondi refused to accept Ryle's conclusion about the distribution of radio galaxies, a growing number of cosmologists could see that the Big Bang model was in the ascendancy and that the Steady State model was looking decidedly unsteady. Worse still, Ryle's radio galaxy measurements were about to deal yet another blow to the Steady Statesmen.

In 1963, the Dutch-American astronomer Maarten Schmidt was studying radio source number 273 from Ryle's 3C survey catalogue, known routinely as 3C 273. By this time, most radio sources were thought to be distant galaxies, but the radio signal from object 3C 273 was so strong that the object was assumed to be a new type of peculiar nearby star within our own Milky Way. Furthermore, 3C 273 could be seen with optical telescopes as a point of light rather than a blur, which reinforced the view that it was a star rather than a galaxy. Schmidt set about trying to measure the wavelengths of light that were being emitted by 3C 273 in order to deduce its composition, but at first he was bemused because the

Figure 95 This poem was written by George Gamow and appeared in his book *Mr Tompkins in Wonderland*. It describes Martin Ryle's research into the distribution of radio galaxies and Fred Hoyle's reaction.

The Steady State Is Out of Date

'Your years of toil'
Said Ryle to Hoyle
'Are wasted years, believe me.
The Steady State
Is out of date
Unless my eyes deceive me.

My telescope
Has dashed your hope;
Your tenets are refuted.
Let me be terse:
Our Universe
Grows daily more diluted!'

Said Hoyle, 'You quote
Lemaître, I note,
And Gamow, well, forget them!
That errant gang
And their Big Bang,
Why aid them and abet them?

You see, my friend,
It has no end
And there was no beginning.
As Bondi, Gold
and I will hold
Until your hair is thinning!'

'Not so!' cried Ryle
With rising bile,
And straining at the tether;
'Far galaxies
Are, as one sees,
More tightly packed together!'

'You make me boil!'
Exploded Hoyle,
His statement rearranging.
'New matter's born
Each night and morn.
The picture is unchanging!'

'Come off it, Hoyle!
I aim to foil
You yet' (the fun commences).
'And in a while,'
Continued Ryle,
'I'll bring you to your senses!'

wavelengths did not seem to correlate with those emitted by any known atoms.

Suddenly he realised what was causing his confusion. He was detecting the well-established wavelengths associated with hydrogen, except that they had been redshifted to an extent never seen before. This was astonishing because 3C 273 was supposed to be a local star, and local stars travel at less than 50 km/s, far too low a speed to account for the redshift observed by Schmidt. In fact, the redshift measurements implied that 3C 273 was receding at 48,000 km/s, roughly 16% of the speed of light. According to Hubble's law, this implied that 3C 273 was the most distant object ever detected, over a billion light years from the Milky Way. Object 3C 273 was not a reasonably bright local star, but a fantastically brilliant far-off galaxy, several hundred times brighter than the brightest galaxies hitherto known. However, its brightness was largely in the form of radio waves rather than visible light.

3C 273 became known as a *quasi-stellar radio object* (or *quasar*), because it was a radio galaxy whose extreme distance and brightness gave it the deceptive appearance of a local star. It was not long before several other radio sources were identified as exceptionally brilliant and far-flung quasar galaxies. Not surprisingly, Gamow celebrated the discovery of quasars with yet another poem, this time stressing the point that astronomers had no idea what was powering these distant quasar galaxies:

Twinkle, twinkle, quasi-star,
Biggest puzzle from afar.
How unlike the other ones,
Brighter than a billion suns.
Twinkle, twinkle, quasi-star,
How I wonder what you are!

Another quasar mystery – one highly relevant to the Big Bang versus Steady State debate – concerned their distribution. Every single quasar seemed to be situated in the far reaches of the cosmos. Proponents of the Big Bang theory were in no doubt about what this meant. They argued that if quasars could be perceived only in the far distance, then it would have taken billions of years for the light to reach us, so we were seeing them as they were billions of years ago – which implied that quasars existed only in an earlier era of the universe. Perhaps the hotter, denser conditions of the early universe were conducive to creating brilliant quasars. According to the Big Bang model, it was quite possible that there were once quasars near to us in the early universe, but in time they would have evolved into ordinary galaxies, which is why we do not see any local quasars today.

However, the quasar distribution was problematic for Hoyle, Gold and Bondi, because the Steady State model claimed that the universe was the same at all times and in all places. If there were quasars far away and in the past, then there should also be quasars right here right now, which did not seem to be the case. The Steady Statesmen tried to save face by suggesting that quasars were rare objects, so perhaps the reason that we did not have any in our neighbourhood was nothing more than bad luck. Also, nobody could explain the true nature of quasars or the power source behind their extraordinary brilliance, so Hoyle, Gold and Bondi argued that their Steady State model could not be overturned by such poorly understood phenomena.

These were weak excuses. The Steady State model was beginning to lose credibility, and an increasing number of cosmologists were moving towards the Big Bang camp. Dennis Sciama, who was one of those who switched sides, called the quasar observations 'the most decisive evidence so far obtained against the Steady State model of the universe'. His change of mind appears to have been a traumatic experience: 'For me the loss of the Steady State theory has been the cause of great sadness. The Steady State theory has a sweep and beauty that for some unaccountable reason the architect of the universe appears to have overlooked. The universe in fact is a botched job, but I suppose we shall have to make the best of it.'

Radio astronomy was opening up a new window onto the universe, discovering entirely new objects and providing critical evidence in the Big Bang versus Steady State debate. Regrettably, Karl Jansky, the father of radio astronomy, received virtually no credit during his lifetime for inadvertently inventing the radio telescope and for making the first radio observations of the sky. He passed away in 1950 at the age of just forty-four. It was only in the decade after his death that radio astronomy would establish itself as a truly major discipline within astronomy.

However, Karl Jansky was eventually immortalised. In 1973 the International Astronomical Union recognised his contribution by naming the unit of radio flux in his honour. This unit, the *jansky*, is used by radio astronomers to indicate the strength of any radio source. A strong quasar might measure 100 janskys, whereas a weak radio object might measure just a few millijanskys.

Bell Laboratories, who had sponsored Jansky's work on radio astronomy, paid their own tribute to him by establishing an ongoing programme of research in radio astronomy. In particular, Bell Labs provided a home for the most famous double act in the history of radio astronomy: an outspoken, ambitious Jewish refugee and a quiet, studious scientist from the oilfields of Texas. Together they would make a discovery that would rock the cosmological establishment.

The Penzias and Wilson Discovery

Arno Penzias was born into a Jewish family in Munich on 26 April 1933, the day that the Gestapo was formed. He first encountered anti-Semitism when he was four years old, while travelling on a trolley-car with his mother:

> When you are the adored eldest son, you sort of get the feeling that you should show off all the time. I said something that made it clear to the other people there that I was Jewish, and that so changed the atmosphere of the trolley-car that my mother had to take us off and wait for the next one. From that incident I learned that I was not supposed to talk about being Jewish in public and that, if you did, you put your family in danger. It was a big shock for me.

Although he was born in Germany, Penzias's father was a Polish citizen, which placed the family under particular pressure. The German authorities had threatened to arrest Poles who refused to leave the country, but the Polish government had cancelled Jewish passports on 1 November 1938, so the Penzias family were unable to cross any borders. It seemed as though they had no prospect of escape from Nazi persecution. However, a campaign was started in America to encourage people to rescue Jewish families by claiming them as relatives, a purely humanitarian ploy that would allow families to obtain permission to leave Germany. With only a month to spare, the Penzias family were informed that an American was willing to sponsor their exit visas, and in the spring of 1939 they fled to Britain. From there they boarded a steamship to New York and started a new life in the Bronx.

Arno's father had run a leather business back in Munich, but now he was forced to take a job as a janitor in an apartment block, stoking the building's furnace and emptying the bins. Arno saw how his father struggled to make a living, and at the same time he noticed that 'people who went to college seemed to dress better and eat more regularly'. Eager for such comfort and security, he worked hard, excelled in his schoolwork, and went on to earn a place at college.

Penzias's passion was physics, but he was concerned that he might not be able to make a living as a physicist, so he asked his father's advice about which subject he should pursue: 'He said that physicists think they can do anything that an engineer can do, and if they can do that they can at least make a living as engineers. In those days, the physics majors were the curve busters. They were the odd, bright kids who didn't fit in. The top bright kids seemed to be attracted to it for aesthetic reasons.'

Earning his first degree at the tuition-free City College of New York, Arno Penzias later embarked on a Ph.D. in radio astronomy at Columbia University's physics department, which by 1956 had already won three Nobel prizes. Penzias's supervisor was Charles Townes, who would become Columbia's fourth Nobel Laureate in physics for his development of the maser, the microwave equivalent of a laser. Penzias's thesis project required him to build an ultra-sensitive radio receiver which incorporated Townes's maser as a key component.

Although the radio receiver performed very well, it did not enable Penzias to achieve his main goal, which was to detect radio waves emitted by the hydrogen gas clouds that were supposed to populate the space in between galaxies. Penzias called his final doctoral thesis 'dreadful', although inconclusive might be a kinder description. Either way, in 1961 he did earn his Ph.D. and left Columbia to take up a research post at Bell Labs, the only industrial laboratory in the world that would employ a budding radio astronomer.

As well as conducting his own pure research, Penzias was also expected to pitch in and help with the more commercial research projects that were being undertaken. For example, Bell Labs had designed Telstar, the first active communications satellite, and after its launch the developers ran into problems trying to point their antenna at the satellite. The new boy Penzias stood up in front of the thirty-strong antenna committee and explained how they could use the known position of a radio galaxy to calibrate the antenna direction and thereby find Telstar. This was a perfect synthesis of pure and commercial research. Penzias's solution was a testament to the Bell Labs ethos of employing pure scientists alongside applied scientists and engineers.

For two years Penzias was the only radio astronomer at Bell Labs, but in 1963 he was joined by Robert Wilson. The young Texan had developed an interest in science while accompanying his father, a chemical engineer, around the local oilfields. He went on to study physics at Rice University in Houston, and after graduating he went to Caltech in 1957 to study for a doctorate. It was there that Wilson took a graduate course on cosmology given by Fred Hoyle, who had become a regular visitor to the California college following his 1953 collaboration with Willy Fowler. Just like Penzias, Wilson's thesis focused on radio astronomy, and after its completion he too abandoned academia and headed for Bell Labs.

Wilson was partly attracted to Bell Labs because of its 6-metre horn radio antenna sited at nearby Crawford Hill, shown in Figure 96. This was originally designed to detect signals from the innovative Echo balloon satellite, which had been launched in 1960. Echo had been squeezed into a 66-centimetre sphere for launching into orbit, but once in space it was inflated into a giant silver globe, 30 metres in diameter, which was capable of passively bouncing signals between an Earth-based transmitter and receiver. However, government intervention in this sector of the communications industry persuaded AT&T to withdraw from the Echo project for economic reasons, leaving the horn antenna free to be transformed into a radio telescope. The horn antenna was doubly suited for radio astronomy: it was largely shielded from local radio interference, and its size meant that it could locate the source of celestial radio signals with good accuracy.

Penzias and Wilson got permission from Bell Labs to spend some of their time scanning the skies to study the various radio sources, but before they could do any serious surveying they first had to fully understand the radio telescope and all its quirks. In particular, they wanted to check that it was picking up a minimal level of 'noise', a technical term used to describe any random interference that might obscure a genuine signal.

This is exactly the same as the noise that you might encounter when you tune your domestic radio to hear a particular radio station. The station's signal might be contaminated with a hiss, which is the noise. There is always a battle between signal and noise, and ideally the signal should be much stronger than the noise. This is generally the case with a domestic radio tuned to a local radio station, because you can usually hear the broadcast very clearly and the noise is insignificant. However, if you tune into a foreign radio station, the signal might well be weaker and the noise level would have a more serious impact on the clarity of the broadcast. In the worst case, the radio signal is completely swamped by the noise and it is impossible to hear anything properly.

In radio astronomy, the signals from a distant galaxy are so feeble that the issue of noise is paramount. To check the noise level, Penzias and Wilson pointed their radio telescope at a part of the sky devoid of radio galaxies, a region where there should be virtually no radio signals from space. Hence, anything that was detected could be attributed to noise. They fully expected the noise to be negligible, so were surprised to discover an unexpected and annoying level of noise. The noise level was disappointing, but not so high that it would seriously affect the measurements that they were intending to make. Indeed, most radio astronomers would have ignored the problem and embarked on their survey. Penzias and Wilson, however, were determined to conduct the most sensitive survey possible, so they immediately set about

Figure 96 Robert Wilson (left) and Arno Penzias posing in front of Bell Laboratories' horn antenna at Crawford Hill, New Jersey. This radio telescope is essentially a giant glorified radio receiver. Its aperture is 6 metres square and the monitoring equipment is housed in a hut at the apex of the cone.

trying to locate the source of the noise and, if possible, reduce it or remove it completely.

Noise sources can be broadly split into two types. First, there is extraneous noise, which is caused by some entity beyond the radio telescope, such as a major city on the horizon or some nearby electrical equipment. Penzias and Wilson surveyed the landscape for any spurious noise sources, and even pointed the telescope towards New York, but there was no increase or decrease in the noise. They also monitored the noise level with time, but again the noise was continuous. In short, the noise was absolutely constant regardless of when and where the telescope was pointed.

This forced the duo to explore the second category of noise, namely noise that is inherent in the equipment. The radio telescope consisted of numerous components, each with the potential to generate its own noise. Exactly the same problem arises with your domestic radio: even if the broadcaster has a strong signal, it can be degraded by noise generated by your radio's amplifier, speaker or wiring. Penzias and Wilson checked every single element of their radio telescope, looking for dodgy contacts, sloppy wiring, faulty electronics, misalignments in the receiver, and so on. Even joints that already seemed to be okay were patched up with aluminium tape just to be sure.

At one point, attention focused on a pair of pigeons that had nested inside the horn antenna. Penzias and Wilson thought that the 'white dielectric material' deposited by the pigeons and smeared on the horn might be the cause of the noise. So they trapped the birds, placed them in a mail van and had them released 50 kilometres away at the Bell Labs site in Whippanny, New Jersey. They scrubbed the antenna until it was bright and shiny, but alas the pigeons obeyed their homing instinct, flew back to the telescope's horn and started depositing white dielectric material all over again. Penzias captured the pigeons once more, but this time he reluctantly decided to get rid of them for good: 'There was a pigeon fancier who was willing to strangle them for us, but I figured the most humane thing was just to open the cage and shoot them.'

After a year of checking, cleaning and rewiring the radio telescope, there was a reduction in the level of noise. Penzias and Wilson could attribute some of the remaining noise to atmospheric effects and some of it to the walls of the radio telescope's horn, and they had to accept that both these noise sources were simply unavoidable. However, this still did not fully account for all of the noise that they were detecting. They had devoted an immense amount of time, effort and money into understanding and minimising the noise of their radio telescope, yet there was one element of noise that was both mysterious and incessant: something, somewhere, somehow was emitting radio waves all the time from all directions.

What the two frustrated radio astronomers had not realised was that they had stumbled into one of the most important discoveries in the history of cosmology. They were completely oblivious to the fact that the omnipresent noise was actually a remnant of the Big Bang: it was the 'echo' from the early expansion phase of the universe. This annoying 'noise' would turn out to be the most convincing evidence yet that the Big Bang model was correct.

If you recall, Gamow, Alpher and Herman had calculated that the universe would undergo a transition roughly 300,000 years after the Big Bang. By this time the universe's temperature would have fallen to roughly 3,000°C, cool enough for the previously free-floating electrons to latch onto nuclei and form stable atoms. The sea of light that filled the universe could no longer interact with either the charged electrons or the charged nuclei, because they had bonded to each other to form neutral atoms. Ever since this moment in the history of the universe, known as recombination, the primordial light has been allowed to pass through the universe completely unchanged – except in one important respect.

Gamow, Alpher and Herman had predicted that, as the universe expanded with time, the wavelength of that primordial light would have been stretched as space itself has been stretched. The light had a wavelength of roughly one-thousandth of a millimetre when it originally emerged from the cosmic fog when the universe was 300,000 years old, but according to the Big Bang model the universe has since expanded by roughly a factor of a thousand. Therefore those light waves should now have a wavelength of roughly 1 millimetre, which would place them in the radio region of the electromagnetic spectrum.

The echo from the Big Bang had transformed itself into radio waves and was being detected as noise by Penzias and Wilson's radio telescope. These waves can be assigned to a sub-category of the radio spectrum known as microwaves, which is why this Big Bang echo came to be known as the cosmic microwave background (CMB) radiation. The existence or non-existence of the CMB radiation was critical to the Big Bang versus Steady State debate, and is listed as the fifth criterion in Table 4, pp. 370–1.

Although the existence of the CMB radiation had been clearly predicted back in the 1940s, the scientific community had largely forgotten about it by the 1960s. That is why Penzias and Wilson failed to make a link between their radio noise and the Big Bang model. However, to their great credit, they refused to ignore the mysterious radio noise and remained distressed and perplexed. They continued to discuss it between themselves and with their colleagues.

Towards the end of 1964, Penzias attended an astronomy conference in Montreal at which he casually mentioned the noise problem to Bernard Burke of the Massachusetts Institute of Technology. A couple of months later, Burke telephoned him excitedly. He had received a preliminary draft of a paper describing the work of cosmologists Robert Dicke and James Peebles at Princeton University. The paper explained that the Princeton team had been studying the Big Bang model and had realised that there ought to be an all-pervasive CMB radiation, which today should reveal itself as a radio signal with a wavelength of a millimetre or so. Dicke and Peebles had no idea that they were walking in the fifteen-year-old footsteps of Gamow, Alpher and Herman. Independently and belatedly, they were re-postulating the CMB radiation. Dicke and Peebles also had no idea that Penzias and Wilson had detected the CMB radiation at Bell Labs.

To summarise, Gamow, Alpher and Herman had predicted the CMB radiation in 1948. But everybody had forgotten this prediction within a decade. Then in 1964 Penzias and Wilson discovered the CMB radiation, but did not realise it. At roughly the same time, Dicke and Peebles predicted the CMB radiation, unaware that the prediction had already been made in 1948. Eventually, Burke told Penzias about the prediction of Dicke and Peebles.

Suddenly, everything fell into place for Penzias. At last, he understood the source of the noise that had plagued his radio telescope and appreciated how highly significant it was. At long last the mystery of the omnipresent noise had been solved. It was nothing to do with pigeons, dodgy wiring or New York, but it had everything to do with the creation of the universe.

Penzias telephoned Dicke and told him that he had detected the CMB radiation described in the Princeton paper. Dicke was stunned, particularly because of the timing of Penzias's phone call. It had interrupted a lunchtime meeting which had been arranged to discuss the construction of a CMB radiation detector at Princeton, because Dicke and Peebles had wanted to test their own prediction. However, such a detector was now pointless, because Penzias and Wilson had already verified the prediction. Dicke put the phone down, turned to his group and exclaimed: 'Boys, we've been scooped!' Wasting no time, Dicke and his team visited Penzias and Wilson the following day. An inspection of the radio telescope and an examination of the data confirmed the scoop. The race to find the CMB radiation was over and the Bell Labs team had unwittingly beaten their Princeton rivals.

In the summer of 1965, Penzias and Wilson published their result in the *Astrophysical Journal.* Their modest 600-word paper conservatively announced exactly what they had detected without offering any personal interpretation. Instead, it was left to Dicke and his team, who published a sister paper in the same journal, to explicitly link Penzias and Wilson's observations with the CMB radiation. They explained how the Bell Labs duo had discovered the predicted echo from the Big Bang. It was a beautiful marriage. Dicke's team had a theory but no observational data, while Penzias and Wilson had observational data but no theory. Putting the Princeton and Bell Labs research together had turned an irritating problem into a tremendous triumph.

The Big Bang model clearly predicted the existence of the CMB radiation and the wavelength it should have today. In contrast, the Steady State model made no mention of the CMB radiation and could not conceive of a scenario in which the universe would be filled with microwaves. Consequently, the discovery of the CMB radiation seemed to be the decisive evidence that proved that the universe started billions of years ago with an almighty Big Bang.

Therefore, the discovery of the CMB radiation also disproved the Steady State model. Wilson's euphoria at establishing the existence of the CMB radiation and the veracity of the Big Bang theory was tinged with sorrow because he had always maintained a certain fondness for the Steady State model: 'I had taken my cosmology from Hoyle at Caltech, and I very much liked the Steady State universe. Philosophically, I still sort of like it.'

His sadness was no doubt moderated by the plaudits that soon flooded in. The NASA astronomer Robert Jastrow said that Penzias and Wilson had 'made one of the greatest discoveries in 500 years of modern astronomy'. And Harvard physicist Edward Purcell was prepared to go even further in his praise of the detection of the CMB radiation: 'It just may be the most important thing anybody has ever seen.'

And yet all this was the result of sheer luck. Penzias and Wilson had been blessed by serendipity. Their primary objective had been to conduct a standard astronomical radio survey, but their greatest distraction turned out to be their greatest discovery. Three decades earlier, Karl Jansky had made a lucky discovery at Bell Labs and had thereby invented radio astronomy; now serendipity had struck again in the same scientific discipline and at the same research establishment. This time the discovery was even more magnificent.

The CMB radiation was just waiting to be discovered by anybody who happened to point a sufficiently sensitive radio antenna at the cosmos, and by chance that turned out to be Penzias and Wilson. However, the serendipitous nature of their discovery was nothing to be ashamed of, because such breakthroughs require not only luck but also considerable experience, knowledge, insight and tenacity. There is strong evidence that the Frenchman Emile La Roux in 1955 and the Ukrainian Tigran Shmaonov in 1957 separately detected the CMB radiation during each of their radio astronomy surveys, but they both shrugged off the apparent noise as a minor defect in their instruments that they were prepared to tolerate. They lacked the determination, persistence and rigour that allowed Penzias and Wilson to discover the CMB radiation.

Even before their paper was published, news of Penzias and Wilson's breakthrough spread rapidly through the cosmological community. The story reached the general public on 21 May 1965 thanks to a lead front-page story in the *New York Times*, which carried the banner headline SIGNALS IMPLY 'BIG BANG' UNIVERSE. Readers became fascinated by the discovery because it was of cosmic significance, yet possessed a certain homely charm. Penzias put it thus:

> When you go out tonight and you take your hat off, you're getting a little bit of warmth from the Big Bang right on your scalp. And if you get a very good FM receiver and if you get between stations, you will hear that sh-sh-sh sound. You've probably heard this kind of rushing sound. It's just sort of soothing. Sometimes it's not much different from the sound of the surf. Of the sound that you're listening to, about one half of one percent of that noise is coming from billions of years ago.

The article in the *New York Times* was an informal endorsement of the Big Bang model of creation. Einstein, Friedmann and Hubble, who had all contributed to the Big Bang model, were no longer alive to see its vindication. The only founding father who survived to witness the conclusion of the greatest cosmological debate in history was Georges Lemaître, who had pioneered the Big Bang's theoretical basis. He was recovering from a heart attack at the University of Louvain's hospital when he heard the news that the CMB radiation had been detected. He would die just one year later, at the age of seventy-one, having lived the life of a loyal priest and devoted cosmologist.

When Gamow, Alpher and Herman heard of the discovery of the CMB radiation, their joy was mixed with some bitterness. It was they who had predicted this echo of the Big Bang well before Dicke and Peebles, but they received virtually no acknowledgement for their pioneering efforts. They were not mentioned in the initial pair of papers announcing the breakthrough in the *Astrophysical Journal*; neither did their names appear in Dicke's subsequent overview in *Scientific American* magazine. Indeed, nearly every academic and popular article failed to mention Gamow, Alpher and Herman in the clamour that followed Penzias and Wilson's discovery.

Instead, Dicke and Peebles were the theorists associated with the prediction of CMB radiation. Dicke and Peebles were undoubtedly brilliant cosmologists, but they had merely retraced the trail already blazed in 1948. The problem was that cosmology had become dominated by a new generation of physicists, who were simply unfamiliar with the work of Gamow, Alpher and Herman.

Gamow tried whenever possible to establish priority for his team's prediction of the Big Bang echo. For example, when the CMB radiation was being discussed at an astrophysics conference in Texas, Gamow was asked if the recently discovered radiation was indeed the phenomenon that he, Alpher and Herman had predicted. Gamow stood at the podium and replied: 'Well, I lost a nickel around here someplace and now a nickel has been found about the place where I lost it. I know all nickels look about the same, but yes, I think it is my nickel.'

When Penzias eventually heard about the original CMB radiation prediction dating back to 1948, he sent Gamow a conciliatory note asking for more information. Gamow offered a detailed description of his earlier research accompanied with the statement: 'Thus, you see, the world did not start with almighty Dicke.'

Ralph Alpher felt even more indignant, because he had been largely responsible for the research programme that predicted the CMB radiation, and yet he received even less recognition than Gamow. He was still a young man when he predicted the CMB radiation and was therefore often overshadowed by Gamow. Worse still, the joky authorship team of Alpher, Bethe, Gamow (Alpha–Beta–Gamma) on the closely related nucleosynthesis paper had pushed him even further down the pecking order.

When a journalist later asked Alpher if he had felt offended by Penzias and Wilson's failure to acknowledge his contribution, he spoke his mind: 'Was I hurt? Yes! How the hell did they think I'd feel? I was miffed at the time that they'd never even invited us down to see the damned radio telescope. It was silly to be annoyed, but I was.'

In *Genesis of the Big Bang*, an account of their work, Alpher and Herman gave a more considered reaction:

> One does science for two reasons: for the thrill of understanding or measuring something for the first time and, having done so, for at least the recognition if not approbation of one's peers. Some colleagues argue that the progress of science is all that matters and that it is of little consequence who does what. Yet we cannot help noticing that these same colleagues are nevertheless pleased with recognition of their work and accept with pleasure and alacrity such approbation as election to prestigious scientific academies.

Meanwhile, recognition for Penzias and Wilson culminated a decade after their discovery with the award of the Nobel Prize for Physics in 1978. In the intervening years, astronomers had refined their measurements of the CMB radiation and accurately checked that all its features matched those that the Big Bang model had predicted. The CMB radiation and therefore the Big Bang model were apparently both genuine.

Penzias used the award ceremony as an opportunity to pay his own quiet tribute to his parents, who had rescued him from Nazi Germany and brought him to New York:

> I wanted, if I can call it that, a Jewish tuxedo, something made in the garment district. My mother worked there, and a whole generation of Jewish immigrants put the next generation through college by

> working there. I didn't want to buy a tuxedo in Princeton, or in a fancy New York store, where it might be sold to you by someone who would make you ashamed of the clothes you were wearing when you came in to buy it. I wanted the tuxedo to be me and not some sort of costume.

He also used the Nobel lecture as an opportunity to set the record straight, explicitly acknowledging and praising the contribution made by Gamow, Alpher and Herman. Penzias gave a historical overview of the development and proof of the Big Bang model, based largely on a lengthy discussion with Alpher that had taken place just a few weeks earlier. It seemed that Alpher had at last found a way of making his peace with the physics community.

Just a month later, however, Alpher suffered a severe heart attack. Perhaps he had become overwhelmed by the stress of fighting for recognition. Perhaps the utter disappointment of not having a share of the Nobel prize proved too much. Alpher gradually recovered, but he would continue to be dogged by ill health.

The Necessary Sprinkling of Wrinkling

The award of the Nobel prize to Penzias and Wilson marked the point at which the Big Bang model became part of the scientific mainstream. In due course, this model of cosmic creation would even find recognition in the Smithsonian's National Air and Space Museum. It was not easy to construct an exhibit which represented the theory and observation that lay behind the development of the Big Bang model, but the curators made some imaginative decisions. The Smithsonian chose to display the Cointreau bottle with which Gamow and Alpher had celebrated their breakthrough in nucleosynthesis, shown in Figure 83 (see p. 335). Ideally the museum would also have installed the six-metre Bell Labs radio telescope used to detect the CMB radiation, but that was impractical. Instead, it displayed the pigeon trap that had been used by Penzias and Wilson during their noise reduction operation, as shown in Figure 97.

The detection of the CMB radiation had given cosmologists a new-found confidence. Not only did the CMB radiation exist, but it existed at the expected wavelength. As well as implying that the Big Bang model was broadly correct, it also meant that cosmologists understood some of the detail of how the temperature and density of the universe had evolved in the wake of the Big Bang.

For the majority of researchers, the CMB radiation was conclusive evidence in favour of a moment of creation and an evolving universe, as opposed to an eternal universe that was essentially steady. As each year passed, increasing numbers of scientists switched their allegiance from the Steady State to the Big Bang model. American astronomers had been polled back in 1959, at the height of the Big Bang versus Steady State controversy, and then again in 1980 after Penzias and Wilson had won their Nobel prize. In 1959 the results showed that 33% of astronomers backed the Big Bang, 24% favoured the Steady State and 43% were unsure. In the 1980 poll, 69% of astronomers supported the Big Bang, only 2% stuck with the Steady State and 29% were unsure.

One of the defectors was the Steady State pioneer Hermann Bondi, who had once said: 'If there was a Big Bang, show me some fossils of it.' He now had to accept that the CMB radiation was a perfect fossil, and ceased to believe in the model that he had helped to create. Thomas Gold, however, kept the faith: 'I can't really see anything wrong with the Steady State theory. I'm not deflected by the numbers of people who believe in one thing or another. Science does not proceed by Gallup poll.'

Similarly, Hoyle continued to mock the Big Bang model and those who believed in it: 'The passionate frenzy with which the Big Bang cosmology is clutched to the corporate scientific bosom

Figure 97 The historic Hav-A-Heart pigeon trap that was used to remove the pigeons from the Bell Labs radio telescope, part of a concerted effort to explain the mysterious source of noise detected by Penzias and Wilson. The trap is now an exhibit at the Smithsonian National Air and Space Museum.

evidently arises from a deep-rooted attachment to the first page of Genesis, religious fundamentalism at its strongest.'

If Hoyle was going to turn the tide of opinion and win the argument, he would have to do more than cast aspersions at the Big Bang supporters. Working with colleagues such as Jayant Narlikar, Chandra Wickramasinghe and Geoffrey Burbidge, he adapted and transformed the original Steady State model into one that began to look more consistent with astronomical observations. The new Quasi-Steady State model required a universe that had regular phases of contraction in between the long-term expansion. And instead of claiming that matter was continually created, the revised model relied on matter being created in intense bursts. Despite these modifications, the Quasi-Steady State model of the universe failed to win widespread support.

Nevertheless, Hoyle continued to defend his model: 'I think that it is fair to say that the theory has demonstrated strong survival qualities, which is what one should properly look for in a theory. There is a close parallel between theory and observation on the one hand, and mutations and natural selection on the other. Theory supplies the mutations, observation provides the natural selection. Theories are never proved right. The best they can do is to survive.' But the Steady State model and its Quasi-Steady State reincarnation were barely surviving. Any unbiased observer could see that they were on the brink of extinction, whereas the Big Bang model was not only surviving, but thriving.

The universe simply made more sense in the context of the Big Bang model. For example, in 1823, when scientists assumed that the universe was infinite and eternal, the German astronomer Wilhelm Olbers wondered why the night sky was not ablaze with starlight. He reasoned that an infinite universe would contain an infinite number of stars, and if the universe was infinitely old then this would have allowed an infinite amount of time for the starlight to have reached us. Hence, our night sky ought to be flooded with an infinite amount of light from all these stars.

The obvious lack of this infinite light from space is known as *Olbers' paradox*. There are various ways to explain why the night sky is not infinitely bright, but the Big Bang explanation is perhaps the most convincing. If the universe was created just a few billion years ago, then the starlight would only have had enough time to reach us from a limited volume of space, because light travels at only 300,000 km/s. In short, a finite age for the universe and a finite speed of light results in a night sky with only a finite amount of light, which is what we observe.

The clearest way to illustrate the superiority of the Big Bang

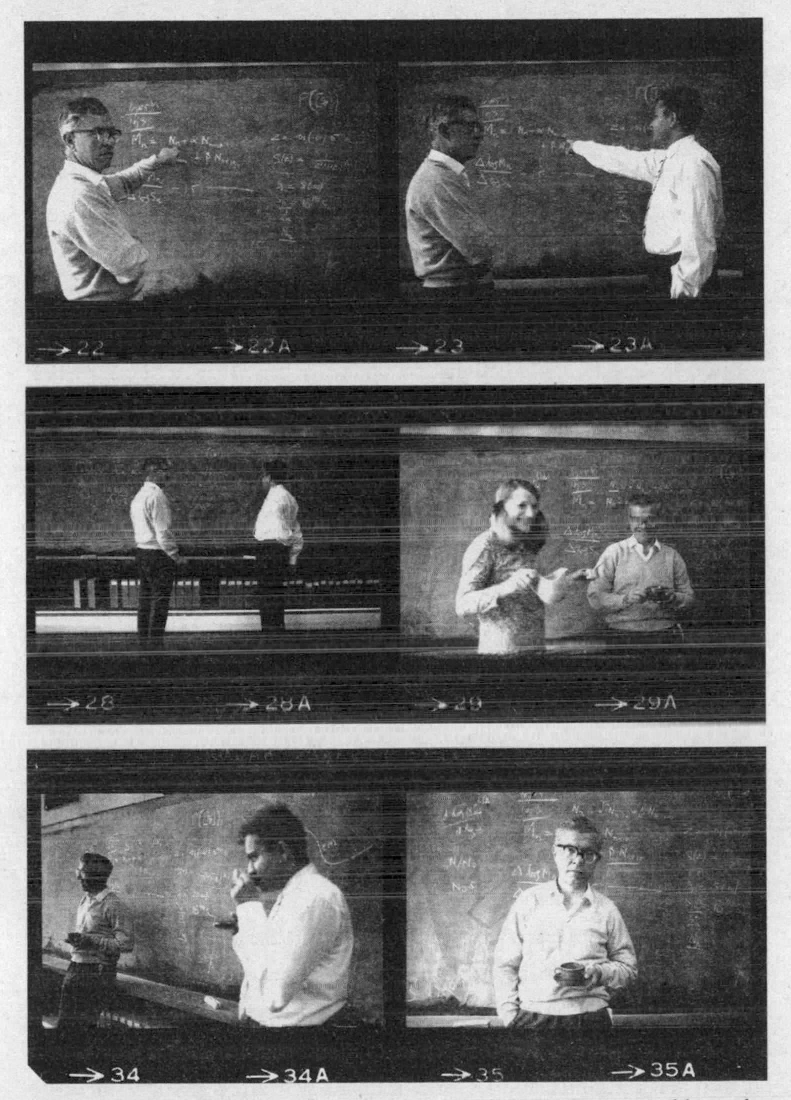

Figure 98 Fred Hoyle with his friend and colleague Jayant Narlikar, who helped him to develop the Quasi-Steady State model of the universe. Their theorising at the blackboard is fuelled by a cup of tea.

over the Steady State is to revisit our table of critical criteria, which appeared at the start of this chapter (Table 4, pp. 370–1). It presented the state of the debate in 1950, with some findings favouring the Big Bang and others supporting the Steady State model. Ever since 1950, however, each new observation seemed to back the Big Bang model and undermine the Steady State model, as demonstrated in Table 6 (pp. 444–5), which shows the state of play in 1978, when Penzias and Wilson won their Nobel prize.

As measured against the seven decisive criteria, the Big Bang model was stronger in four cases. The three remaining criteria could be judged as one outright victory for the Steady State model, one that was a success for both models and one that was a failure for both models.

Leaving aside the issue of creation, which remained a difficulty for both models, cosmologists focused their attention on the only other issue that was problematic for the Big Bang model. It was not clear how a universe created from a Big Bang could evolve to form galaxies. As Hoyle had once pointed out: 'If you postulate an explosion of sufficient violence to explain the expansion of the Universe, condensations looking at all like the galaxies could never have been formed.' In other words, Hoyle was complaining that the Big Bang was absurd because it would have blown apart all the existing matter to create a universe containing a thin and even smattering of substance, as opposed to a universe with its matter concentrated in galaxies.

Big Bang supporters were forced to agree that the Big Bang would have resulted, at least initially, in an even soup of matter that would indeed have been blown apart by the cosmic expansion. The challenge to the Big Bang model was clear – how could a universe created with a landscape of unparalleled blandness evolve into one populated by massive galaxies separated by vast empty voids?

Big Bang cosmologists comforted themselves with the hope that the early universe, although very uniform, could not have been perfectly uniform. They were optimistic that somehow the early universe must have had its homogeneity disrupted in some small way. If this were the case, then they believed that these tiniest imaginable variations in density would have been sufficient to trigger the necessary evolution of the universe.

Slightly denser regions would have attracted matter by the force of gravity, making these regions even denser, thereby attracting even more matter, and so on until the first galaxies were formed. In other words, if cosmologists hypothesised the slightest variations in density, then it was not too hard to imagine how gravity could have goaded the universe into forming rich and complicated structures and sub-structures.

If this was the mechanism by which the Big Bang model formed galaxies, then the earliest fluctuations in density would have seeded an extraordinary cosmic condensation. Today's universe is full of objects that have an average density of roughly 1 g/cm^3, the same as water. For example, the Sun is slightly denser than water, at 1.4 g/cm^3, while Saturn is a little less dense, at 0.7 g/cm^3. On the other hand, there are huge empty reaches of the cosmos, vast voids of near-nothingness. Consequently, the overall average density of the universe, taking into account everything from the galaxies to the empty voids, is roughly 0.000000000000000000000000000001 g/cm^3. This means that there are regions of the universe, in particular those that we inhabit, that are a million million million million million times denser than the average density.

So, the Big Bang narrative was that the early universe consisted of the most uniform, harmonised, consistent, smooth soup of matter conceivable; tiny variations in this almost homogeneous sea then sparked a chain of events that led, within a few billion years, to a universe in which there were massive discrepancies between high-density galaxies and near-zero density voids.

Table 6

This table lists various criteria against which the Big Bang and Steady State models could be judged. It shows how the two models fared on the basis of the data available in 1978, and it is an updated version of Table 4 (pp. 370–1).

Criterion	Big Bang Model	Success
1. Redshift and the expanding universe	Expected from a universe that is created in a dense state and then expands	✓
2. Abundances of the atoms	The observed proportions of light atoms (e.g. hydrogen, helium) are very close to the Big Bang prediction by Gamow and colleagues; heavier atoms are produced in the stars	✓
3. Formation of galaxies	The Big Bang expansion might have pulled apart baby galaxies before they could grow; nevertheless, galaxies did evolve, but nobody could explain how	✗
4. Distribution of galaxies	The distribution of galaxies varies with distance, as shown by Ryle; young galaxies (e.g. quasars) are observed, but only at great distances, as they would have existed only just after the Big Bang	✓
5. Cosmic microwave background (CMB) radiation	This echo of the Big Bang was predicted by Gamow, Alpher and Herman, and was found by Penzias and Wilson	✓
6. Age of the universe	Recent age measurements show that the objects in the universe are younger than the universe itself, so everything is consistent	✓
7. Creation	There is still no explanation of the creation of the universe	?

The ticks and crosses give a crude indication of how well each model fared in relation to each criterion, and a question mark indicates a lack of data or a mixture of agreement and disagreement.

Criterion	Steady State Model	Success
1. Redshift and the expanding universe	Expected from an eternal universe that expands, with new matter being created in the gaps	✓
2. Abundances of the atoms	Cannot really explain the observed abundances of light atoms; heavier atoms are produced in the stars	✗
3. Formation of galaxies	There is more time and no initial violent expansion; this allows galaxies to develop and die, to be replaced by new galaxies built from created matter	✓
4. Distribution of galaxies	Young galaxies should be evenly distributed, because they can be born anywhere and at any time out of the matter created in between old galaxies, but this is not backed by observation	✗
5. Cosmic microwave background (CMB) radiation	Cannot explain the observed CMB radiation	✗
6. Age of the universe	There is no evidence for anything older than 20 billion years, yet the universe is supposedly infinitely old	?
7. Creation	There is still no explanation of the continuous creation of matter	?

To prove that such a tremendous transition really took place, Big Bang cosmologists would have to find evidence for the variations in density that initiated the formation of the galaxies. Otherwise, with no hard evidence for these fluctuations, the Big Bang model would remain open to criticism from the few remaining Steady Statesmen such as Hoyle.

The obvious place to look for hints of fluctuations in the early universe was in the universe's oldest relic, namely the CMB radiation. This radiation was released at a specific moment in the history of the universe, so it now serves as a fossil, indicating the state of the universe as it was when atoms first formed roughly 300,000 years after the moment of creation. In their detection of this CMB radiation, radio astronomers were therefore effectively looking back in time and seeing the universe at a very early stage of its evolution. The Big Bang model estimated the universe to be at least 10 billion years old, so being able to see the universe as it was when it was just 300,000 years old was equivalent to seeing the universe when it was just 0.003% of its current age. Let us give the universe a more human timescale by imagining that it is now a mature seventy-year-old person – the CMB radiation would have been created when the universe was a newborn baby, just a few hours old.

It might not be immediately obvious that observing the CMB radiation is equivalent to looking back in time, but exactly the same thing happens when astronomers observe a distant star. If a star is 100 light years away, then its light will have taken 100 years to reach us, so we can only see that star as it was 100 years ago. Similarly, if the CMB radiation was released billions of years ago and has taken billions of years to reach us, then when astronomers eventually detect it they are effectively sensing the universe as it was billions of years ago, when it was only 300,000 years old.

If there were density variations at this moment in the history of the universe, then they ought to have been imprinted on the CMB radiation we see today. This is because a patch of universe that had a slightly higher density than average – a clump – would have a well-defined effect on the CMB radiation that emerged from it. The radiation from such a region would have experienced a slightly greater struggle as it escaped the extra gravitational attraction caused by the above-average density of the clump. Therefore the emerging radiation would have lost some energy as it emerged from the clump, giving it a slightly longer wavelength.

So, by examining the CMB radiation coming from different directions in the universe, astronomers hoped to detect slight variations in wavelength. Radiation arriving from a particular direction with a slightly longer wavelength would indicate that it had emerged from a part of the ancient universe that was slightly more dense, whereas radiation from a different direction with a slightly shorter wavelength would indicate that it had emerged from a part of the ancient universe that was slightly less dense. If astronomers could find these wavelength variations in the CMB radiation, then they would be able to prove that there were density variations in the early universe which would indeed have seeded the galaxies. The Big Bang model would then become even more compelling.

Penzias and Wilson had proved that the CMB radiation existed and that it had roughly the right wavelength, but now astronomers began to measure it with increasing precision, trying to show that the radiation coming from one part of the universe had a slightly different wavelength from the radiation coming from another part. Unfortunately, the CMB radiation appeared to be the same wherever they looked. It was supposed to be roughly uniform because the early universe had been very similar at every point in space, but the measurements showed the radiation from every direction to be not just similar, but identical. There was no sign of the most minuscule increase or decrease in wavelength.

The Steady Statesmen seized upon this negative result as damaging for the Big Bang model, because no variation in the observed wavelength of today's CMB radiation meant no density variations in the early universe, which meant no explanation for the galaxies we see today.

But the majority of cosmologists did not panic. They argued that the variations must indeed be present, but were too small to be detected because the available observational technology was too crude. This seemed like a reasonable argument. For example, the paper on which this page is printed looks perfectly smooth, but with sufficiently sensitive equipment its surface variations become apparent, as shown in Figure 99. Perhaps the same would prove to be true of the CMB radiation, and the variations would be revealed upon closer inspection.

By the 1970s, the latest equipment was sensitive enough to detect potential differences in the CMB radiation down to 1 part in 100, but still there was no sign of any variation. This still left open the possibility of variations at less than 1 part in 100, but detecting such small variations seemed to be impossible from the surface of the Earth. The problem was that the CMB radiation is in the microwave region of the electromagnetic spectrum, and the moisture in the atmosphere continuously emits microwaves which, although very weak, would be sufficient to overwhelm any minuscule variations in the CMB radiation that might exist.

One innovative solution was to design a CMB radiation detector that could be hoisted into the air by a giant, helium-filled, high-altitude balloon, capable of rising tens of kilometres above the Earth, virtually to the edge of space. A balloon-borne detector would have the advantage of floating in a region of the atmosphere that contained virtually no moisture, which would therefore contain very few atmospheric microwaves.

However, balloon experiments were fraught with difficulty. The

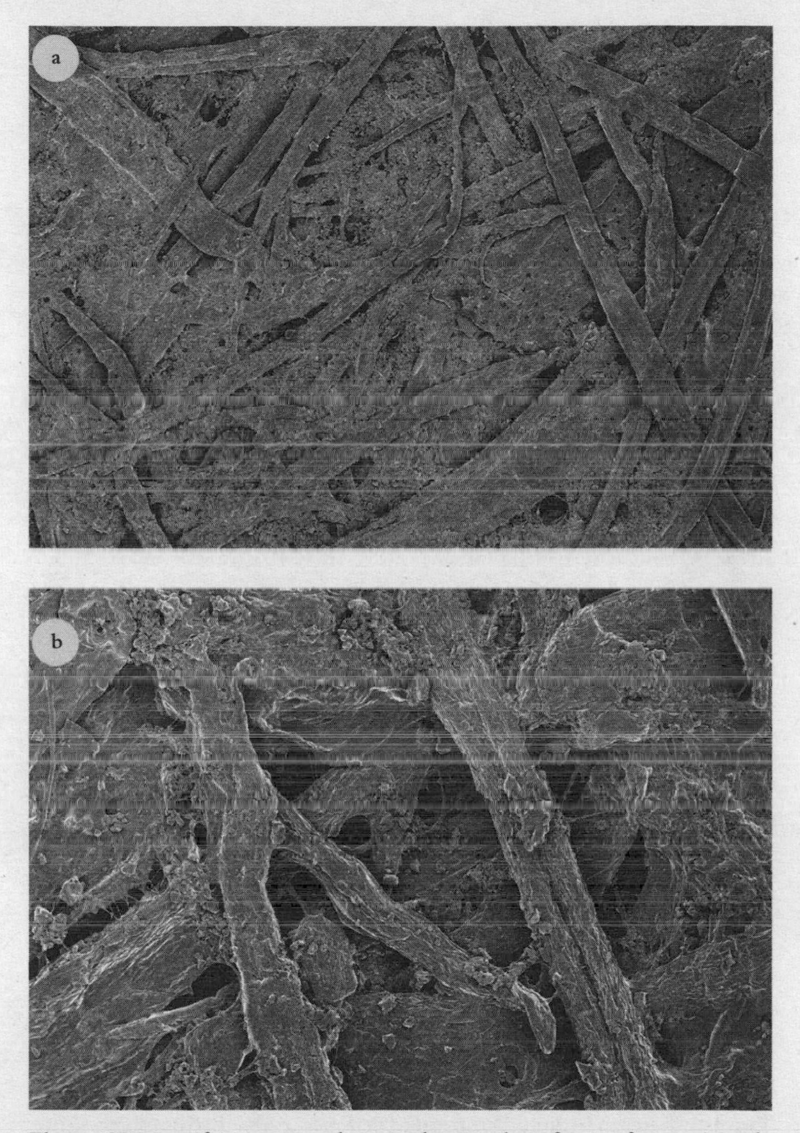

Figure 99 Magnifying apparently smooth paper by a factor of approximately 250, as in diagram (a), reveals its structure and variation. A magnification of approximately 1,000, as in diagram (b), reveals even more detail.

sheer cold could shatter glue and cause detectors to fall apart. Also, if a fault developed in the equipment then the astronomers were helpless. Even if the equipment functioned normally, the detector could operate for only a few hours before the balloon descended. Worst of all, the gondola containing the detector might crash to the ground, ending up lost or destroyed, with years of careful preparation going to waste.

George Smoot of the University of California at Berkeley, who had become obsessed with the search for variations in the CMB radiation, took part in several balloon experiments, but he became disillusioned in the mid-1970s. His balloon experiments regularly ended in disaster, and those that landed in one piece still failed to reveal any variations in the CMB radiation. Instead, Smoot adopted a new strategy. He planned to install a microwave detector on board an aeroplane, so that he could make observations over a longer period with greater reliability. It had to be better than perilously suspending an experiment below a flimsy balloon.

Smoot attempted to identify an aeroplane with a high-altitude capability which was also able to stay aloft for long periods of time, both necessary criteria for effective CMB radiation measurements. In the end, he decided that the ideal vehicle was the Lockheed Martin U-2 reconnaissance plane, legendary for its Cold War spying missions. He made a formal approach to the US Air Force and, to his amazement, they responded positively. They were delighted by the thought of taking part in a research project that might crack what was becoming the biggest mystery in cosmology. Senior military figures were so co-operative that they even told Smoot about a top-secret upper viewing hatch that could be installed on the U-2, which would give his experiment a clear view of the sky. The hatch was originally intended for use only during the testing of intercontinental ballistic missiles, when the U-2's task was to monitor their re-entry.

Previous balloon-borne experiments had used quite crude detectors, because nobody was prepared to invest lots of money in equipment that might end up crashing to the ground and being destroyed. Now that Smoot had a more reliable airborne platform, he constructed a CMB radiation detector using the very latest technology. It was capable of comparing the CMB radiation arriving from two different directions with greater sensitivity than ever before.

The experiment was launched on board a U-2 in 1976, and within just a few months Smoot and his colleagues had discovered a staggering variation in the CMB radiation. The radiation coming from one half of the sky had a wavelength that was 1 part in 1,000 longer than the radiation coming from the opposite half of the sky. It was an important result, but *not* the one that Smoot had really been looking for.

The variations that would have seeded the galaxies in the early universe would have been very irregular, and so would have shown themselves as a patchwork of haphazard regions across the sky. However, Smoot had detected a very simple two-part variation. The difference between what was observed and what cosmologists really wanted to see is shown in Figure 100.

There was a relatively obvious explanation for Smoot's measurements. The broad hemispherical variation was caused simply by the Earth's own motion and the resulting Doppler effect. As the Earth swept through space, if the detector was looking forwards it perceived the incoming CMB radiation to have a slightly shorter wavelength; if the detector was looking backwards then the wavelength appeared to be slightly longer. By measuring the difference in wavelengths, Smoot could actually measure the speed of the Earth through the cosmos. This speed was the combined effect of the Earth moving around the Sun, the Sun moving within our Milky Way galaxy, and the Milky Way's own movement. The result was announced on the front page of the *New York Times* on 14 November 1977: GALAXY'S SPEED THROUGH UNIVERSE FOUND TO EXCEED A MILLION M.P.H.

Although this was an interesting result, it contributed nothing to the big question: where were the variations in the CMB radiation that had seeded the universe? Even when the contribution from the Doppler effect was removed, there was still no sign of the Big Bang variations. They had to be present if the Big Bang model was correct, but nobody could find them. Smoot's equipment was very sensitive, so his failure to see the telltale patchwork pattern told him

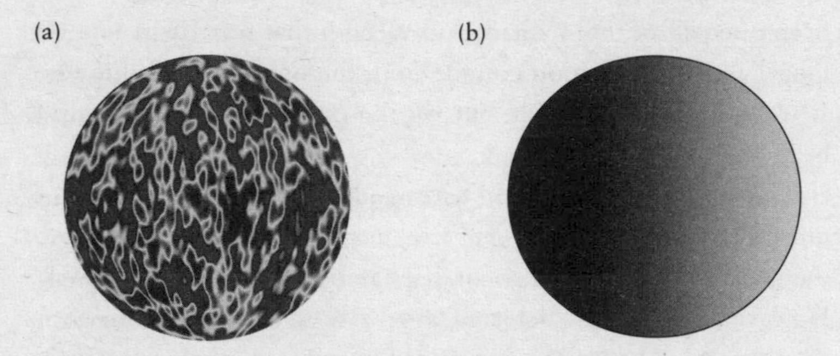

Figure 100 These two spheres represent two different maps of the CMB radiation. From our Earth-based view, at the centre of the spheres, we are looking out into space, and the shading represents the average wavelength of the CMB radiation we see arriving from different directions. Darker shading represents a slightly longer average wavelength for the radiation, whereas lighter shading represents a slightly shorter average wavelength.

Map (a) shows a patchwork of variation, the sort of pattern that cosmologists desperately needed to find. A region of longer average wavelength would indicate that it was slightly denser in the early universe and could therefore have seeded galaxy formation. Cosmologists were unsure of what the exact pattern in the CMB radiation might be, but they knew that it ought to be fairly complex to explain the modern arrangement of galaxies. Map (b) shows a simple structure, with shorter wavelengths in one hemisphere and longer wavelengths in the other. This sort of variation was detected by Smoot's U-2 experiment. It had nothing to do with the complex variation that would explain galaxy formation in the Big Bang model of the universe.

that the variations had to be less than 1 part in 1,000. Such small variations would be hard to detect even in an airborne experiment, because there was still a thin atmosphere above the detector that would obscure very fine measurements.

Astronomers gradually began to recognise that their only hope for finding the elusive variations (if they existed) was to get right above the Earth's atmosphere with a CMB radiation detector mounted on board a space satellite. A satellite-borne experiment would be properly isolated from atmospheric microwaves, it would be perfectly stable, it would be capable of scanning the entire sky and it would be able to operate day after day.

Even while Smoot had been working on his spyplane-based experiment, he had suspected that a satellite might be the only way to detect variations in the CMB radiation. For this reason he had already become involved in planning a more ambitious experiment. Back in 1974, NASA had asked scientists to submit ideas for its latest round of Explorer satellites, which were a series of relatively inexpensive projects aimed at supporting the astronomical community. A Berkeley team, which included George Smoot, submitted a proposal for a satellite-borne CMB radiation detector, but they were not alone. A group from the Jet Propulsion Laboratory in Pasadena, California, had put forward a similar proposal, as did John Mather, an ambitious twenty-eight-year-old NASA astrophysicist.

NASA, keen to back an experiment of such cosmological significance, unified the three proposals and funded a detailed feasibility study of what was to be named the Cosmic Background Explorer satellite, known by the acronym COBE (pronounced to rhyme with Toby). The collaboration began designing the experiment in 1976, while Smoot was still heavily involved in his U-2 spyplane measurements, but this was still only a preliminary phase, so Smoot's divided loyalty was not a problem. The team of scientists and engineers would spend the next six years working out how to build a detector that could achieve the cosmological goal of finding variations in the CMB radiation, and which was also small enough and strong enough to be launched into space.

The final design included three separate detectors, each one measuring a different aspect of the CMB radiation. Mike Hauser of the Goddard Space Flight Center, where the entire project was based, led the team responsible for the Diffuse Infrared Background Experiment (DIRBE). John Mather was responsible for the second detector, the Far Infrared Absolute Spectrophotometer (FIRAS). George Smoot was in charge of the third detector, the Differential Microwave Radiometer (DMR), and it was this detector that had been designed specifically to find variations in the CMB radiation. The DMR detector, as its name suggests, was intended to simultaneously detect the CMB radiation coming from two directions and measure the difference in the two sets of microwave radiation.

The COBE project finally got the green light in 1982, eight years after it was proposed. Construction could begin at last, and COBE was scheduled to be launched on board a space shuttle in 1988. However, after four years of building the satellite, the entire project was thrown into jeopardy. On 28 January 1986 the space shuttle *Challenger* exploded soon after lift-off, killing all seven of its crew.

'I was stunned,' recalled Smoot. 'We all were. We grieved for the astronauts. The tragedy of the accident was uppermost, but slowly the probable implications for COBE began to dawn . . . With one shuttle lost and three grounded, NASA's schedule had gone to hell. Nothing was flying. There was no telling how long the COBE launch might be delayed; maybe years.'

The astronomers and engineers had spent over a decade designing and building the COBE satellite, and its future now seemed bleak. All shuttle flights were abandoned, and the backlog of shuttle payloads rapidly built up. Even if launches were resumed, it was clear that there would be other priorities that would push COBE to the back of the queue. In fact, before the end of 1986, NASA officially announced that COBE had been dropped from the space shuttle launch programme.

The COBE team began to search for a substitute launch vehicle, and the only serious option was an old-fashioned disposable rocket. The best available launcher was the European Ariane rocket, but NASA had funded COBE and was not prepared to see a foreign rival steal the glory of launching the satellite. One COBE team member noted: 'We had two or three discussions with the French, but when NASA headquarters found out about it, they ordered us to cease and desist and threatened us with bodily harm if we didn't.' Not surprisingly, talking to the Russians was also completely out of the question.

The rocket business was largely in decline, so there were very few other alternatives. For instance, the COBE team approached McDonnell-Douglas, but the company had halted its Delta rocket production line. They had only a few spare rockets left, which had all been earmarked as targets in weapons tests for the new Strategic Defense Initiative (nicknamed the Stars Wars programme). However, when the Delta engineers heard about the plight of COBE, they were delighted that one of their beautifully crafted vehicles might be used for something more constructive than target practice. They immediately offered their services, but there was still one outstanding problem that had to be overcome.

The complete COBE satellite would weigh almost 5 tonnes, but the Delta rocket could cope with a payload only half that weight, so COBE would have to slim down considerably. The COBE team was forced to completely redesign the satellite, drastically reducing its size and making huge sacrifices that wasted years of previous work. At the same time, the team had somehow to ensure that the satellite's scientific capabilities remained intact – that it could still probe the CMB radiation and test the Big Bang model. Even worse, the entire redesign and build had to be completed within just three years, because there would be a launch opportunity in 1989, and missing this deadline would have led to further severe delays.

Hundreds of scientists and engineers worked weekends and around the clock to meet one of the most demanding deadlines in the history of space adventure. At last, on the morning of 18 November 1989, fifteen years after the proposal was originally submitted to NASA, the COBE satellite was ready for launch. Others had continued to search for the variations in the CMB radiation throughout this period, using detectors on the ground and carried aloft by balloons and aeroplanes, but the CMB radiation continued to look perfectly smooth. It was not too late for the COBE satellite to make a name for itself.

The COBE team had gone out of its way to make sure that Ralph Alpher and Robert Herman, who had originally predicted the CMB radiation in 1948, were not forgotten, and invited them to Vandenberg Air Force Base in California to witness the launch. The two theorists were even permitted to ascend the gantry and pat the rocket's nose cone just before lift-off. Smoot was also among the hundreds of people who had gathered for the launch. All his ambitions depended on COBE and the Delta rocket: 'On an earlier trip I had seen the rocket up close, and had been aghast at how decrepit it looked, rusting here and there, patched here and there, spot repairs made with Glyptal. Our professional life's work was on top of that thing. We didn't say a word, only silent prayers.'

When the countdown reached zero, the Delta rocket rose from the launch pad. Within thirty seconds it had broken the sound barrier, and within eleven minutes COBE was successfully in orbit. A final booster stage lifted the satellite to an altitude of 900 km, and thereafter it followed a polar orbit, circumnavigating the Earth fourteen times a day.

From the very first batch of data that was beamed back to Earth, it was clear that COBE was operating perfectly and that each detector had survived the physical stress of the rocket launch. However, Smoot and his colleagues were unable to make any announcements relating to the main objective of their mission.

Proving, or disproving, the existence of the variations in the CMB radiation would require a very subtle and long-term analysis of data from the DMR detector, and even accumulating these measurements was a slow process. The detector could simultaneously measure and compare the CMB radiation from two small patches of the sky 60° apart, but in order to measure the radiation across the entire sky the satellite first had to orbit the Earth hundreds of times. The DMR detector eventually completed its first rough survey of the entire sky in April 1990.

The first analysis revealed no sign of any variation in the CMB radiation at a level of 1 part in 3,000. After the second trawl there was no sign of any variation at a level of 1 part in 10,000. Science writer Marcus Chown described the measurements as 'unbroken blandness'.

COBE had been sent into space to find the variations that seeded today's galaxies. Perhaps they were just proving difficult to find. Or perhaps they did not exist at all, which would be disastrous for the Big Bang model because then there would be no explanation for the creation of the galaxies. And without galaxies, there would be no stars, no planets and no life. The situation was becoming distressing. As John Mather put it: 'We haven't ruled out our own existence yet. But I'm completely mystified as to how the present day structure exists without having left some signature on the background radiation.'

Optimists hoped that more data and closer scrutiny would reveal the CMB radiation variations. Pessimists worried that a more detailed inspection would prove that the CMB radiation was perfectly smooth and that the Big Bang model was flawed. As each

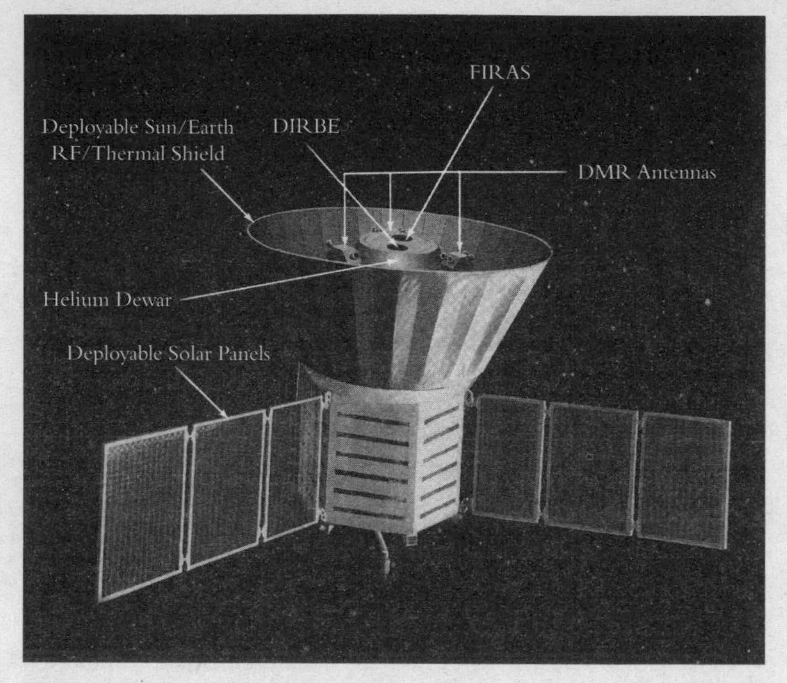

Figure 101 The COBE satellite was launched in 1989. The three detectors are partly obscured by a shield that protects them from heat and microwaves from the Sun and Earth. The Dewar flask at the centre of the shield contains liquid helium, which cools the satellite components to reduce the emission of microwave radiation by the satellite itself.

So far, I have given the impression that the CMB radiation from any given direction has a single wavelength, but in reality the CMB radiation from any one direction has a range of wavelengths. However, the characteristic of this wavelength distribution can be described by noting just the dominant or peak wavelength, which is why the CMB radiation has been treated as though it consists of a single wavelength.

The fate of the Big Bang model depended on measurements made by the DMR detector. It could compare the incoming CMB radiation from two different directions and look for differences in the peak wavelength. Such differences would be indicative of density variations in the early universe, and higher-density regions would have seeded today's galaxies.

The FIRAS detector and the DIRBE detector were designed to analyse other aspects of the CMB radiation.

month passed, with no statement on the existence or non-existence of variations, rumours began to circulate around the cosmological community and in the scientific press. Theorists began to develop ad hoc variants of the Big Bang model that did not necessarily require variations in the CMB radiation. *Sky & Telescope* magazine summarised the mood when it ran the headline THE BIG BANG: DEAD OR ALIVE? The small Steady State community took heart and began to criticise the Big Bang model anew.

What nobody outside the COBE team realised was that the long-awaited variations were gradually beginning to emerge. Signs of the variations were so tentative that the research was kept a closely guarded secret.

The COBE DMR detector had continued gathering more data throughout 1990 and 1991, and it had completed its first thorough mapping of the entire sky by December 1991, taking 70 million measurements along the way. At last, a variation had appeared at the level of just 1 part in 100,000. In other words, the peak wavelength of the CMB radiation varied by 0.001% depending on where COBE was looking. The CMB radiation showed only tiny variations across the sky, but crucially they did exist. And they were just about big enough to indicate density fluctuations in the early universe that were sufficient to seed the subsequent development of galaxies.

Some COBE scientists were anxious to publish quickly, but others were more cautious and it was they who prevailed. The COBE team embarked on a thorough review to reassure themselves that the apparent variations were not down to a glitch in the detector or a fault in the analysis. To engender an atmosphere of caution and self-criticism, Smoot offered a free plane ticket to anywhere in the world to anyone who could find a mistake in the analysis. He realised that he was involved in making one of the most sensitive measurements in the history of science, and it would be all too easy for a well-camouflaged error to contaminate the results. He once likened the challenge of finding the tiny variations in the CMB radiation to 'listening for a whisper during a noisy beach party while radios blare, waves crash, people yell, dogs bark, and dune buggies roar'. In such a situation it is easy to hear the wrong thing or even imagine hearing something that is not really there.

After almost three months of further analysis and argument, there was a consensus among the COBE team that the variations were genuine. It was time to go public. A paper was submitted to the *Astrophysical Journal*, and it was agreed that the discovery would be announced at a conference organised by the American Physical Society in Washington on 23 April 1992.

Smoot, spokesman for the team that had constructed the DMR detector, had the honour of addressing the assembled crowd and presenting a truly momentous result. It had been a quarter of a century since Penzias and Wilson had discovered the CMB radiation and now, at long last, the anticipated variations had been identified. The result was still a closely guarded secret, so not even the conference organisers were aware that Smoot would be making such a significant announcement. Hence, he had been allotted only the standard twelve minutes, but that was long enough to present one of the most important discoveries in the history of science. The assembled audience watched in awe as the cosmological landscape dramatically fell into place. The Big Bang could, indeed, explain the formation of the galaxies.

At noon there was a major press conference. With the accompanying press release were COBE's maps of the universe, each a mix of reds, pinks, blues and mauves, which would take on iconic status. Black and white versions of the maps are shown in Figure 102. Each lozenge-shaped map represents the entire sky, unwrapped and reformatted for the flat page, just as a map of the spherical Earth is distorted to fit on the page of an atlas.

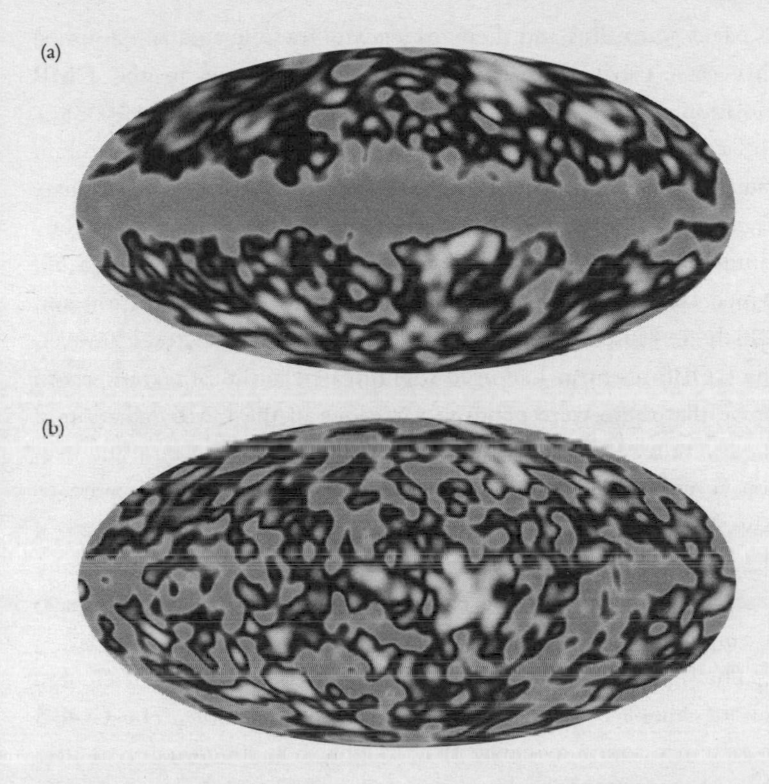

Figure 102 As COBE looked into space, it saw the CMB radiation arriving from all directions. The variations in the radiation were mapped onto the surface of a sphere, as if COBE were positioned at the centre of the sphere and looking out. COBE created several spherical maps, and two of these have been unwrapped and shown here as two-dimensional maps. The maps were originally colour-coded, but are shown here in black, white and grey. The shading reflects variations in the intensity of the CMB radiation as measured by COBE's DMR detector.

Map (a) is dominated by the radiation from the stars in our own Milky Way, which appears streaked across the equator. This image was nicknamed 'the hamburger'.

Map (b) has had the Milky Way contribution removed. It is a better indication of variations in the CMB radiation across the universe. Most of the map is still dominated by random noise, but a statistical analysis shows genuine variations in the CMB radiation at the level of 1 part in 100,000.

Many journalists and their readers saw these images and assumed that each patch represented a genuine variation in the CMB radiation, one of the much vaunted differences of 1 part in 100,000. In fact, the COBE measurements were severely affected by random radiation emitted by the DMR detector itself, so the critical map, Figure 102(b), contains a significant random contribution. This contamination is so severe that by sight alone it is impossible to tell which blotches are genuine variations in the CMB radiation and which are caused by chance fluctuations in the detector. However, the COBE scientists had used sophisticated statistical techniques to prove that there were genuine variations in the CMB radiation at the level they had declared, so their result was valid, even if the map was somewhat misleading. It would have been more accurate to have handed journalists a statistical analysis of the data instead of the images, but no news journalist would have understood it. In any case, the picture editors were certainly grateful for a striking image to run alongside the articles that would appear the following day.

The statistical analysis might have been complicated, but George Smoot's message to the rest of the world was simple. The COBE satellite had found evidence that, roughly 300,000 years or so after the moment of creation, there were tiny density variations across the universe at the level of 1 part in 100,000, which grew with time and ultimately resulted in the galaxies that we see today. Having spent the previous evening thinking up snappy sentences he could deliver at the press conference, Smoot told the assembled journalists: 'We have observed the oldest and largest structures ever seen in the early universe. These were the primordial seeds of modern-day structures such as galaxies, clusters of galaxies, and so on.' Smoot also gave journalists a more memorable and snappier quotation: 'Well, if you're religious, it's like seeing the face of God.'

The press responded by devoting entire front pages to the COBE result. *Newsweek* magazine typified the dramatic coverage with the headline THE HANDWRITING OF GOD. Although slightly embarrassed by the fervour that his words had inspired, Smoot nonetheless claimed to have no regrets: 'If my comment got people interested in cosmology, then that's good, that's positive. Anyhow it's done now. I can't take it back.'

The mention of God, the striking images and the sheer scientific importance of the COBE breakthrough guaranteed that this was without doubt the biggest astronomy story of the decade. Even more fuel was added to the fire by Stephen Hawking, who said: 'It's the discovery of the century, if not of all time.'

At last, the challenge to prove the Big Bang model was over. Generations of physicists, astronomers and cosmologists – Einstein Friedmann, Lemaître, Hubble, Gamow, Alpher, Baade, Penzias, Wilson, the entire COBE team, and many others – had succeeded in addressing the ultimate question of creation. It was clear that the universe was dynamic, expanding and evolving, and that everything we see today emerged from a hot, dense, compact Big Bang over 10 billion years ago. There had been a revolution in cosmology, and the Big Bang model was now accepted. The paradigm shift was complete.

CHAPTER 5 - PARADIGM SHIFT
SUMMARY NOTES

① 1950 - THE COSMOLOGICAL COMMUNITY WAS DIVIDED BETWEEN THE STEADY STATE MODEL AND THE BIG BANG MODEL.

QUESTIONS HAD TO BE ANSWERED AND CONFLICTS HAD TO BE RESOLVED BEFORE ONE MODEL COULD CLAIM TO BE THE TRUE DESCRIPTION OF THE UNIVERSE.

FOR EXAMPLE: IF THERE WAS A BIG BANG, THEN:

- WHY WAS THE UNIVERSE YOUNGER THAN THE STARS?
- HOW WERE THE HEAVY ELEMENTS FORMED?
- WHERE WAS THE CMB RADIATION?
- AND HOW DID THE GALAXIES FORM?

② FIRST BAADE AND THEN SANDAGE RECALIBRATED THE DISTANCE SCALE TO THE GALAXIES AND SHOWED THAT THE BIG BANG ACTUALLY PREDICTED A MUCH OLDER UNIVERSE, COMPATIBLE WITH THE AGES OF THE STARS AND GALAXIES WITHIN IT.

③ HOYLE SET OUT TO EXPLAIN THE FORMATION OF HEAVY ELEMENTS AND SHOWED HOW THEY WERE FORMED BY FUSION WITHIN THE HEARTS OF AGEING STARS.

THE PROBLEM OF NUCLEOSYNTHESIS HAD BEEN SOLVED:

- HEAVY ELEMENTS FORMED IN DYING STARS
- LIGHT ELEMENTS FORMED SOON AFTER THE BIG BANG

④ 1960S ASTRONOMERS USED RADIO ASTRONOMY AND DISCOVERED NEW GALAXIES (eg YOUNG GALAXIES AND QUASARS) THAT TENDED TO EXIST IN THE FAR REACHES OF THE UNIVERSE.

THIS UNEVEN DISTRIBUTION OF GALAXIES WENT AGAINST THE STEADY STATE MODEL WHICH CLAIMED THAT THE UNIVERSE WAS ROUGHLY THE SAME EVERYWHERE.

HOWEVER, THIS OBSERVATION WAS WHOLLY COMPATIBLE WITH THE BIG BANG MODEL.

⑥ MID 1960s – PENZIAS AND WILSON ACCIDENTALLY DISCOVERED THE CMB RADIATION PREDICTED BY ALPHER, GAMOW AND HERMAN BACK IN 1948, PROVIDING COMPELLING EVIDENCE IN FAVOUR OF THE BIG BANG.

THIS PIECE OF SERENDIPITY WON THE NOBEL PRIZE IN 1978.

NEARLY ALL COSMOLOGISTS MOVED TO THE BIG BANG CAMP.

⑦ 1992 –

THE COBE SATELLITE

DISCOVERED TINY VARIATIONS IN THE CMB RADIATION COMING FROM DIFFERENT PARTS OF THE SKY, WHICH INDICATED TINY VARIATIONS IN DENSITY IN THE EARLY UNIVERSE, WHICH WOULD HAVE SEEDED THE FORMATION OF THE GALAXIES.

THE PARADIGM SHIFT FROM AN ETERNAL UNIVERSE TO A BIG BANG UNIVERSE WAS COMPLETE.

THE BIG BANG MODEL WAS PROVED TO BE TRUE!

THE END?

EPILOGUE

If you want to make an apple pie from scratch, you must first create the universe.

CARL SAGAN

What continues to amaze me is that human beings have had the audacity to conceive a theory of creation and that now we are able to test that theory.

GEORGE SMOOT

We argue for the Big Bang model as the most persuasive and inclusive physical theory of the cosmos at this time because the model has predictive ability (i.e. it encompasses simultaneously many and diverse astronomical observations) and in particular because, as any viable theory must, it continues to survive the challenges of observational falsification . . . In the case of the Big Bang, not only has the model survived over a number of decades, but the case for it has become progressively stronger.

RALPH ALPHER and ROBERT HERMAN

Ten or twenty billion years ago, something happened – the Big Bang, the event that began our universe. Why it happened is the greatest mystery we know. That it happened is reasonably clear.

CARL SAGAN

The Big Bang model of the universe is arguably the most important and glorious scientific achievement of the twentieth century. However, the Big Bang model can also be considered to be very ordinary in the way it was conceived, developed, explored, tested, proved and eventually accepted. In these respects it has much in common with ideas in many less glamorous areas of science. The development of the Big Bang model was an archetypal example of the scientific method in action.

Just like many other areas of science, cosmology started by attempting to explain things that had previously been in the domain of myth or religion. The earliest cosmological models were useful, but not perfect, and very soon inconsistencies and inaccuracies began to emerge. A new generation of cosmologists offered an alternative model and campaigned in favour of their view of the universe, while the scientific establishment defended the existing model. Both the establishment and the rebel camp argued their cases, drawing on theory, experiment and observation, sometimes working for decades before making a breakthrough, occasionally changing the scientific landscape overnight with a serendipitous discovery. Both sides made the most of the latest technology – everything from lenses to satellites – in an effort to find the pivotal piece of evidence that would prove their model. Eventually, the case in favour of the new model became overwhelming, and cosmology underwent a revolution as the establishment discarded its old model in favour of the new model. Most former critics of the new model became convinced and switched their allegiance, and the paradigm shift was complete.

Importantly, in the majority of scientific battles there is no paradigm shift. Typically, a newly proposed scientific model is soon found to be flawed, and the establishment model remains in place as the best explanation of reality. It is reassuring that this is the case, otherwise science would be constantly revising its position and it would be an unreliable framework for exploring and understanding the universe. However, when a paradigm shift does occur, it is one of the most extraordinary moments in the history of science.

The path from an old paradigm to a new one may be several decades long and may require contributions from dozens of scientists. This gives rise to an interesting question: who deserves the credit for the new paradigm? This issue was neatly explored in the play *Oxygen* by Roald Hoffmann and Carl Djerassi. The play is based around the retro-Nobel, a fictional award given in recognition of a discovery that took place before the founding of the Nobel Academy. A committee meets and soon agrees that the award should be given for the discovery of oxygen. Unfortunately, the members cannot agree who deserves credit for the discovery. Was it the Swedish apothecary Carl Wilhelm Scheele, because he was the first to synthesise and isolate the gas? Or was it the English Unitarian minister Joseph Priestley, because he was the first to publish the discovery and provide details of his research? Or was it the French chemist Antoine Lavoisier, because he properly understood that oxygen was not merely a version of air ('dephlogisticated air'), but an entirely new element? The play discusses the question of priority at length, travelling back in time to allow each man to argue his case, which serves only to reveal the complexities of attributing credit.

If the question of who deserves the credit for discovering oxygen is hard to answer, then the question of who invented the Big Bang model is virtually impossible. Developing, testing, revising and proving the complete Big Bang model required a number of theoretical, experimental and observational stages, and each one has its own heroes. Einstein deserves some credit for explaining gravity through his theory of general relativity, without which no serious cosmological model could have been developed. However, at first he fought against the idea of an evolving universe, so it was left to Lemaître and Friedmann to develop the theory of the Big Bang. Their work would not have been taken seriously had it not been for the observations by Hubble, who demonstrated that the universe was expanding. But Hubble's claim to the Big Bang crown is tempered by the fact that he was reluctant to draw any cosmological conclusions from his own research. The Big Bang model would have remained in the doldrums were it not for the theoretical contributions of Gamow, Alpher and Herman and the observational work of Ryle, Penzias, Wilson and the COBE team. Even Fred Hoyle, the Steady State protagonist, made theoretical contributions to nucleosynthesis, inadvertently helping to bolster the Big Bang. Clearly, the Big Bang model cannot be attributed to any single individual.

In fact, this book mentions only a small fraction of those who contributed to the development of the Big Bang model, because it would be impossible to give a complete and definitive account of the Steady State versus Big Bang debate in just a few hundred pages. Each subsection of each chapter of this book would need to be expanded into its own dedicated volume to do justice to everyone who has contributed to the development of the Big Bang model.

In addition to the limitations of space, this account of the history of the Big Bang model has also been constrained by an effort to minimise the number of mathematical equations. Mathematics is the language of science, and in many cases a full and accurate explanation of a scientific concept is possible only by presenting a detailed mathematical exposition. However, it is usually possible to give a general description of a scientific concept by using mere words and a few pictures to illustrate the key points. Indeed, the mathematician Carl Friedrich Gauss once stressed the value of 'notions, not notations'.

Evidence that the Big Bang theory can be explained in words and pictures appeared on Friday 24 April, 1992. This was the day after the COBE press conference, when the front page of the *Independent* newspaper summarised all the essential elements of the Big Bang model of the universe in a single, simple diagram, shown here in Figure 103. Some of the time and temperature values in the diagram differ from those quoted in earlier chapters, because of improvements in theory and observation up to 1992. The numbers are still only approximate, but to a large extent they continue to represent the consensus among today's cosmologists.

The *Independent*'s diagram neatly sums up our current understanding of the Big Bang universe. First, as it points out, 'all matter and energy were condensed to a point' and then there was an almighty Big Bang. The term 'Big Bang' implies some sort of explosion, which is not a wholly inappropriate analogy, except that the Big Bang was not an explosion *in* space, but an explosion *of* space. Similarly, the Big Bang was not an explosion *in* time, but an explosion *of* time. Both space and time were created at the moment of the Big Bang.

Within a second, the super-hot universe expanded and cooled dramatically, its temperature falling from a few trillion to a few billion degrees. The universe contained mainly protons, neutrons and electrons, all bathed in a sea of light. The protons, equivalent to hydrogen nuclei, reacted with other particles in the next few minutes to form light nuclei such as helium. The ratio of hydrogen to helium in the universe was largely fixed within these first few minutes, and is consistent with what we see today.

The universe continued to expand and cool. It now consisted of simple nuclei, energetic electrons and vast amounts of light, with everything scattering off everything else. After roughly 300,000 years, the temperature of the universe had cooled sufficiently to allow the electrons to slow down, latch onto the nuclei and form fully fledged atoms. This effectively prevented any further scattering of the light, which ever since has been sailing through the universe largely unhindered. This light has become known as the cosmic microwave background (CMB) radiation, a sort of luminous echo of the Big Bang, which was predicted by Gamow, Alpher, and Herman, and detected by Penzias and Wilson.

Thanks to the COBE satellite's detailed measurements of the CMB radiation, we know that the universe contained regions of slightly higher-than-average density when it was 300,000 years old. These regions gradually attracted more matter and grew denser, so that the first stars and galaxies had formed by the time that the universe was roughly a billion years old. The nuclear reactions that were initiated within stars went on to form the medium-weight elements, while the heaviest elements would be created in the intense conditions of a star's violent death throes. It is thanks to the stellar formation of elements such as carbon, oxygen, nitrogen, phosphorus and potassium that it was ultimately possible for life to evolve.

And here we are today, 15 billion years later (give or take a couple of billion years). The uppermost section of the newspaper illustration, which contains humans, is somewhat flattering, as it exaggerates the role we have played in the history of the universe. Although life has existed on Earth for a few billion years, humans have existed for only a hundred thousand years or so. To put this

Figure 103 In Britain, the COBE discovery dominated the front page of the *Independent* newspaper on Friday 24 April, 1992. The newspaper heralded the variations in the CMB radiation as the ultimate endorsement of the Big Bang model of the universe, which it explained with the aid of this bold diagram.

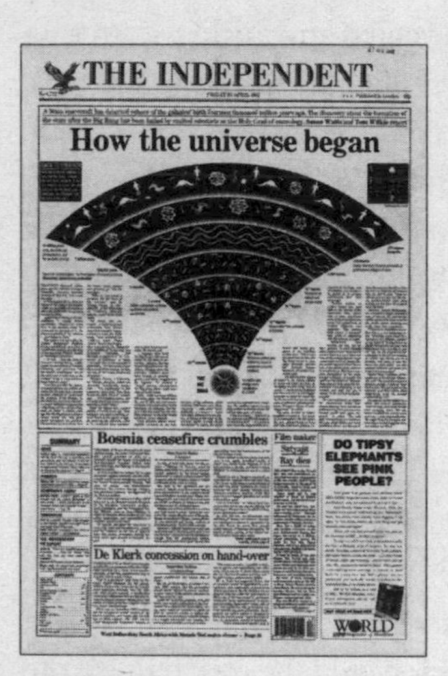

THE INDEPENDENT

How the universe began

Bosnia ceasefire crumbles

DO TIPSY ELEPHANTS SEE PINK PEOPLE?

De Klerk concession on hand-over

WORLD

How th

FOURTEEN thousand million years ago the universe hiccuped. Yesterday, American scientists announced that they have heard the echo.

A Nasa spacecraft has detected ripples at the edge of the Cosmos which are the fossilised imprint of the birth of the stars and galaxies around us today.

According to Michael Rowan-Robinson, a leading British cosmologist, "What we are seeing here is the moment when the structures we are part of — the stars and galaxies of the universe — first began to form."

The ripples were spotted by the Cosmic Background Explorer (Cobe) satellite and presented to excited astronomers at a meeting of the American Physical Society in Washington yesterday.

"Oh wow . . . you can have no idea how exciting this is," Carlos Frenk, an astronomer at Durham University, said yesterday. "All the world's cosmologists are on the telephone to each other at the moment trying to work out what these numbers mean."

Cobe has provided the answer to a question that has baffled scientists for the past three decades in their attempts to understand the structure of the Cosmos. In the 1960s two American researchers found definitive evidence that a Big Bang had started the whole thing off about 15 billion years ago. But the Big Bang would have spread matter like thin gruel evenly throughout the universe. The problem was to work out how the lumps (stars, planets and galaxies) got into the porridge.

"What we have found is evidence for the birth of the universe," said Dr George Smoot, an astrophysicist at the University of California, Berkeley, and the leader of the Cobe team.

Dr Smoot and colleagues at Berkeley joined researchers from several American research organisations to form the Cobe team. These included the Goddard Space Flight Center, Nasa's Jet Propulsion Laboratory, the Massachusetts Institute of Technology and Princeton University. Joel Primack, a physicist at the University of California at Santa Cruz, said that if the research is confirmed, "it's one of the major discoveries of the century. In fact, it's one of the major discoveries of science."

Michael Turner, a University of Chicago physicist, called the discovery "unbelievably important . . . The significance of this cannot be overstated. They have found the Holy Grail of cosmology . . . if it is indeed correct, this certainly would have to be considered for a Nobel Prize."

Since the ripples were created almost 15 billion years ago, their radiation has been travelling toward Earth at the speed of light. By detecting the radiation, Cobe is "a wonderful time machine"

able to view the young
verse, Dr Smoot said.

A remnant glow
the Big Bang is still ar
today, in the form o
crowave radiation tha
bathed the universe fo
billions of years since
explosion. Galaxies
have formed by gro
gravitational forces br
ter together. To
"lumpy" universe, rad
the Big Bang should
signs of being lumpy.

Cobe, which has be
500 miles above the
the end of 1989, has i
on board that are sens
extremely old radiatio
ples Cobe has found a
hard evidence of the l
lumpiness in the radia

Cobe detected alm
ceptible variations in

niverse began

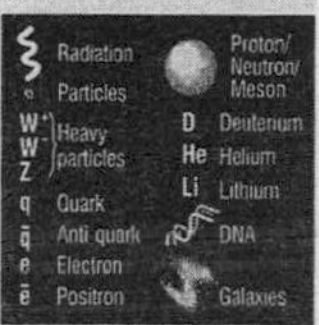

GRAPHIC: MICHAEL ROSCOE

-270 degrees Centigrade

-255 degrees
Heavy chemical elements produced in gravitational collapse of stars

6,000 degrees

10^9 degrees
Formation of helium and lithium nuclei

10^{10} degrees

10^{15} degrees
More matter than antimatter in Cosmos

10^{27} degrees

10^{32} degrees
All forces unified and violent increase in expansion (cosmic inflation)

conds

THE
IG
ANG

All matter and energy were condensed to a point

ne radiation, which
C below zero. Those
only about thirty-
a degree – repre-
ferences in the den-
at the edge of the
les of wispy clouds
slightly less dense

matter, the scientists said yesterday. The smallest ripples the satellite picked up stretch across 500 million light years of space.

Cobe has taken a snapshot of the universe just 300,000 years after Big Bang itself – at a point in time when the foggy fireball of radiation and matter produced by the explosion cooled down. "The results also show that the idea of a Big Bang model is once again brilliantly successful," Professor Rowan-Robinson, of London University, said.

He described the ripples as similar to the chaotic pattern of waves you might see from an aeroplane window flying over an ocean. "I can be pretty confident now that if we had an even bigger telescope in space we could see the fluctuations that are the early signs of individual galaxies themselves. It's just a matter of technology now," he added.

The point in time of Cobe's snapshot is known as "the epoch of recombination". At this point, the early galaxies began to form and light from these galaxies, released from the foggy soup of radiation, was set free to be picked up by modern astronomers with their telescopes.

"Further analysis of Cobe's results will shed light on the identity of the mysterious dark matter that we know contributes most of the mass of the universe," Dr Carlos Frenk, of Durham University, said yesterday. This mystery dark matter is scientists' best guess at explaining why the universe is lumpy.

Astronomers have worked out that, for today's galaxies to have formed, there ought to be far more matter around than they have observed. One of the leading theories to get round this is the Dark Matter theory, which says that about 99 per cent of the matter of the universe is invisible to us. This theory predicts fluctuations in the background radiation of exactly the size Cobe has observed. "Because these had not been seen, the theoreticians were beginning to get worried that they had got it wrong," Professor Rowan-Robinson said.

"If Cobe had found no ripples the theoreticians would have been in disarray; their best shot at understanding how galaxies were formed would have been disproved," he added. "The cold dark matter theory is a very beautiful one which makes very exact predictions about what the size of these fluctuations should be. How big they are depends on how fast they are able to grow. These results are just the size that the theory predicts. People have been looking for this kind of variation since the 1950s."

However, Arnold Wolfendale, the Astronomer Royal, sounded a note of caution. He said the scientific community must examine the results before shouting too loudly about their importance.

"There is no doubt that, if verified, this is a very important result. Detecting these small fluctuations is very difficult. Another group reported having picked up similar fluctuations last year, then later found they were due to cosmic rays. At the frequencies our colleagues in the US are working at, cosmic rays should not be a problem, but there is dust between the stars which can also produce radiation and make you think it is cosmological."

Martin Rees, Professor of Astrophysics at Cambridge University, said: "We needed equipment sensitive enough to pick up these fluctuations. We can expect in the next year or so there will be other observations from the ground corroborating this."

He said the results opened up a whole new area of astronomy. "Now we have seen them we can start analysing them. We can learn a lot about the history of the universe – what happened when. We might find, for example, that there was a second foggy era after the original fog lifted."

into context, if the history of the universe were represented as a timeline running between the fingertips of two outstretched arms, then a nail file could extinguish human existence with a single scrape.

It is important to remember that this history of creation and evolution is backed up with concrete evidence. Physicists such as Gamow, Alpher and Herman performed detailed calculations, estimated the conditions of the early universe and made predictions about how the early universe would leave its mark on the current universe, namely in terms of the ratio of hydrogen to helium and the CMB radiation. These predictions have turned out to be uncannily accurate. As pointed out by the Nobel prize-winning physicist Steven Weinberg, the Big Bang model is more than just idle speculation: 'Our mistake is not that we take our theories too seriously, but that we do not take them seriously enough. It is always hard to realise that these numbers and equations we play with at our desks have something to do with the real world. Even worse, there often seems to be a general agreement that certain phenomena are just not fit subjects for respectable theoretical and experimental effort. Gamow, Alpher and Herman deserve tremendous credit above all for being willing to take the early universe seriously, for working out what the known physical laws have to say about the first three minutes.'

When a newspaper is prepared to splash a broad-brush exposition of a cosmological model across its front page, then it is a strong indication that, as Arthur Eddington would have put it, the Big Bang model has moved from the theoretical workshop into the scientific showroom. Yet this does not mean that the model is polished and complete, because there will always be some outstanding issues and some details that need to be filled in. The rest of this epilogue is a brief dip into some of those still to be resolved issues and details. A few paragraphs cannot hope to convey the subtlety, depth and true significance of any of these problems, but what follows should demonstrate that, while the broad concept of the Big Bang model has been proved to be correct, it will be a long time before the dole queues are full of redundant cosmologists.

For example, we know that today's galaxies were seeded by variations in density that existed in the universe roughly 300,000 years after the Big Bang, but what was responsible for these density variations? Also, according to Einstein's general theory of relativity, space can be either flat, or curved inwards, or curved outwards. In a flat universe a ray of light can keep on travelling in a straight line for ever, just like a ball rolling along a flat, frictionless surface, but in a curved universe the ray could follow a circular path and return to where it started, just like an aeroplane flying around the equator of the curved Earth. Our universe seems to be flat according to astronomical observations, so the question is this: why is our universe flat, when it could have been curved?

One possible explanation for both the origin of the variations and the apparent flatness of the universe is provided by the theory of *inflation*, which was developed towards the end of 1979 by Alan Guth. When he first conceived of cosmic inflation, Guth was so amazed that he scribbled 'SPECTACULAR REALIZATION' in his notebook. This was not an understatement, as inflation looks to be a valuable addition to the Big Bang model. There are various versions of inflation, but in essence the theory proposes a brief and gargantuan phase of expansion in the very earliest moments of the universe, perhaps ending after just 10^{-35} seconds. During this inflationary era, the universe doubled in size every 10^{-37} seconds, which means roughly a hundred doublings. This might not sound a lot, but a famous fable shows the power of doubling.

The fable explains how a Persian vizier once asked his sultan if he could be paid in grains of rice, such that there was 1 grain on the first square of a chessboard, then 2 on the second square, then 4,

8, 16, and so on. The sultan agreed, thinking that the final amount of rice would be trifling, but in fact he was bankrupted because the final square of the chessboard contained 9,223,372,036,854,775,808 grains. The combined total for all the squares would have been almost twice this number, which far outstrips today's worldwide annual production of rice.

So inflation would have vastly expanded the universe in an instant, before giving way to the more leisurely expansion that we see today. Crucially, in 0.00000000000000000000000000000000001 seconds, inflation would have had a major influence on the development of the universe. Primarily, the newborn universe would have had only insignificant variations in density, but inflation would have blown up and exaggerated these minor variations, thereby leading to the significant variations that astronomers know existed after 300,000 years. These variations, in particular the higher-density clumps, then went on to seed the formation of the galaxies.

Another consequence of inflation is that a universe that was not flat before inflation would have become very flat afterwards. The surface of a billiard ball is clearly not flat, but if it repeatedly doubled in size twenty-seven times then it would be as big as the Earth. The Earth still has a curved surface, but it is much less curved than a billiard ball, and on the human scale it gives the appearance of being flat. Similarly, an inflated universe would tend to give the impression of being flat, which is what astronomers see today.

As well as addressing the problems of generating variations and flatness, inflation could potentially throw light on another mystery. When astronomers compare their views of the universe in opposite directions, there seems to be a strong similarity between patches of the cosmos that are more than 20 billion light years apart. Cosmologists expected much more divergence between such distant regions of the universe, but inflation could explain why this is not so. Two parts of the universe might have been very close to each other before inflation, so they would have been very similar because of their proximity to each other. Then, after the fantastic expansion associated with inflation, they would suddenly have been separated by a relatively large distance, yet they would retain their initial similarity because the separation had occurred so rapidly.

Guth's inflation theory is still in the workshop, but many cosmologists think that in due course it will be incorporated into the Big Bang model. Jim Peebles once said: 'If inflation is wrong, God missed a good trick! Inflation is a beautiful idea. However, there are many other beautiful ideas that nature has decided not to use so we shouldn't complain too much if it's wrong.'

Something else that keeps Big Bang cosmologists awake at night is *dark matter*. Observations show that stars orbiting the periphery of galaxies have tremendous speeds, yet the gravitational pull of all the stars closer to the heart of the galaxy is not enough to prevent these peripheral stars from flying off into the cosmos. Therefore, cosmologists believe that there must be vast quantities of dark matter in a galaxy, namely matter that does not shine but which exerts enough of a gravitational pull to keep the stars in their orbits. Although the idea of dark matter dates back to Fritz Zwicky at Mount Wilson in the 1930s, cosmologists are still unsure of its true nature, which is rather embarrassing as calculations imply that the universe has more dark matter than ordinary stellar matter.

Some candidates for dark matter are so-called *massive compact halo objects* (MACHOs), a category which includes black holes, asteroids and giant Jupiter-like planets. We would not see such objects in a galaxy, because they do not shine, but they would all contribute to the gravitational attraction within a galaxy. Other candidates for dark matter come under the heading of *weakly interacting massive particles* (WIMPs), which includes various types of particles that do not form objects like MACHOS, but which might permeate the entire universe, hardly making their presence felt, except through the force of gravity.

As yet, there are only vague clues to the nature and amount of dark matter in the universe, which is rather frustrating because cosmologists need a respectable understanding of dark matter before they can fill in some gaps in the Big Bang model. For example, the gravitational influence of dark matter would have played a major role in attracting more ordinary matter in the early stages of the universe, thereby helping to form galaxies.

And, at the other end of the timeline, dark matter might play a decisive role in the ultimate fate of the universe. The universe has been expanding ever since the Big Bang, but all of the mass of the universe should have been pulling the matter inwards and gradually slowing down the expansion. This leads to three possible futures, which were first proposed by Alexander Friedmann in the 1920s. First, the universe might expand for ever, but at an ever-decreasing rate. Second, the universe might gradually slow down to the point where the expansion grinds to a halt. Third, the universe might slow down, stop, and then begin to contract towards what is now known as the Big Crunch or Big Squeeze. So the future of the universe depends on the gravitational pull within the universe, which depends on the mass of the universe, which in turn depends on the amount of dark matter in the universe.

In fact, a fourth potential future is now seriously being considered. In the late 1990s astronomers focused their telescopes on a particular variety of supernova known as a Type Ia supernova. These supernovae are very bright and so can be seen even if they erupt in remote galaxies. Type Ia supernovae also have the advantage of having a telltale brightness variation that can be used to gauge their distances and thus the distances to the galaxies that contain them. And, by using spectroscopy, it is possible to measure their recessional velocity. As astronomers studied more and more

Type Ia supernovae, the measurements seemed to be implying that the universe was actually expanding at an ever-increasing rate. So, instead of the expansion of the universe slowing down, it seems to be speeding up. The universe is apparently blowing itself apart. The repulsive driving force for this runaway universe is still a mystery, and has been labelled *dark energy*.

With a momentarily violent period of inflation, peculiar dark matter and weird dark energy, the new Big Bang universe of the twenty-first century is a strange place indeed. It seems that the eminent scientist J.B.S. Haldane had tremendous foresight when he wrote in 1937: 'My suspicion is that the universe is not only queerer than we suppose, but queerer than we can suppose.'

Completely solving the remaining mysteries of the Big Bang will require a three-pronged attack, involving further theoretical developments, laboratory experiments and, most important of all, even clearer observations of the cosmos. For example, the COBE satellite completed its scientific mission on 23 December 1993 and has been superseded by satellites with improved detectors, such as WMAP, whose results appear in Figure 104. Even better satellites are already being designed, and on the Earth's surface there will be more sensitive radio telescopes, more powerful optical telescopes and experiments on the lookout for signs of dark matter.

Future observations will challenge, test and stretch the Big Bang model. They may lead to a revision of the estimate of the age of the universe, diminish the influence of dark matter in the universe or fill in some gaps in our knowledge, but cosmologists generally agree that these will only be tweaks to the overall scheme of the Big Bang model, rather than a paradigm shift to a completely new model. This is a view endorsed by Big Bang pioneers Ralph Alpher and Robert Herman in *Genesis of the Big Bang*, published in 2001: 'Although many questions about cosmological modeling are still unanswered, the Big Bang model is in reasonably good shape. We

Figure 104 The WMAP (Wilkinson Microwave Anisotropy Probe) satellite was designed to measure the CMB radiation at thirty-five times better resolution than the COBE satellite could manage. Its observations were turned into the maps shown here, released in 2003. The lozenge-shaped format is equivalent to the projection of the COBE maps shown in Figure 102 (p. 461). This map can be rolled up to form a sphere, and the two opposite sides of the sphere are also shown. You can imagine the WMAP satellite in the centre of the sphere looking out at variations in the CMB radiation across the sky.

The WMAP data allowed various parameters of the universe to be measured with greater accuracy than ever before. The WMAP team estimated that the universe is 13.7 billion years old, to within an error of just 0.2 billion years. They also calculated that the universe is 23% dark matter, 73% dark energy and 4% ordinary matter. Furthermore, the size of the variations is compatible with what astronomers would expect to see if there was an inflationary phase in the early universe.

are certain that future theoretical and observational work will at the very least fine-tune it, but we do not anticipate that, after more than 50 years, the model will turn out to be basically inadequate. Would that we could come back after another 50 years and see how it all came out.'

Although the majority of cosmologists would agree with Alpher and Herman, it is important to note that the Big Bang still has a few staunch critics, who continue to prefer the notion of an eternal universe. When the Steady State model became untenable, a few of its supporters switched to the modified version, the Quasi-Steady State model. Cosmologists who continue to support this minority view are fiercely proud of their role in challenging the Big Bang orthodoxy. Indeed, Fred Hoyle, who died in 2001, went to his grave in the firm belief that the Quasi-Steady State model was correct and that the Big Bang model was wrong. In his autobiography he wrote: 'To claim, however, as many supporters of Big Bang cosmology do, to have arrived at the correct theory verges, it seems to me, on arrogance. If I have ever fallen into this trap myself, it has been in short spells of hubris, inevitably to be followed by nemesis.' Such healthy defiance is an inherent part of science and should never be discouraged. After all, the Big Bang model itself was a consequence of a rebellion against the establishment.

And Hoyle's hatred of the Big Bang model was probably compounded by the fact that it was his naming of it that helped to establish it in the public consciousness. 'Big Bang' turned out to be a short, punchy and memorable title for the theory of creation, yet it was invented by the theory's greatest critic. While some cosmologists like the tabloid tone of the phrase Big Bang, others complain that it seems inappropriate for a concept of such magnificent grandeur. Even the cartoon characters Calvin and Hobbes pointed out the problem in a comic strip by Bill Watterson that appeared on 21 June 1992. Calvin says to Hobbes: 'I've been reading about the beginning of the universe. They call it "The Big Bang". Isn't it weird how scientists can imagine all the matter of the universe exploding out of a dot smaller than the head of a pin, but they can't come up with a more evocative name for it than "The Big Bang"? That's the whole problem with science. You've got a bunch of empiricists trying to describe things of unimaginable wonder.' Calvin goes on to suggest 'The Horrendous Space Kablooie!' as an alternative title, which some cosmologists actually used for a while, sometimes abbreviating it to the HSK.

The following year, *Sky & Telescope* magazine ran a competition to replace the Big Bang label, but the esteemed panel of judges, Carl Sagan, Hugh Downs and Timothy Ferris, were not impressed by the entries. Suggestions for new titles included 'Hubble Bubble', 'Bertha D. Universe' and 'SAGAN' ('Scientists Awestruck by God's Awesome Nature'). They concluded that none of the 13,099 suggestions from forty-one countries was any better than Hoyle's original derisive 'Big Bang' tag.

It seems as though this is a testament to the fact that the Big Bang model is now part of our culture. An entire generation has grown up with the Big Bang as the model that explains the creation, evolution and history of the universe, and we could not imagine this theory by any other name.

Even the Church has grown to love the Big Bang model. Ever since Pope Pius XII endorsed the Big Bang, the Catholic Church has largely tolerated this scientific view of creation. It has effectively abandoned any pretence that Scripture gives a literal explanation for the universe. This has proved to be a very pragmatic change of attitude. In the past, God provided the guiding hand behind all the mysteries of the universe, from volcanic eruptions to the setting of the Sun, but one by one science has provided rational and natural explanations for these phenomena. The chemist Charles Coulson coined the term 'God of the gaps' to point out that a deity who was supposedly responsible for everything beyond our comprehension would have his power diminished as each gap in knowledge was filled by science. But now the Catholic Church concentrates on the spiritual world and leaves the job of explaining the natural world to science, which means that it can remain secure in the knowledge that any future scientific discoveries cannot diminish the status of God. Science and religion can live independently, side by side.

In 1988, as if to reinforce this independence, Pope John Paul II declared: 'Christianity possesses the source of its justification within itself and does not expect science to constitute its primary apologetic.' Then, in 1992, the Vatican even admitted that it had been wrong to persecute Galileo. Advocating a Sun-centred view of the universe had been considered heresy because, according to the Bible: 'God fixed the Earth upon its foundation, not to be moved for ever.' However, after an inquiry that lasted thirteen years, Cardinal Paul Poupard reported that theologians at the time of Galileo's trial 'failed to grasp the profound non-literal meaning of the Scriptures when they describe the physical structure of the universe'. And in 1999 the Pope symbolically put an end to the centuries-old conflict between religion and cosmology when he toured his Polish homeland and visited the birthplace of Nicholas Copernicus, specifically praising Copernicus's scientific achievements.

Perhaps encouraged by the Church's newfound tolerance, some cosmologists have decided to delve into the philosophical implications of the Big Bang model. For example, the model describes how the universe started from a hot, dense, primordial soup and then evolved into the vast array of galaxies, stars, planets and life forms that exist today – was this inevitable, or could the universe have been different? The Astronomer Royal, Martin Rees, addresses this issue in his book *Just Six Numbers*. In it he explains how the structure of the universe ultimately depends on just six parameters, such as the strength of gravity. Scientists can measure the value of each of these parameters, which gives the eponymous six numbers. Rees wonders how things might have been different if these numbers had taken on other values when the universe was created. For example, if the number assigned to gravity had been larger, then the force of gravity would have been stronger, which would have resulted in stars that formed more quickly.

One number, which Rees labels ε, reflects the strength of the strong nuclear force, which glues together the protons and neutrons in the nucleus of an atom. The bigger the value of ε, the stronger the glue. Measurements show that $\varepsilon = 0.007$, which is incredibly fortunate, because if it were much different then the consequences would have been catastrophic. If $\varepsilon = 0.006$, the nuclear glue would have been slightly weaker, and it would have been impossible to fuse hydrogen into deuterium. This is the first step on the road to forming helium and all the heavier elements. In fact, if $\varepsilon = 0.006$, then the entire universe would be filled with nothing but bland hydrogen, so there would be no chance of any life. Instead, if $\varepsilon = 0.008$, the nuclear glue would have been slightly stronger, and hydrogen would have all too readily transformed itself into deuterium and helium – so much so, that all the hydrogen would have disappeared in the early phase of the Big Bang and there would be none left to fuel the stars. Again, there would be absolutely no chance of life.

Rees examines the other five numbers that define the universe and explains how changing any of them would have severely affected the evolution of the universe. In fact, some of these five numbers are even more sensitive to change than ε. Had they been even very slightly different from the values we measure, then the universe could easily have been sterile, or it could have destroyed itself as soon as it was born.

Consequently, it seems that these six numbers have been tuned for life. It is as though the six dials that dictated the evolution of the universe had been carefully set in order to create the conditions necessary for us to exist. The eminent physicist Freeman Dyson wrote: 'The more I examine the universe and the details of its architecture, the more evidence I find that the universe in some sense must have known we were coming.'

This harks back to the anthropic principle mentioned in Chapter 5, which Fred Hoyle exploited to work out how carbon is created within stars. The anthropic principle states that any cosmological theory must take into account the fact that the universe has evolved to contain us. It implies that this should be a significant element in cosmological research.

The Canadian philosopher John Leslie devised a firing squad scenario to elucidate the anthropic principle. Imagine that you have been accused of treason and are awaiting execution in front of a firing squad of twenty soldiers. You hear the command to fire, you see the twenty guns fire – and then you realise that none of the bullets has hit you. The law says that you are allowed to walk free in such a situation, but as you head for freedom you begin to wonder why you are still alive. Did the bullets all miss by chance? Does this sort of thing happen once in every ten thousand executions, and did you just happen to be lucky? Or was there a reason why you survived? Perhaps all twenty members of the firing squad deliberately missed because they believed you to be innocent? Or when the rifle sights were calibrated the previous night, were they all mistakenly set so the rifles all fired 10° to the right of the intended target? You could live the rest of your life assuming that the failed execution was nothing more than chance, but it would be hard not to read some deeper significance into your survival.

Similarly, it seems to defy the odds that the six numbers that characterise the universe have very special values that allow life to flourish. So do we ignore this and just count ourselves extremely lucky, or do we look for special meaning in our extraordinarily good fortune?

According to the extreme version of the anthropic principle, the fine-tuning of the universe which has allowed life to evolve is indicative of a tuner. In other words, the anthropic principle can be interpreted as evidence for the existence of a God. However, an alternative view is that our universe is part of a *multiverse*. The dictionary definition of the universe is that it encompasses everything, but cosmologists tend to define the universe as the collection of only those things that we can perceive or that can influence us. By this definition, there could be many other separate and isolated universes, each defined by its own set of six numbers. The multiverse would thus consist of numerous diverse universes, perhaps an infinity of universes. The overwhelming majority of them would be either sterile or short-lived, or both, but by chance just a few will contain the sort of environment that is capable of evolving and sustaining life. Of course, we happen to live in one of the universes that is conducive to life.

'The cosmos maybe has something in common with an off-the-shelf clothes shop,' says Rees. 'If the shop has a large stock, we're not surprised to find one suit that fits. Likewise, if our universe is selected from a multiverse, its seemingly designed or fine tuned features wouldn't be surprising.'

This question – was our universe designed for life or is it the lucky universe in a generally unlucky multiverse? – is at the very edge of scientific speculation and the subject of heated debate among cosmologists. The only question that surpasses it in metaphysical magnitude is the biggest question of all: what came before the Big Bang?

So far, the capacity of the Big Bang model has been limited to describing how the vast cosmos observable today emerged and evolved from a dense, hot state billions of years ago. Exactly how far back you are prepared to extend the Big Bang model depends on whether you include features such as the early inflationary phase or the latest theories in particle physics, which purport to describe the universe when it had a temperature of 10^{32} degrees Celsius and was only 10^{-43} seconds old.

That still leaves the outstanding issue of the actual moment of creation and what caused it. This was something that George Gamow rapidly shied away from when critics questioned him about the scope of his research. He added a disclaimer in the second printing of his popular treatise *Creation of the Universe*:

> In view of the objections raised by some reviews concerning the use of the word 'creation', it should be explained that the author understands this term, not in the sense of 'making something out of nothing', but rather as 'making something shapely out of shapelessness', as, for example, in the phrase 'the latest creation of Parisian fashion'.

Failure to address what happened before the Big Bang would be a disappointment, but not a ruinous failure for cosmology. At worst, the Big Bang model would remain valid but incomplete, which would put it on a par with many other scientific theories. Biologists are a long way from explaining how life was created, but this does not bring into question the validity of their theory of evolution by natural selection, or the concepts of genes and DNA. Cosmologists, though, have to admit that they are probably in a worse position than biologists. There is every reason to believe that the standard laws of chemistry as we understand them were behind the construction of the first cell and the first piece of DNA, whereas it is not clear that the known laws of physics were valid in the moment of cosmic creation. As we run the clock backwards and the universe approaches the moment of zero time, it seems that all matter and energy was concentrated at one point, which causes a major problem for the laws of physics. At the moment of creation, the universe seems to enter an unphysical state known as a *singularity*.

Even if cosmologists could cope with the physics of a singularity, many of them claim that the question 'What came before the Big Bang?' is impossible to answer because it is invalid. After all, the model states that the Big Bang gave rise not only to matter and radiation but also to space and time. So if time was created during the Big Bang, then time did not exist before the Big Bang, and it is therefore impossible to use the phrase 'before the Big Bang' in any meaningful way. Another way to think of this is in terms of the word 'north', which can be used sensibly in the questions 'What is north of London?' or 'What is north of Edinburgh?', but makes no sense in the context of 'What is north of the North Pole?'

Critics may feel that if this is the best that cosmologists can offer, then 'What came before the Big Bang?' is a puzzle that has to be relegated to the realm of myth or religion, a gap for God which will forever remain beyond the reach of science. In his book *God and the Astronomers*, the American astronomer Robert Jastrow was pessimistic about the ambition of the Big Bang theorist: 'He has scaled the mountains of ignorance; he is about to conquer the highest peak; as he pulls himself over the final rock, he is greeted by a band of theologians who have been sitting there for centuries.'

One way to finesse the problem of creation is to consider a slightly overweight universe. The universe would expand, but the extra mass would result in a greater gravitational pull that would halt the expansion and then reverse it so that the universe actually began to contract. The universe would look to be heading for a Big Crunch, as mentioned earlier, but instead there is a Big Bounce. As the matter and energy become concentrated, the universe might reach a critical stage at which the pressure and energy counteract gravity and begin to push the universe back outwards. This leads to another Big Bang and another expansion phase, until gravity halts the expansion, resulting in a contraction followed by another Big Crunch, and another Big Bang, and so on.

This rebounding, oscillating, eco-friendly, recyclable, phoenix universe would be eternal, but it could not be considered as being in a Steady State. This is not a version of the Steady State model, but rather a multiple Big Bang model. It has been seriously discussed by several cosmologists, including Friedmann, Gamow and Dicke.

Others, such as Eddington, detested this vision of a recycled universe: 'I would feel more content that the universe should accomplish some great scheme of evolution and, having achieved whatever might be achieved, lapse back into chaotic changelessness, than that its purpose should be banalized by continual repetition.' In other words, an ever-expanding universe will eventually become cold and barren because its stars will run out of hydrogen fuel and stop shining, and Eddington preferred this 'Big Freeze' (or 'heat death') scenario to an infinitely repetitive and tedious universe.

In addition to Eddington's subjective criticism, the rebounding Big Bang faces a range of practical problems. For example, no cosmologist has yet been able to give a full account of the forces that would be required to cause a cosmic rebound. In any case, the latest observations indicate that the universe's expansion is accelerating, which reduces the likelihood of the current expansion turning into a contraction.

Despite its flaws, the rebounding universe scenario does allow the collapse of the universe to trigger the next Big Bang, which at least addresses the issue of cause-and-effect that lies at the heart of our desire to find out what came before the Big Bang. But perhaps cause-and-effect is a common-sense prejudice that should be set aside in this cosmological context. After all, the Big Bang expansion started on a miniature scale, and common sense does not really apply in this extreme realm. Instead, it is the weird rules of quantum physics that hold sway.

Quantum physics is the most successful and utterly bizarre theory in the whole of physics. As Niels Bohr, one of the founders of quantum physics, famously said: 'Anyone who is not shocked by quantum theory has not understood it.'

Although cause-and-effect is a valid principle in the everyday macroscopic world, it is the so-called *uncertainty principle* that rules the sub-microscopic quantum domain. This principle dictates that events can happen spontaneously, which has been shown to be the case experimentally. It also allows matter to appear from nowhere, even if only temporarily. At the everyday level the world seems deterministic and the laws of conservation hold true, but at the microscopic level determinism and conservation can both be violated.

Hence, *quantum cosmology* offers various hypotheses that allow for the universe to have started from nothing for no reason. For example, a baby universe could have spontaneously emerged from nothing, possibly alongside a multitude of other universes, making it part of a multiverse. As Alan Guth, the father of inflation theory, put it: 'It's often said there is no such thing as a free lunch. But the universe itself may be a free lunch.'

Unfortunately, the scientific community has to admit that all these possible answers, from rebounding universes to spontaneous quantum creation, are highly speculative and do not yet properly address the ultimate question of where the universe came from. Nevertheless, the current generation of cosmologists should not be downhearted. They should rejoice in the fact that the Big Bang model is a coherent and consistent description of our universe. They should be proud that the Big Bang model is a pinnacle of human achievement, because it explains so much of the universe's present by revealing its past. They should go out and tell the world that the Big Bang model is a tribute to human curiosity and our intellect. And if a member of the public should ask the toughest question of all, 'What came before the Big Bang?', then they might consider following St Augustine's example.

In his autobiography, *Confessions*, written in about AD 400, the philosopher and theologian St Augustine quotes an answer he has heard to the theological equivalent of 'What came before the Big Bang?':

> What was God doing before He created the Universe?
> Before He created Heaven and Earth, God created hell to be used for people such as you who ask this kind of question.

WHAT IS SCIENCE?

The words 'science' and 'scientist' are surprisingly modern inventions. In fact, the word 'scientist' was coined by the Victorian polymath William Whewell, who used it in the *Quarterly Review* in March 1834. The Americans took to the word almost immediately, and by the end of the century it was also popular in Britain. The word is based on the Latin *scientia*, which means 'knowledge', and it supplanted other terms such as 'natural philosopher'.

This book is a history of the Big Bang model, but at the same time it attempts to provide an insight into what science is and how it works. The Big Bang model is a good example of how a scientific idea is created, tested, verified and accepted. Nevertheless, science is such a broad activity that this book's description of it is incomplete. So, in an attempt to fill in some of the gaps, here is a selection of quotations about science.

Science is organized knowledge.
HERBERT SPENCER (1820–1903), English philosopher

Science is the great antidote to the poison of enthusiasm and superstition.
ADAM SMITH (1723–90), Scottish economist

Science is what you know. Philosophy is what you don't know.
BERTRAND RUSSELL (1872–1970), English philosopher

[Science is] a series of judgements, revised without ceasing.
PIERRE EMILE DUCLAUX (1840–1904), French bacteriologist

[Science is] the desire to know causes.
WILLIAM HAZLITT (1778–1830), English essayist

[Science is] the knowledge of consequences, and dependence of one fact upon another.
THOMAS HOBBES (1588–1679), English philosopher

[Science is] an imaginative adventure of the mind seeking truth in a world of mystery.
CYRIL HERMAN HINSHELWOOD (1897–1967), English chemist

[Science is] a great game. It is inspiring and refreshing. The playing field is the universe itself.
ISIDOR ISAAC RABI (1898–1988), American physicist

Man masters nature not by force but by understanding. This is why science has succeeded where magic failed: because it has looked for no spell to cast over nature.
JACOB BRONOWSKI (1908 –74), British scientist and author

That is the essence of science: ask an impertinent question, and you are on the way to a pertinent answer.
JACOB BRONOWSKI (1908–74), British scientist and author

It is a good morning exercise for a research scientist to discard a pet hypothesis every day before breakfast. It keeps him young.
KONRAD LORENZ (1903–89), Austrian zoologist

Truth in science can best be defined as the working hypothesis best suited to open the way to the next better one.
KONRAD LORENZ (1903–89), Austrian zoologist

In essence, science is a perpetual search for an intelligent and integrated comprehension of the world we live in.
CORNELIUS VAN NEIL (1897–1985), American microbiologist

The scientist is not a person who gives the right answers, he is one who asks the right questions.
CLAUDE LEVI-STRAUSS (1908–), French anthropologist

Science can only ascertain what *is*, but not what *should be*, and outside of its domain value judgements of all kinds remain necessary.
ALBERT EINSTEIN (1879–1955), German-born physicist

Science is the disinterested search for the objective truth about the material world.
RICHARD DAWKINS (1941–), English biologist

Science is nothing but trained and organised common sense differing from the latter only as a veteran may differ from a raw recruit; and its methods differ from those of common sense only as far as the guardsman's cut and thrust differ from the manner in which a savage wields his club.
THOMAS HENRY HUXLEY (1825–95), English biologist

The sciences do not try to explain, they hardly even try to interpret, they mainly make models. By a model is meant a mathematical construct which, with the addition of certain verbal interpretations, describes observed phenomena. The justification of such a mathematical construct is solely and precisely that it is expected to work.
JOHN VON NEUMANN (1903–57), Hungarian-born mathematician

The science of today is the technology of tomorrow.
EDWARD TELLER (1908–2003), American physicist

Every great advance in science has issued from a new audacity of imagination.
JOHN DEWEY (1859–1952), American philosopher

Four stages of acceptance:
i) this is worthless nonsense,
ii) this is an interesting, but perverse, point of view,
iii) this is true, but quite unimportant,
iv) I always said so.
J.B.S. HALDANE (1892–1964), English geneticist

Philosophy of science is about as useful to scientists as ornithology is to birds.
RICHARD FEYNMAN (1918–88), American physicist

A man ceases to be a beginner in any given science and becomes a master in that science when he has learned that he is going to be a beginner all his life.
ROBIN G. COLLINGWOOD (1889–1943), English philosopher

GLOSSARY

Terms in italics have their own entry in the glossary.

absorption The process by which *atoms* absorb light at specific *wavelengths*, allowing their presence to be detected by *spectroscopy* by identifying the 'missing' wavelengths.

alpha particle A subatomic particle ejected during certain kinds of *radioactive decay*. The particle, consisting of two *protons* and two *neutrons*, is identical to the *nucleus* of a *helium atom*.

anthropic principle The principle that states that, since humans are known to exist, the laws of physics must be such that life can exist. In its extreme form, the anthropic principle states that the universe has been designed to allow life.

arcminute A unit used in the measurement of very small angles, equal to 1/60 of 1°.

arcsecond A unit used in the measurement of very small angles, equal to 1/60 of an *arcminute* or 1/3,600 of 1°.

atom The smallest component of an *element*, comprising a positively charged *nucleus* surrounded by negatively charged *electrons*. The number of positively charged *protons* in the nucleus uniquely determines which chemical element the atom belongs to. For example, every atom containing a single proton is an atom of *hydrogen*, while every atom containing 79 protons is an atom of gold.

Big Bang model The currently accepted *model* of the universe, according to which time and space emerged from a hot, dense, compact region between 10 and 20 billion years ago.

Cepheid variable star A type of *star* whose brightness varies over a precise, regular period, usually between 1 and 100 days. The period of variation is directly linked to the star's average luminosity, which can therefore be calculated. By comparing the star's luminosity to the apparent brightness as seen from the Earth, its distance can be accurately determined. These stars therefore play an important role in determining the cosmic distance scale.

CMB radiation See *cosmic microwave background radiation*.

COBE (Cosmic Background Explorer) A satellite launched in 1989 to make accurate measurements of the *cosmic microwave background* (CMB) *radiation*. Its DMR detector provided the first evidence for variations in the CMB radiation, indicative of regions in the early universe that led to *galaxy* formation.

Copernican model The Sun-centred *model* of the universe, proposed by Nicholas Copernicus in the sixteenth century.

cosmic microwave background (CMB) radiation A pervasive 'sea' of *microwave radiation* emanating almost uniformly from every direction in the universe, which dates back to the moment of *recombination*. This radiation is the 'echo' of the *Big Bang*, predicted by Gamow, Alpher and Herman in 1948, and detected by Penzias and Wilson in 1965. Originating in the heat of the Big Bang, it has since been stretched from *infrared* to microwave *wavelengths* by the expansion of the universe. The *COBE* satellite measured variations in the CMB radiation.

cosmological constant An extra parameter incorporated by Einstein into the equations of his *general theory of relativity* when it became clear that his equations implied either a growing or a shrinking universe. By effectively introducing anti-gravity, the equations then permitted a static universe.

cosmological principle The principle that no location in the universe is preferred over any other, and that the overall features of the universe appear to be the same in all directions (*isotropic*) and no matter where the observer is located (*homogeneous*).

cosmology The study of the origin and evolution of the universe.

creation field (C-field) A theoretical concept introduced as part of the *Steady State model*. The C-field maintained the overall density of the universe by creating matter to fill the gaps resulting from the expansion of the universe.

cross-section A quantity used by particle physicists to assess the likelihood of two particles colliding.

dark energy A postulated form of energy that could account for recent observations which imply that the expansion of the universe is accelerating. Although calculations suggest that it may make a dominant contribution to the mass-energy in the universe, there is no agreement on its nature.

dark matter A postulated form of matter, believed to make up a significant fraction of the matter in the universe. It makes its presence felt via its *gravity*, but emits little or no *visible light*.

deferent The large circle used to describe the motion of a celestial body around the Earth in the *Ptolemaic model*. When combined with a smaller *epicycle*, the observed planetary motions could be approximately replicated.

deuterium An *isotope* of *hydrogen* containing one *proton* and one *neutron* in the *nucleus*.

Doppler effect The change in *wavelength* of sound or *electromagnetic waves* emitted by a moving source. The same effect occurs if it is the observer (not the source) that is moving. Waves are compressed ahead of the source and stretched behind it, producing, for example, the familiar change in the pitch of a siren from high to low as an ambulance passes by at speed. A similar effect causes the *redshift* in the spectrum of a receding *galaxy*.

electromagnetic radiation A travelling form of energy, including *visible light*, *radio waves* and X-rays. Electromagnetic radiation moves through space as *electromagnetic waves* at the *speed of light*. The *wavelength* of the radiation determines its qualities.

electromagnetic spectrum The complete range of *wavelengths* of *electromagnetic radiation*, from short-wavelength (high-energy) gamma rays and X-rays, through *ultraviolet*, *visible light* and *infrared*, to long-wavelength (low-energy) *radio waves*.

electromagnetic waves A harmonised vibration of electric and magnetic fields, each sustaining the other and propagating together through space as *electromagnetic radiation*.

electron A subatomic particle with negative charge. Electrons can exist independently or in orbit around the positively charged *nucleus* of an *atom*.

element One of the basic materials of the universe, as listed in the periodic table. The smallest quantity of an element is an *atom*, and the number of *protons* in the atom determines the type of element.

emission The process by which *atoms* are excited (for example by heating) and emit light at specific *wavelengths*, allowing their presence to be detected by *spectroscopy*.

epicycle The small circle used in the Earth-centred *Ptolemaic model* of the universe, in addition to the *deferent*, to account for the looping *retrograde motions* of some planets as they moved in their supposed orbits around the Earth.

ether The all-pervading substance through which light was once believed to propagate. Its existence was disproved by the *Michelson–Morley experiment.*

exponential notation A convenient method of abbreviating very large or very small numbers. For example 1,200 can be written as 1.2×10^3 because it equals $1.2 \times (10 \times 10 \times 10)$, and 0.0005 can be written as 5×10^{-4} because it equals $5 \div (10 \times 10 \times 10 \times 10)$.

fission The process by which a large atomic *nucleus* is broken apart to produce two smaller nuclei, generally releasing energy as a result. *Radioactive decay* is a fission process that occurs spontaneously.

fusion The process by which two small atomic *nuclei* join together to make a single larger nucleus, generally releasing energy as a result. For example, *hydrogen* nuclei can fuse via a multi-step process to form *helium.*

galaxy A collection of *stars*, gas and dust held together by *gravity*, usually separated from neighbouring galaxies, and often spiral or elliptical in shape. Galaxies range in size from around a million stars to several billion.

general theory of relativity Einstein's theory of *gravity*, which underpins the science of *cosmology*. General relativity describes gravity as a curvature in four-dimensional *spacetime.*

gravity An attractive force experienced between any pair of massive bodies. Gravity was first described by Newton, but Einstein produced a more accurate description in his *general theory of relativity* that depended on the curvature of *spacetime.*

helium The second most common and second-lightest *element* in the universe, after *hydrogen*. Its *nucleus* contains two *protons* and (usually) two *neutrons*. The pressures and temperatures inside *stars* can force helium to undergo nuclear *fusion* to form heavier nuclei.

homogeneous Having similar properties at all locations.

Hubble constant (H_0) A measurable parameter of the universe, describing its rate of expansion. It is believed to have a value of 50–100 km/s/Mpc, meaning that a *galaxy* 1 megaparsec away will be receding at between 50 and 100 km/s. The Hubble constant emerges from the definition of *Hubble's law.*

Hubble's law The empirically determined law stating that the recessional velocity of a *galaxy* is proportional to its distance: $v = H_0 \times d$. The constant of proportionality in the equation (H_0) is the *Hubble constant.*

hydrogen The simplest and most abundant *element* in the universe, containing one *proton* in its *nucleus*, orbited by one *electron*. See also *deuterium*.

inflation The phase of extremely rapid expansion during the first 10^{-35} seconds of the universe. Although inflation is hypothetical, it would explain several features of the universe.

infrared The portion of the *electromagnetic spectrum* with slightly longer *wavelengths* than *visible light*.

isotope A variant of a single *element*, distinguished by having a different number of *neutrons* in its *nucleus*. For example, *hydrogen* has three isotopes, possessing zero, one and two neutrons respectively, but all contain just one *proton*.

isotropic Similar in all directions.

light wave See *electromagnetic wave*.

light year The distance travelled by light in one calendar year, approximately 9,460,000,000,000 km.

Michelson–Morley experiment An experiment conducted in the late nineteenth century to detect the Earth's motion through the *ether* by measuring the *speed of light* both in the direction of and perpendicular to the Earth's direction of travel. The experiment disproved the existence of the ether.

microwave radiation A portion of the *electromagnetic spectrum* with *wavelengths* of a few millimetres or centimetres. It is usually regarded as a subdivision of *radio waves*.

Milky Way A name given to the *galaxy* in which our Solar System resides. The Milky Way is a spiral galaxy containing around 200 billion *stars*, and the Sun is located in one of its spiral arms.

model A self-consistent set of rules and parameters intended to describe mathematically some aspect of the real world.

multiverse An alternative *model* to the single universe, in which many different universes co-exist, each accommodating a different set of physical laws, and each isolated from all the others.

nebula A cloud of gas and, often, dust in the *Milky Way* galaxy seen as an indistinct patch of light in the night sky, in contrast to the point-like *stars*. In the twentieth century, following the resolution of the Great Debate, many objects labelled as nebulae before 1900 were recognised to be separate *galaxies*.

neutron A particle found inside atomic *nuclei*. A neutron has almost the same mass as a *proton*, but carries no electric charge.

nova A *star* that rapidly becomes, typically, 50,000 times brighter in a few days, and then returns to its former brightness over the course of a few months. A nova is fuelled by material flowing from a nearby companion star.

nuclear physics The study of atomic *nuclei*, their interactions and their structure.

nucleon A generic term for *protons* and *neutrons*, the two particles found in atomic *nuclei*.

nucleosynthesis The formation of the *elements* via nuclear *fusion*, particularly in *stars* and in *supernova* explosions. The nucleosynthesis of the lightest atomic *nuclei* took place in the moments after the *Big Bang*.

nucleus The compact structure at the centre of an *atom*, containing *protons* and *neutrons*, and accounting for at least 99.95% of the mass of any atom.

Occam's razor A rule of thumb stating that in the presence of alternative, adequate explanations for a phenomenon, the simpler one is the more likely to be correct.

parallax The apparent shift in location of an object when an observer changes position. *Stellar parallax* is used in astronomy to measure the distance to the closest *stars*.

parsec A unit of distance used in astronomy, equal to about 3.26 *light years*. Short for 'parallax second', it is the distance at which an object would show a *stellar parallax* of one *arcsecond*. A distance of 1 million parsecs is known as 1 megaparsec (Mpc).

perfect cosmological principle An extension of the *cosmological principle* which states that the universe is not only *homogeneous* and *isotropic*, but also unchanging with time. This principle is the basis of the *Steady State model*.

plasma A high-temperature state of matter in which atomic *nuclei* become separated from their *electrons*.

primeval atom theory Georges Lemaître's early version of the *Big Bang model* in which all the *atoms* in the universe were originally contained in one compact 'primeval atom'. The explosion of the primeval atom initiated the universe.

proper motion The apparent motion of a *star* across the sky, caused by its actual motion with respect to the Sun. The effect is so slight that it was not detected until 1718.

proton A positively charged subatomic particle found in the *nucleus* of an *atom*.

Ptolemaic model The flawed Earth-centred model of the universe, in which all other cosmic bodies followed orbits around the Earth. These orbits were constructed from perfect circles called *deferents* and *epicycles.*

quasar An intensely bright object, appearing like a *star* ('quasi-stellar'), but now known to be a highly luminous young *galaxy* that existed early in the universe. Quasars are observable today only in the most distant reaches of the universe, as the light reaching us from the far universe started its journey when the universe was much younger.

Quasi-Steady State model A modified version of the *Steady State model* which attempts to make up for some of the inconsistencies in the original model.

radial velocity The velocity of a *star* or *galaxy* towards or away from the Earth. This component of a star's motion can be determined from the *Doppler effect* on light or other *electromagnetic waves* emitted by the star or galaxy.

radioactive decay The process by which an atomic *nucleus* spontaneously transforms itself and releases energy. Typically it will change into a lighter, more stable nucleus.

radioactivity The tendency of certain *atoms* (e.g. uranium) to undergo *radioactive decay*.

radio astronomy The study of *radio waves* emitted by astronomical objects, using *radio telescopes* rather than optical telescopes.

radio galaxy A type of *galaxy* notable for its strong emission of *radio waves*. The radio emission from such a galaxy is about a million times as strong as that from a normal galaxy such as the *Milky Way*. Only about one in a million galaxies falls into this category.

radio telescope An instrument designed to detect *radio waves* from celestial radio sources. Radio telescopes are highly sensitive radio receivers and have the form of an antenna or a dish.

radio waves *Electromagnetic radiation* with *wavelengths* longer than a few millimetres, including microwaves. The study of radio waves emitted by celestial objects is called *radio astronomy*.

recombination The moment when the universe had cooled sufficiently to allow *electrons* to bind to *nuclei*, transforming matter from *plasma* to *atoms* with no overall electric charge. This occurred when the universe was roughly 300,000 years old and at a temperature of about 3,000°C. From that moment, *electromagnetic radiation* was able to travel through the universe almost unhindered; today we detect it as the *cosmic microwave background radiation*.

redshift An increase in the *wavelength* of emitted light caused by the emitter's recessional velocity and the resulting *Doppler effect*. In cosmology this term is usually associated with the stretching of *light waves* from a distant *galaxy* as the universe expands. The galaxy is not receding through space, but the expansion of space itself is causing the redshift.

relativity See *general theory of relativity* and *special theory of relativity*.

retrograde motion The temporary change in the apparent direction of motion of Mars, Jupiter and Saturn. It is a consequence of observing these planets from the Earth, which has a higher orbital speed around the Sun.

RR Lyrae star A type of variable *star* less luminous than a *Cepheid variable*, and with a period between 9 and 17 hours. The inability to detect any RR Lyrae stars in the Andromeda Galaxy in the 1940s was an important clue that the *galaxy* was more distant than previously thought.

similar triangles Any pair of triangles of the same shape but different sizes. The two triangles have all three angles in common, and corresponding edges scale in the same proportion.

Solvay Conferences A series of prestigious invitation-only conferences held every few years to discuss particular problems at the forefront of physics.

spacetime The unified construct in which the three dimensions of space combine with the fourth dimension of time to produce the underlying framework of our universe. The concept of spacetime is an integral part of Einstein's *special* and *general theories of relativity*. The curvature of spacetime leads to the force that we interpret as *gravity*.

special theory of relativity Einstein's theory based on the premise that the *speed of light* is the same for all observers regardless of their own motion. Its most famous consequence is the equivalence of energy and matter, expressed by the equation $E = mc^2$. It also implies that the perception of time and space depends on the observer. The theory is a 'special' case because it does not deal with objects which are accelerating or experiencing *gravity*, for which Einstein later developed the *general theory of relativity*.

spectroscope An instrument that separates *light waves* into their component *wavelengths* for analysis. It can be used to identify the *atoms* that emitted the light or to measure the amount of *redshift*.

spectroscopy The study of light by splitting it into its component *wavelengths* to learn about the nature of its source.

speed of light (*c*) A constant equal to exactly 299,792,458 m/s. According to the *special theory of relativity*, the speed of light is the same for all observers, regardless of their own motion.

star A ball of predominantly *hydrogen*, pulled together under the effect of its own *gravity*, with sufficient mass for the temperatures and pressures inside to initiate nuclear *fusion*. Stars tend to occur in formations called *galaxies*.

Steady State model A largely discredited *model* of the universe in which the universe expands and new matter is created in the growing gaps between *galaxies*. The universe would thus maintain a similar density at all times, and would last for eternity.

stellar parallax The apparent shift in position of a nearby *star* against the background of distant stars, caused by the observer's shift in position as the Earth orbits the Sun.

supernova The catastrophic explosion of a *star* that has exhausted its *hydrogen* fuel source. Heavier elements vital for life are generated in the events leading up to and during a supernova.

thought experiment An experiment conducted by thinking through a logical chain of events. This is useful when the conditions required to carry out a real experiment are prohibitive.

ultraviolet (UV) *Electromagnetic radiation* with a slightly shorter *wavelength* than that of *visible light*.

visible light A region of the *electromagnetic spectrum* containing the *electromagnetic radiation* which humans can see. The *wavelengths* range from 0.0004 mm (violet) to 0.0007 mm (red).

wavelength The distance between two successive peaks (or troughs) of a wave. The wavelength of *electromagnetic radiation* determines which part of the *electromagnetic spectrum* it belongs to and its overall properties.

FURTHER READING

This book has tried to explain a big subject in a relatively small amount of space. For readers who would like to explore some of the topics in more detail, the following list of books (and a few articles) might be of interest. They range from popular science to more technical texts, and the books are arranged under the chapter heading that is most appropriate to their contents. Many of them were used in the research and writing of *Big Bang*, but others go beyond the scope of this book, particularly those that relate to material covered in the Epilogue.

Chapter 1

Allan Chapman, *Gods in the Sky* (Channel 4 Books, 2002)

The Oxford historian of science discusses the development of ancient astronomy and the overlap with religion and mythology.

Andrew Gregory, *Eureka!* (Icon, 2001)

The development of science, mathematics, engineering and medicine in ancient Greece.

Lucio Russo, *The Forgotten Revolution* (Springer-Verlag, 2004)

An exploration of the rise of science in ancient Greece, and a discussion of why it came to an end and how it influenced Copernicus, Kepler, Galileo and Newton.

Michael Hoskin (editor), *The Cambridge Illustrated History of Astronomy* (CUP, 1996)

An excellent introduction to the history of astronomy.

John North, *The Fontana History of Astronomy and Cosmology* (Fontana, 1994)

A detailed overview of the history of astronomy, stressing its development as a science from ancient times.

Arthur Koestler, *The Sleepwalkers* (Arkana, 1989)
An account of the development of cosmology from ancient Greece through to the seventeenth century.
Kitty Ferguson, *The Nobleman and His Housedog* (Review, 2002)
A highly accessible account of the partnership between Tycho Brahe and Johannes Kepler.
Martin Gorst, *Aeons* (Fourth Estate, 2001)
A history of humankind's attempts to measure the age of the universe, from Bishop Ussher to Hubble's law.
Dava Sobel, *Galileo's Daughter* (Fourth Estate, 2000)
An account of the life of Galileo, which includes letters sent to him by his daughter, who lived in a convent from the age of thirteen.
Carl Sagan, *Cosmos* (Abacus, 1995)
The book based on the famous television series, which must have been the inspiration for numerous careers in astronomy.

Chapter 2

James Gleick, *Isaac Newton* (Fourth Estate, 2003)
An accessible and concise account of the life of Isaac Newton.
Hans Reichenbach, *From Copernicus to Einstein* (Dover, 1980)
A short history of the ideas that contributed to relativity theory.
David Bodanis, $E = mc^2$ (Walker, 2001)
The biography of an equation, inspired by Cameron Diaz, who once asked if somebody could explain the meaning of Einstein's famous formula.
Clifford Will, *Was Einstein Right?* (Basic Books, 1999)
An examination of the various tests that have been applied to Einstein's theories, including the measurement of Mercury's anomalous orbit and Eddington's eclipse expedition.
Jeremy Bernstein, *Albert Einstein and the Frontiers of Science* (OUP, 1998)
A popular biography with clear explanations of Einstein's work.
John Stachel, *Einstein's Miraculous Year* (Princeton University Press, 2001)
A moderately technical discussion of the remarkable papers that established Einstein's reputation in 1905.
Michio Kaku, *Einstein's Cosmos* (Weidenfeld & Nicolson, 2004)
A fresh account of Einstein's work on special relativity and general relativity, which also discusses his attempts to unify the laws of physics.

Russell Stannard, *The Time and Space of Uncle Albert* (Faber & Faber, 1990)

Uncle Albert and his niece Gedanken explore the relativistic world in a book aimed at young people from the age of eleven.

Edwin A. Abbott, *Flatland* (Penguin Classics, 1999)

Subtitled *A Romance of Many Dimensions*, this quirky, thought-provoking novella gives a useful insight into a multi-dimensional universe.

Melvyn Bragg, *On Giants' Shoulders* (Sceptre, 1999)

Twelve of history's greatest scientists are profiled, including several who played a role in the development of cosmology.

Arthur Eddington, *The Expanding Universe* (CUP, 1988)

This entertaining and popular essay about the expanding universe hypothesis was written in 1933, when the concept of the Big Bang was being developed.

E. Tropp, V. Frenkel and A. Chernin, *Alexander A. Friedmann: The Man Who Made the Universe Expand* (CUP, 1993)

A short but excellent biography of Friedmann, focusing on his professional life. It includes some semi-technical explanations of his cosmological ideas.

Chapter 3

Richard Panek, *Seeing and Believing* (Fourth Estate, 2000)

A history of the telescope and how it has changed our view of the universe.

Kitty Ferguson, *Measuring the Universe* (Walker, 2000)

A history of humankind's attempts to measure the cosmos, from the ancient Greeks to modern cosmology.

Alan Hirshfeld, *Parallax* (Owl Books, 2002)

A detailed, popular account of the heroic attempts to measure the distances to the stars.

Tom Standage, *The Neptune File* (Walker, 2000)

The discovery of Neptune is not relevant to the big questions in cosmology, but this excellent book covers a fascinating period in the history of astronomy.

Michael Hoskin, *William Herschel and the Construction of the Heavens* (Oldbourne, 1963)

An account of William Herschel's work to elucidate the structure of the Milky Way, with some of his original papers.

Solon I. Bailey, *History and Work of the Harvard Observatory 1839–1927* (McGraw Hill, 1931)

An interesting and largely non-technical (if somewhat dry) account of the research projects pursued at the Harvard College Observatory from its founding until the mid-1920s. It covers the work of Henrietta Leavitt and Annie Jump Canon, and explains the techniques and instruments they employed.

Harry G. Lang, *Silence of the Spheres* (Greenwood Press, 1994)

Subtitled *The Deaf Experience in the History of Science*, this book includes sections on John Goodricke and Henrietta Leavitt.

Edwin Powell Hubble, *The Realm of the Nebulae* (Yale University Press, 1982)

A somewhat technical book, based on the 1935 Silliman Lectures delivered by Hubble at Yale University. It is an interesting snapshot of cosmology soon after Hubble's major breakthroughs.

Gale E. Christianson, *Edwin Hubble: Mariner of the Nebulae* (Institute of Physics Publishing, 1997)

A non-technical and highly readable biography of Edwin Hubble.

Michael J. Crowe, *Modern Theories of the Universe from Herschel to Hubble* (Dover, 1994)

A good mix of history and science, including extracts from original writings by astronomers and cosmologists.

W. Patrick McCray, *Giant Telescopes* (Harvard U P, 2004)

An up-to-date history of the development of the telescope following the Edwin Hubble era.

Chapter 4

Helge Kragh, *Cosmology and Controversy* (Princeton University Press, 1999)

This book is a definitive yet largely accessible account of the entire Big Bang versus Steady State debate. The book focuses on the historical development of the debate and the personalities involved, and the relevant science is clearly explained along the way. This is probably the single most important book about the development of the Big Bang model.

F. Close, M. Marten and C. Sutton, *The Particle Odyssey: A Journey to the Heart of the Matter* (OUP, 2004)

An excellent guide to the history of atomic, nuclear and sub-nuclear physics, including its links to cosmology.

Brian Cathcart, *The Fly in the Cathedral* (Viking, 2004)

The story of Ernest Rutherford, his protégés and the Cavendish

Laboratory. A popular account of how physicists transformed our understanding of the atomic nucleus and split the atom.

George Gamow, *My World Line* (Viking Press, 1970)

Gamow's 'informal autobiography' gives a delightful insight into the life of one the twentieth century's most charismatic physicists.

George Gamow, *The New World of Mr Tompkins* (CUP, 2001)

An enchanting and light-hearted introduction into the weird world of quantum and relativistic physics by one of the great practitioners.

Joseph D'Agnese, 'The Last Big Bang Man Left Standing', *Discover* (July 1999, pp. 60–67)

An article which gave Ralph Alpher an important opportunity to describe to a general readership his role in the development of the Big Bang.

R. Alpher and R. Herman, *Genesis of the Big Bang* (OUP, 2001)

An excellent and not too technical account of the origin of the Big Bang model and its development up to the present day.

Iosif B. Khriplovich, 'The Eventful Life of Fritz Houtermans', *Physics Today* (July 1992, pp. 29–37)

An article documenting the life of Fritz Houtermans, written with affection and illustrated with charming photographs.

Fred Hoyle, *The Nature of the Universe* (Basil Blackwell, 1950)

Based on the BBC radio series that inadvertently christened the Big Bang model, this book gives an overview of cosmology in 1950.

Fred Hoyle, *Home Is Where the Wind Blows* (University Science Books, 1994)

An engaging autobiography which details Hoyle's numerous exploits as a mathematician, radar researcher, physicist, cosmologist and all-round maverick.

Thomas Gold, *Getting the Back off the Watch* (OUP, 2005)

Thomas Gold had just finished writing his memoirs when he passed away in 2004. The book is due to be published in 2005, and this is only its working title.

Chapter 5

J.S. Hey, *The Evolution of Radio Astronomy* (Science History Publications, 1973)

A concise overview of the development of radio astronomy from Jansky to the present day, written by one of its first practitioners.

Stanley Hey, *The Secret Man* (Care Press, 1992)
A short memoir.

Nigel Henbest, 'Radio Days', *New Scientist* (28 October 2000, pp. 46–7)
An interesting article about the early days of radio astronomy and Stanley Hey's contribution to the field.

Marcus Chown, *The Magic Furnace* (Vintage, 2000)
An excellent account of how physicists and cosmologists explained the mystery of nucleosynthesis.

Jeremy Bernstein, *Three Degrees above Zero* (CUP, 1984)
A history of scientific research conducted at Bell Labs, including interviews with Arno Penzias and Robert Wilson.

G. Smoot and K. Davidson, *Wrinkles in Time* (Little Brown, 1993)
The story of COBE by the head of the Differential Microwave Radiometer team.

John C. Mather, *The Very First Light* (Penguin, 1998)
The story of COBE by the head of the Far Infrared Absolute Spectrophotometer team.

M.D. Lemonick, *Echo of the Big Bang* (Princeton University Press, 2003)
The story of the cosmic microwave background radiation and the WMAP satellite.

F. Hoyle, G.R. Burbidge and J.V. Narlikar, *A Different Approach to Cosmology* (CUP, 2000)
The authors, who remain unconvinced by the Big Bang model, put forward their own arguments and challenge the interpretations of various observations.

Epilogue

Karl Popper, *The Logic of Scientific Discovery* (Routledge, 2002)
First published in 1959, Popper presents an academic and revolutionary view of the philosophy of science.

Thomas S. Kuhn, *The Structure of Scientific Revolutions* (University of Chicago Press, 1996)
First published in 1962, this is Kuhn's alternative view of the nature of scientific progress.

Steve Fuller, *Kuhn vs Popper* (Icon, 2003)
A re-examination of the Kuhn versus Popper debate on the philosophy of science, which is more accessible than their original publications cited above.

Lewis Wolpert, *The Unnatural Nature of Science* (Faber & Faber, 1993)
A discussion of what science is, what it can do, what it cannot do, and how it works.

Alan H. Guth, *The Inflationary Universe* (Vintage, 1998)
The father of inflation theory explains how it came to be and what it says about our universe.

F. Tipler and J. Barrow, *The Anthropic Cosmological Principle* (OUP, 1996)
An exploration of the relationship between the existence of our universe and the existence of life within it.

Mario Livio, *The Accelerating Universe* (Wiley, 2000)
A discussion of one of the most important discoveries in cosmology in the 1990s, namely that the universe appears to be expanding at an ever-increasing rate.

Lee Smolin, *Three Roads to Quantum Gravity* (Perseus, 2002)
A discussion of the relationship between the theories of quantum physics and general relativity. How might these theories be unified, and what are the implications for cosmology?

Brian Greene, *The Elegant Universe* (Random House, 2000)
A weighty, chart-topping volume that explains general relativity and string theory.

Martin Rees, *Just Six Numbers* (Basic Books, 2001)
The Astronomer Royal describes how six numbers, the constants of nature, define the qualities of the universe, and asks why the numbers appear to be just right for the evolution of life.

John Gribbin, *In Search of the Big Bang* (Penguin Books, 1998)
The story of the Big Bang, the evolution of the universe and the creation of galaxies, stars, planets and life, updated since its initial publication in 1986.

Steven Weinberg, *The First Three Minutes* (Basic Books, 1994)
Although slightly dated, this is still one of the best popular accounts of the Big Bang and the earliest moments of the universe.

Paul Davies, *The Last Three Minutes* (Basic Books, 1997)
Part of the Science Masters series, this book examines the ultimate fate of the universe.

Janna Levin, *How the Universe Got Its Spots* (Phoenix, 2003)
Written as a series of letters to her mother, Janna Levin's intensely personal account gives a unique perspective on cosmology and what it is like being a cosmologist.

'Four Keys to Cosmology', *Scientific American* (February 2004, pp. 30–63)
A set of four excellent articles that give details of the latest measurements of the CMB radiation and their implications for cosmology: 'The Cosmic Symphony', by Wayne Hu and Martin White, 'Reading the Blueprints of Creation' by Michael A. Strauss, 'From Slowdown to Speedup' by Adam G. Riess and Michael S. Turner, and 'Out of the Darkness' by Georgi Dvali.

Stephen Hawking, *The Universe in a Nutshell* (Bantam, 2002)
A richly illustrated book by the most famous cosmologist in the world. It won the 2002 Aventis prize for science books and is far more comprehensible than Hawking's *A Brief History of Time*.

Guy Consolmagno, *Brother Astronomer* (Schaum, 2001)
How religion and science can live together, by an astronomer at the Vatican Observatory.

R. Brawer and A. Lightman, *Origins* (Harvard U P, 1990)
Interviews with twenty-seven leading cosmologists, including Hoyle, Sandage, Sciama, Rees, Dicke, Peebles, Hawking, Penrose, Weinberg and Guth.

Andrew Liddle, *Introduction to Modern Cosmology* (Wiley, 2003)
A textbook covering all aspects of cosmology, which acts as a good introduction for readers with a moderate background in science.

Carl Gaither and Alma E. Cavazos-Gaither, *Astronomically Speaking* (Institute of Physics, 2003)
An excellent collection of quotations on the subject of astronomy. Part of a series that includes *Mathematically Speaking*, *Scientifically Speaking* and *Chemically Speaking*.

ACKNOWLEDGEMENTS

Over the last couple of years I have relied on numerous people to help me write this book. I am enormously grateful to Ralph Alpher, Allan Sandage, Arno Penzias and the late Thomas Gold, who all took time to tell me about their contributions to the development of cosmology. Their patience and good nature were much appreciated. Helge Kragh at the University of Aarhus and Ian Morrison at Jodrell Bank were also very supportive, as were Nancy Wilson and Don Nicholson at Mount Wilson Observatory. I was also able to visit Bell Laboratories, and I would like to thank all the people who took me around the various facilities there, in particular Saswato Das.

Thanks also go to Arthur Miller at University College (London), who introduced me to the work of Fritz Houtermans, and to Nigel Henbest, who pointed me towards the important contributions of Stanley Hey. I was able to interview Arno Penzias and heard the original Third Programme recordings of Fred Hoyle while I was working on the BBC radio programmes 'The Serendipity of Science' and 'Material World'; I am grateful to the producers of those programmes, Amanda Hargreaves, Monise Durrani and Andrew Luckbaker, who unwittingly helped to re-ignite my interest in cosmology.

Various people gave me valuable feedback on the manuscript as it developed, including Martin Rees and David Bodanis, who managed to find time to help me even though they were ridiculously busy on their own projects. Emma King, Alex Seeley, Amarendra Swarup and Mina Varsani all helped me during the various stages of this project, and I am grateful for all their contributions. In particular, my assistant Debbie Pearson helped me to research several parts of this book, arranging for me to visit the Mullard Radio Astronomy Observatory in Cambridge, and tracking down many of the photographs in the book.

Various archives and libraries are mentioned in the picture credits section, but the following people and institutions went beyond the call of duty in their efforts to help: Peter D. Hingley (Royal Astronomical Society), Heather Lindsay (Emilio Segré Visual Archives), Dan Lewis (The Huntington Library), John Grula (Observatories of the Carnegie Institution of Washington), Jonathan Harrison (St John's College Library), Iosif Khriplovich (University of Novosibirsk), Cheryl Dandridge (Lick Observatory Archives), Lewis Wyman (Library of Congress), Liliane Moens (Lemaître Archives, Catholic University, Louvain), and Mark Hurn, Sarah Bridle and Jochen Weller (Institute of Astronomy, University of Cambridge).

I should like to thank Iolo ap Gwynn at University of Wales Bio-Imaging Laboratory, who specially created the magnified images of paper shown in Figure 99, and Alison Doane at Harvard College Observatory, who changed her plans at short notice to show me the photographic stacks that house the work of Henrietta Leavitt and her colleagues. I have also been able to include several remarkable photographs of Fred Hoyle in this book, and I am enormously grateful to Barbara Hoyle and the Master and Fellows of St John's College for permission to use these images from the Hoyle Collection at St John's College, Cambridge.

Numerous friends and colleagues have kept me cheerful by adding a mix of interesting and fun diversions to my daily diet of reading, learning and writing about cosmology (which is already interesting and fun in itself). Hugh Mason, Ravi Kapur, Sharon Herkes and Valerie Burke-Ward have all worked with me on developing UAS, a project that encourages undergraduates to spend time in schools. Being involved in this project has kept me continually aware of the various issues relating to science education. Claire Ellis and Claire Greer have run code-breaking workshops in schools on my behalf, demonstrating the relevance of mathematics to young people. It is their hard work and enthusiasm that have taken the Enigma Project to tens of thousands of students. And I should also like to thank Nick Mee, who turned the idea of *The Code Book* on CD-ROM into a reality. He has also allowed me to have regular glimpses of the night sky through his telescope, which is always a great experience.

Over the last couple of years I have been involved with the National Museum of Science and Industry, the Science Media Centre and the National Endowment for Science, Technology and the Arts (NESTA), which have all stretched my brain in new ways. Thanks go to everybody at all three institutions for putting up with my ideas and interference. Suzanne Stevenson is deserving of special thanks because she and her colleagues help to keep me on track, providing me with constant advice, support and encouragement. Without her rock-solid presence, I would not be able to develop new ideas and projects.

Raj and Francesca Persaud have done a terrific job of keeping me sane and focused, while Roger Highfield, Holly, Rory, Asha and Sachin have given me a unique insight into the parks and museums of Greenwich, which has been a refreshing highlight of my week. Richard Wiseman has played a valuable role in keeping my obsessions in check, and he also introduced me to the wonders of bubble magic, for which I am truly grateful. Shyama Perera went out of her way to drag me away from my desk, which was always much needed and appreciated. And I am also grateful to my two nieces, Anna and Rachael, who continue to act as my fashion police and who have been eradicating my cardigan fixation. I am also glad that Rachael provided me with an excuse to travel to southern India last year, where she was teaching at Teddy School in Tirumangalam and making the most of her gap year. I know that the pupils, teachers and staff were grateful for her enthusiasm and cool head, and I know that she learned a lot from the devoted staff at this inspirational institution, which is helping to transform the local community. I am also grateful to Fiona Burtt, whose friendly words of advice have been much appreciated over the last couple of years.

While writing *Big Bang* I have received considerable support in turning my scribblings into a real book. Raymond Turvey drew all the illustrations, and Terence Caven was responsible for designing the overall look and layout of the book. John Woodruff, who was involved in *The Code Book*, played a central role in transforming a rough and ready draft into a polished manuscript. Indeed, he has played a quiet, behind-the-scenes role in editing and correcting dozens of science books over the last twenty years. He is one of the unsung heroes of science publishing.

Christopher Potter has been a guiding influence ever since I started writing, and I am glad that he has been able to be involved in the development of *Big Bang*. Mitzi Angel has been my new editor and has been absolutely brilliant, acting as a perfect sounding board and source of friendly advice. I should also like to thank my overseas editors from Italy to Japan, from France to Brazil, from Sweden to Israel and from Germany to Greece, who continue to support my writing across the globe. In turn, they work with some of the world's best translators, who take on the challenge of translating a book that contains both narrative prose and scientific explanation. There are very few translators capable of working in such a specialised area of publishing, and I am grateful to those who make my books accessible beyond the English-speaking world.

Finally, everybody at Conville & Walsh Literary Agency has been professional and charming during the writing of this book, even though I am sure that I am one of their more awkward authors. In particular, Patrick Walsh, who has been my literary agent ever since I started writing, almost a decade ago, works tirelessly on my behalf. He provides me with frank comments on my work and is always there during times of crisis, and I cannot imagine there are many literary agents who would accompany their authors to Zambia to witness a solar eclipse. In short, Patrick has been the best friend that any author could wish for.

Simon Singh
London
June 2004

PICTURE CREDITS

Illustrations by Raymond Turvey. All remaining images in the book were obtained courtesy of the following sources.

Figures 10–12	Royal Astronomical Society
Figures 15, 16	Royal Astronomical Society
Figure 18	(Copernicus) AIP Emilio Segrè Visual Archives, T.J.J. See Collection
Figure 18	(Tycho) Royal Astronomical Society
Figure 18	(Kepler) Institute of Astronomy, University of Cambridge
Figure 18	(Galileo) AIP Emilio Segrè Visual Archives, R. Galleria Uffizi
Figure 21	Getty Images (Hulton Archives)
Figure 28	Institute of Astronomy, University of Cambridge
Figure 29	AIP Emilio Segrè Visual Archives, Soviet Physics Uspekhi
Figure 31	Archives Lemaître', Université Catholique de Louvain.
Figure 32	Royal Astronomical Society
Figure 33	Institute of Astronomy, Cambridge University
Figure 35	Owen Gingerich, Harvard University
Figures 36, 37	The Birr Castle Archives and Father Browne Collection
Figure 37	Julia Muir, Glasgow University
Figure 38	Edwin Hubble Papers, Huntington Library
Figure 39	(Curtis) The Mary Lea Shane Archives/Lick Observatory
Figure 39	(Shapley) Harvard College Observatory
Figure 42	Julia Margaret Cameron Trust
Figure 42	Royal Astronomical Society
Figures 43, 44	Harvard College Observatory
Figures 46–48	Edwin Hubble Papers, Huntington Library

Figure 49	Observatories of the Carnegie Institution of Washington
Figure 50	Julia Muir, Glasgow University
Figure 56	Royal Astronomical Society
Figure 57	(1950) Palomar Digital Sky Survey
Figure 57	(1997) Jack Schmidling, Marengo, Illinois
Figure 63	Astronomical Society of the Pacific
Figure 65	Brown Brothers
Figure 66	California Institute of Technology Archives
Figure 68	Cavendish Laboratory, University of Cambridge
Figure 75	Library of Congress
Figure 76	Cavendish Laboratory, University of Cambridge & AIP Emilio Segrè Visual Archives
Figure 77	AIP Emilio Segrè Visual Archives, Institut International Physique Solvay
Figure 78	Herb Block Foundation
Figure 79	George Gamow, *The Creation of the Universe*
Figure 83	Ralph Alpher
Figures 84, 85	St John's College Library, Cambridge
Figure 87	St John's College Library, Cambridge
Figure 88	AIP Emilio Segrè Visual Archives, Institut International Physique Solvay
Figure 92	Lucent Technologies Inc/Bell Labs
Figure 94	Mirror Group News
Figure 96	Lucent Technologies Inc/Bell Labs
Figure 97	Smithsonian National Air and Space Museum
Figure 98	St John's College Library, Cambridge
Figure 99	University of Wales Bio-Imaging Laboratory
Figure 101	NASA (COBE group)
Figure 102	NASA (DMR group)
Figure 103	The *Independent* newspaper, 24 April 1992
Figure 104	NASA (WMAP group)

INDEX

Page numbers in *italic* refer to Figures

P.S.

Ideas, interviews & features . . .

About the author

About the book

Read on

About the author

From Somerset to Space

Simon Singh talks to Louise Tucker

Tell me about your own universe: where were you born and where did you grow up?
I grew up and went to school in Somerset – I belong to the West Country Singh clan. For some strange reason my grandfather and others from our village, Thakarki in the Punjab, chose to settle in Somerset and Devon when they arrived in England in the 1930s. Virtually all of my family's close relatives are now settled in Britain, America or Canada, but we still go back to India every couple of years.

What did you want to be when you grew up?
I have a strong recollection of being ten years old and my sister asking me what I wanted to be when I grew up, and I replied that I would become 'a nuclear physicist'. In fact my PhD was in particle physics, which is more like sub-nuclear physics, but it was a fairly accurate prediction.

What is your first memory of science?
I remember my father once showed me something called a dozel when I was maybe seven or eight. It was a stick with a propeller on the end, which spun freely. When a rod was scraped along the stick's ridged back nothing would happen . . . until my father said the magic word 'dozel', whereupon the propeller would spin. Repetition of the word would cause it to spin the other way, and a third mention of dozel would bring it to a halt. It turns out that it is the way you hold your hand that determines whether or not it

spins and its direction, but at the time I was baffled. I have only ever seen a dozel on one other occasion, which was in America and it was called a Hooey stick. It now sits on my mantelpiece.

Your dedication credits many science TV presenters, such as Magnus Pyke and James Burke, for inspiring your interest in science. In an era in which television is sometimes blamed for not being educational enough, do you think it still has a role in inspiring children?
Absolutely – I was inspired by TV boffins and I would hope that TV can help inspire the next generation of scientists. Children's TV seems to be doing a great job, but I am not sure that the major channels are giving science enough prominence in prime time or nurturing new on-screen talent. Unless it's dinosaurs or medicine, TV execs seem to think science is not a mainstream subject.

You are very involved in various educational projects promoting science. What do you think of current science education?
I am worried that science and mathematics education is in a major crisis, and it comes down to one thing – a lack of qualified schoolteachers. Teachers inspire students, so a shortage means fewer students study science at university, so fewer become teachers, and then you enter a downward spiral. My job is easy compared to a schoolteacher's because I can choose my subjects and I am writing for and broadcasting to a willing audience, ►

TOP TEN

Simon Singh's Top Ten Favourite Books:

Surely You're Joking, Mr Feynman!
Richard Feynman

The Making of the Atomic Bomb
Richard Rhodes

Chaos
James Gleick

Flatland
Edwin A. Abbott

A Mathematician's Apology
G.H. Hardy

Our Final Century
Martin Rees

The Surgeon of Crowthorne
Simon Winchester

Dr Tatiana's Sex Advice to All Creation
Olivia Judson

Interpreter of Maladies
Jhumpa Lahiri

The Phantom Tollbooth
Norton Juster

Simon Singh was born in Somerset in 1964. He studied physics at Imperial College, London, and then completed his PhD in particle physics at Cambridge University and CERN, Geneva. Having failed to discover the top quark, he abandoned any pretence at being a serious scientist and joined the BBC in 1991, working on programmes such as *Tomorrow's World* and *Horizon*. His documentary about Fermat's Last Theorem won a BAFTA, and this notorious mathematics problem also became the subject of his first book.

Simon's second book, *The Code Book*, examined the history of cryptography from Ancient Greece to the internet. He then presented a TV series on Channel 4 entitled *The Science of Secrecy* based on the book. He has also hosted *Mind Games* on

From Somerset to Space *(continued)*

whereas teachers have to make even the dull parts of the curriculum interesting to students who might require convincing that science is important. I have a huge respect for those people who do go into teaching. It requires both knowledge and the ability to communicate, and if you have that combination of talents then there are many other job opportunities, particularly in this age of technology. I wish I could offer some solutions, but I am not sure how we turn the tide and encourage more people to go into the classroom.

You've written books, produced and directed TV and radio programmes and lectured all over the world. Of what are you most proud?
The BBC *Horizon* documentary that I made in 1996 with John Lynch still means a great deal to me. It proved that mathematics could be exciting and passionate, and because of the power of TV it has been seen by tens of millions of people around the world. It showed older mathphobes that numbers could be thrilling and it showed younger people that mathematics was worth studying. I still meet people who say that it encouraged them to pursue mathematics at university, and that is incredibly satisfying for me.

As a writer who prides himself on communicating difficult ideas, who are the writers that you admire most and who have influenced you?
I have a particular respect for scientists who take the time to write for the general reader.

There is no reason why Martin Rees, Steve Jones or Marcus du Sautoy should take a break from doing top-notch research, and instead write books. Presumably they realise that their books have a huge impact on the public and budding scientists.

Is there any book that you wish you had written?
I admire clever ideas for books. So although I could not have written *Dr Tatiana's Sex Advice to All Creation* by Olivia Judson, I wish I had thought of writing a book about animal behaviour in terms of letters from various species to a zoological agony aunt. Similarly, I wish I had thought of writing about the development of the telegraph, which puts the current information revolution into context. This is the subject of *The Victorian Internet* by Tom Standage.

What do you read when you're writing?
Books directly relevant to whatever I am writing about and nothing else. I write about subjects that are generally new to me, so I have to completely immerse myself in these subjects – mathematics, cryptography or cosmology – in order to get up to speed.

What motivates you to write?
There are two reasons. First, I love learning about science and mathematics, so it is great fun and a real privilege to spend a couple of years reading about a particular subject and meeting the leading thinkers. Second, I want to get other people excited about science and mathematics. I find them ►

LIFE ***at a Glance***
(continued)

BBC4 and *Five Numbers* on Radio 4. In 2002 he worked alongside Richard Wiseman on a live show entitled *Theatre of Science*, which was performed at the Soho Theatre in London and at the Edinburgh Fringe.

Simon lives in London. He loves rattlebacks, non-transitive dice, magic tricks, gambling, electrocuting gherkins, trash TV and the Violent Femmes. His website is at **www.simonsingh.net**

From Somerset to Space *(continued)*

amazing and fascinating and either I have to grab people on the street and tell them about what I have just learned or I write books about it, and the latter is less likely to get me into trouble.

In an interview a few years ago you mentioned that Big Bang *would be your last book. Is that still the case?*
Bearing in mind the answer to my last question, this is a tough one to answer, but I do indeed think that this is likely to be my last major book. I have always flitted between careers, and I have enjoyed not quite knowing where I will go next. I spent a few years doing my PhD, a few years directing TV, a few years broadcasting and a few years writing, so now it seems like time for a change. The biggest problem with writing a book is that it completely takes over my life, and there are always lots of other things that I would like to do which get ignored. Even if I stopped writing for a decade, I could begin to explore some of these other interests. And many of these interests still relate to communicating science and mathematics, so to a large extent my answer to the previous question still holds true. ■

A Critical Eye

It is difficult to summarise the breadth of reviewers' responses to *Big Bang*. Descriptions ranged from 'epic' and ' wonderful' to 'beautiful' and 'entertaining'. Many reviews emphasised the lucidity of Singh's writing: **Nature** said that 'this very well-written book conveys the ideas underpinning cosmological theory with great clarity', and the **Economist** called it 'a model of clarity'.

Singh's ability to tell a complex story engagingly was also widely praised: 'Singh is a very gifted storyteller who never misses the chance to make his subject clearer or more entertaining,' wrote Scarlett Thomas in the **Independent on Sunday**; the **Guardian** commented that 'Singh tells his tale well, with chatty anecdotes leavening the astrophysics'; and the **Daily Telegraph** described the book as 'an epic tale brilliantly told, packed with courage and tragedy, heroes and martyrs'.

But, most of all, the book won plaudits for its tackling of a subject that still strikes fear into many otherwise very intelligent adults: mathematics. 'Even the most mathematically hobbled of us,' said the **Sunday Telegraph**, could be enabled to understand 'the history of man's intellectual engagement with the dark spaces around him' by reading *Big Bang*. And in the **Daily Mail** the maths-phobic were entreated not to worry since 'Simon Singh spares us most of the maths, and he juggles big ideas with tact and care'. For anyone who struggles to understand science, suggested the reviewers, *Big Bang* was a good place to begin: 'Even ►

A Critical Eye *(continued)*

if the cosmologists don't know where the universe is going, at least they have found out where it has come from. Anybody who wants to understand this wonderful achievement will not do better than start with Singh's book,' declared the **Mail on Sunday.** ■

The Missing Pages

Simon Singh

At roughly 550 pages, *Big Bang* is substantially longer than originally planned. In fact my original book proposal was entitled *The Little Book of the Big Bang* and the entire story of the greatest scientific theory in history was supposed to have been encapsulated in a couple of hundred pages. I soon realised that it was impossible to do justice to the Big Bang without running to several hundred pages, and before long there was a serious risk of breaking the 1,000-page barrier. My main problem was that I felt compelled to pass on everything I found during my research, from mind boggling scientific notions to yet another forgotten hero of cosmology.

Hence, after completing my first draft of *Big Bang*, I embarked on the painful process of cutting back the content to make the book less intimidating and more accessible, less meandering and more engaging. Of course some sections were harder to cull than others, simply because I had become particularly fond of them.

> 'I soon realised that it was impossible to do justice to the Big Bang without running to several hundred pages.'

For example, *Big Bang* used to contain the tale of the Heike crab, a species that is famous because its shell often looks like a samurai mask. The traditional explanation is that the crabs contain the souls of samurai soldiers belonging to the Heike clan, who drowned in a sea battle in 1185. For this reason, today's fishermen always throw back to the sea any crabs exhibiting the distinctive samurai shell, as it would be unthinkable to eat a creature with the soul of a samurai. In fact a tiny fraction of these crabs have always exhibited ▶

The Missing Pages *(continued)*

a vaguely samurai appearance, but their numbers increased dramatically and their samurai appearance was enhanced after the battle of 1185 when fishermen started to take more notice of them and the practice of not eating such crabs began. Suddenly there was a huge survival advantage for a crab to look like a samurai, so this property was exaggerated and promulgated to create the large population of samurai-looking crabs that we have today.

I have wanted to retell this piece of evolutionary biology ever since reading about it in Carl Sagan's *Cosmos*, so I included it in my first draft of Chapter 1. I still think it is a fascinating story, but in the context of the Big Bang it was a self-indulgent detour. I will not explain how I originally managed to fit the Heike crab into Chapter 1, but I will admit that it required clumsy crowbarring of the worst kind.

Similarly, the first incarnation of Chapter 2 had a section about the Jesuit priest Giovanni Riccioli who published the *New Almagest* in 1651, which included meticulous drawings of the lunar surface and labelled features such as the lowland plains. He labelled each so-called sea according to a state of mind, such as the Sea of Tranquillity or the Sea of Madness, and this rule for naming lunar plains has remained ever since. In 1959, however, a Soviet probe sent back images of the previously unseen far side of the Moon and the Russians broke Riccioli's protocol by calling one region the Sea of Moscow. The International Astronomical Union rejected the name, until Soviet astronomers convinced the IAU that there was such a thing as a 'Moscow state of mind'. I still

> 'Some sections were harder to cull than others, simply because I had become particularly fond of them.'

think this is a wonderfully quirky story, but reluctantly I had to discard it, because Riccioli and his lunar seas were not directly relevant to the real story that I ought to have been telling in Chapter 2.

Exactly the same problem has arisen with my previous books – several quirky asides had to be lost because they were not directly relevant to the history of Fermat's Last Theorem or the development of cryptography. Perhaps my favourite lost episode concerns a magic square that appeared in an early draft of *The Code Book*.

S	A	T	O	R
A	R	E	P	O
T	E	N	E	T
O	P	E	R	A
R	O	T	A	S

This square carries the Latin motto *sator arepo tenet opera rotas*, which means something along the lines of 'Arepo the sower holds the wheels at work.' It is a magic square because the same message can be read from left to right, right to left, up the columns and down the columns. However, the letters only gain their true meaning when they are anagrammed into a new arrangement. ►

```
         P
         A
    A    T    O
         E
         R
PATERNOSTER
         O
         S
    O    T    A
         E
         R
```

The Missing Pages *(continued)*

The word *paternoster* (Our Father) can be created twice. This is only possible if the *N* is shared, so the letters are formed into a cross. Moreover, the only letters not required to construct *paternoster* twice are a pair of *A*s and *O*s, which symbolise alphas and omegas, beginnings and ends. In other words, the magic square is clever as it stands, but it also has hidden cleverness because encoded within it are three elements (*paternoster*, the cross, alpha/omega) of deep significance to Christians. Historians are divided as to the origins and true significance of this magic square, but it seems to be a beautiful illustration of a steganography, which means hiding an important message within an apparently innocuous text or image. This cryptographic oddity has been lying in my files ever since it was cut from *The Code Book*.

It seems a shame to have cut the magic square from *The Code Book*, just as it was disappointing to have omitted the tales of the Heike crab and the Sea of Madness from *Big Bang*, but ultimately I have to remember that my books are not random collections of vaguely related episodes, but rather they are supposed to be coherent histories of science. I am just grateful that this postscript essay has at least allowed me the opportunity to get these three curiosities out of my system. ■

Have You Read?

Other titles by Simon Singh

Fermat's Last Theorem
In 1963 a schoolboy browsing in his local library stumbled across the world's greatest mathematical problem: Fermat's Last Theorem, a puzzle that has baffled mathematicians for over 300 years. Aged just ten, Andrew Wiles dreamed that he would crack it. Wiles's lifelong obsession with a seemingly simple challenge set by a long-dead Frenchman is an emotional tale of sacrifice and extraordinary determination. In the end Wiles was forced to work in secrecy and isolation for seven years, harnessing all the power of modern maths to achieve his childhood dream. Many before him had tried and failed, including an eighteenth-century philanderer who was killed in a duel. At roughly the same time a Frenchwoman made a major breakthrough in solving the riddle, but she had to attend maths lectures at the Ecole Polytechnique disguised as a man since women were forbidden entry to the school. A remarkable story of human endeavour and intellectual brilliance over three centuries, *Fermat's Last Theorem* will fascinate both specialist and general readers.

The Code Book
Since humans began writing, they have been writing in code. This quest for secrecy has often changed the course of history. In *The Code Book*, Simon Singh offers a sweeping view of the subject of encryption as well as its more dramatic effects on the outcome of wars, monarchies and individual lives. ►

Have You Read? *(continued)*

Included in this fascinating book is the story of Mary Queen of Scots, trapped by her own code and put to death by Queen Elizabeth. Also recounted is the history of the Beale Ciphers, created in the early nineteenth century to obscure the location of a treasure of gold, buried somewhere in Virginia. Singh also traces the monumental improvements in code-making and -breaking brought on by the First and Second World Wars, including the development of the German Enigma cipher machine, which was cracked by the brilliant Allied code-breakers at Bletchley Park. Now, in the Information Age, the possibility of a truly unbreakable code looms large, and information security has become one of the major debates of our times. Simon Singh investigates how technology and the ways we communicate will affect our personal privacy and our everyday lives. Dramatic, compelling and remarkably far-reaching, this book will forever alter your view of history, what drives it, and how private that e-mail you just sent really is.

The Cracking Code Book
Simon Singh brings life to an amazing story of puzzles, codes, lost languages and riddles in this abridged and adapted version of *The Code Book* aimed at younger readers. Children and teenagers will be drawn into the history of codes and code-breaking from Ancient Egypt to the internet. This history of cryptography involves tales of heroism and villainy, and at the same time it explains the science and mathematics behind the world's cleverest codes. ■

If You Loved This, You Might Like . . .

A Beautiful Mind
Sylvia Nasar
Made into a Hollywood film, this is the biography of the games theorist John Forbes Nash.

The Man who Loved Only Numbers
Paul Hoffman
Another biography of a fascinating mathematical life, namely Paul Erdös.

The Victorian Internet
Tom Standage
The history of the telegraph, a communications revolution that puts the growth of the internet into context.

Fingerprints
Colin Beavan
An account, as gripping as any crime novel, of how fingerprinting became central to crime detection.

Can Reindeer Fly?
Roger Highfield
A quirky and thought-provoking examination of the science of Christmas.

Mendeleyev's Dream
Paul Strathern
The story of the chemist who pioneered the Periodic Table, an icon of modern science. ►

If You Loved This *(continued)*

Snowball Earth
Gabrielle Walker
A great detective story about a hitherto unknown period of Earth's history.

The Fifth Miracle
Paul Davies
How did life get started? Davies looks at the latest ideas and breakthroughs.

Strange Beauty
George Johnson
A highly readable biography of Murray Gell-Mann, one of the great physicists of the twentieth century.

The Cogwheel Brain
Doron Swade
The heroic tale of Charles Babbage, a Victorian genius who tried to build a mechanical computer. ■

Find Out More

Websites selected by Simon Singh

http://map.gsfc.nasa.gov/m_uni.html
This NASA website offers a tutorial, Cosmology 101, that covers everything from the development of the core Big Bang model to the latest ideas and observations.

http://www.nasa.gov/multimedia/imagegallery/index.html
This NASA website is dedicated to showcasing stunning space-related images, from the Apollo Moon landings to remote spiral galaxies. The Image of the Day is always worth a look.

http://hubblesite.org/
The Hubble Space Telescope has an excellent website. The Gallery shows spectacular images, Discoveries explains the science behind the images, and the Newscenter carries the latest observations.

http://galileo.rice.edu/
Based at Rice University, Houston, the Galileo Project is the definitive Galileo website.

http://www.newtonproject.ic.ac.uk/
Based at Imperial College, London, the Newton Project is the definitive Isaac Newton website.

http://www.alberteinstein.info/
http://www.albert-einstein.org/
http://www.aip.org/history/einstein/
These three sites tell you everything you could ever want to know about Einstein, ranging from an academic archive website ►

Find Out More *(continued)*

to a more engaging website aimed at the general visitor.

http://www.mtwilson.edu/
The Mount Wilson Observatory website contains a section on its history and its current research activities. It also contains information for those who want to visit the observatory.

http://www.thechromatics.com/
The Chromatics write and perform songs about astronomy. They are unique.

http://www.badastronomy.com/
Phil Plait devotes his life to debunking 'bad' astronomy, which covers everything from minor misunderstandings to claims that the Moon landings were a hoax! ■